AF615628

GROWTH FACTORS AND SIGNAL TRANSDUCTION IN DEVELOPMENT

MODERN CELL BIOLOGY

RECENT VOLUMES PUBLISHED IN THE SERIES

Volume 11 — Antisense RNA and DNA
James A.H. Murray, Editor

Volume 12 — Genes in Mammalian Reproduction
Ralph B.L. Gwatkin, Editor

Volume 13 — Microtubules
Jeremy S. Hyams and Clive W. Lloyd, Editors

Volume 14 — Growth Factors and Signal Transduction in Development
Marit Nilsen-Hamilton, Editor

GROWTH FACTORS AND SIGNAL TRANSDUCTION IN DEVELOPMENT

Marit Nilsen-Hamilton, Editor
Department of Biochemistry and Biophysics
Iowa State University
Ames, Iowa

WILEY-LISS
A JOHN WILEY & SONS, INC., PUBLICATION
New York • Chichester • Brisbane • Toronto • Singapore

Address All Inquiries to the Publisher
Wiley-Liss, Inc., 605 Third Avenue, New York, NY 10158-0012

Printed in United States of America

Library of Congress Cataloging-in-Publication Data

Growth factors and signal transduction in development / editor, Marit
Nilsen-Hamilton.
p. cm. — (Modern cell biology ; 14)
Includes bibliographical references and index.
ISBN 0-471-30539-1
1. Growth factors. 2. Cellular signal transduction.
3. Developmental biology. I. Nilsen-Hamilton, Marit. II. Series.
QH573.M63 vol. 14
[QP552.G76]
574.87 s—dc20
[574.87′61] 94-6541
CIP

The text of this book is printed on acid-free paper.

10 9 8 7 6 5 4 3 2 1

Contents

Contributors

Rosemary J. Akhurst, Department of Medical Genetics, University of Glasgow, Yorkhill, Glasgow G3 8SJ, Scotland **[97]**

Daniel F. Bowen-Pope, Department of Pathology, SJ-60, University of Washington, Seattle, WA 98195 **[51]**

Igor B. Dawid, Laboratory of Molecular Genetics, National Institute of Child Health and Human Development, National Institutes of Health, Bethesda, MD 20892 **[199]**

Ravindra N. Dhir, Department of Obstetrics and Gynecology, University of Pennsylvania Medical Center, Philadelphia, PA 19104 **[1]**

Richard A. Franklin, Division of Basic Sciences, Department of Pediatrics, National Jewish Center for Immunology and Respiratory Medicine, Denver, CO 80206 **[75]**

William T. Garside, Department of Obstetrics and Gynecology, University of Pennsylvania Medical Center, Philadelphia, PA 19104 **[1]**

Erwin W. Gelfand, Division of Basic Sciences, Department of Pediatrics, National Jewish Center for Immunology and Respiratory Medicine, Denver, CO 80206 **[75]**

Iva Greenwald, Department of Biochemistry and Molecular Biophysics, College of Physicians and Surgeons, Columbia University, New York, NY 10032 **[179]**

Richard T. Hamilton, Department of Zoology and Genetics, Iowa State University, Ames, IA 50011-3223 **[123]**

Susan Heyner, Department of Obstetrics and Gynecology, University of Pennsylvania Medical Center, Philadelphia, PA 19104 **[1]**

Keli Hippen, Department of Zoology and Genetics, Iowa State University, Ames, IA 50011-3223 **[139]**

Kelaginamane T. Hiriyanna, Department of Zoology and Genetics, Iowa State University, Ames, IA 50011-3223 **[139]**

F. Michael Hoffman, McArdle Laboratory for Cancer Research and Laboratory of Genetics, University of Wisconsin–Madison, Madison, WI 53706 **[165]**

Thomas S. Ingebritsen, Department of Zoology and Genetics, Iowa State University, Ames, IA 50011-3223 **[139]**

Eric C. Liebl, McArdle Laboratory for Cancer Research, University of Wisconsin–Madison, Madison, WI 53706 **[165]**

Joseph J. Lucas, Division of Basic Sciences, Department of Pediatrics, National Jewish Center for Immunology and Respiratory Medicine, Denver, CO 80206 **[75]**

Keith Miller, Eppley Institute for Research in Cancer and Allied Diseases, Department of Pathology and Microbiology, University of Nebraska Medical Center, Omaha, NE 68198-6805 **[19]**

Marit Nilsen-Hamilton, Department of Biochemistry and Biophysics, Iowa State University, Ames, IA 50011 **[xiii]**

Angie Rizzino, Eppley Institute for Research in Cancer and Allied Diseases, Department of Pathology and Microbiology, University of Nebraska Medical Center, Omaha, NE 68198-6805 **[19]**

The numbers in brackets are the opening page numbers of the contributors' articles.

Ronald A. Seifert, Department of Pathology, SJ-60, University of Washington, Seattle, WA 98195 **[51]**

Naohiro Terada, Division of Basic Sciences, Department of Pediatrics, National Jewish Center for Immunology and Respiratory Medicine, Denver, CO 80206 **[75]**

Simon Tuck, Department of Biochemistry and Molecular Biophysics, College of Physicians and Surgeons, Columbia University, New York, NY 10032 **[179]**

Cynthia R. Ward, Department of Medicine, School of Veterinary Medicine, University of Pennsylvania, Philadelphia, PA 19104 **[1]**

Preface

Over the past few years studies of growth factors and their mechanisms of action have been exceptionally productive and have led to the almost complete elucidation of a signal transduction pathway from tyrosine kinase receptors through Ras to the activation of specific genes. Other pathways will soon be known. These enormous advances in our understanding of the molecular aspects of the regulation of cell function by growth factors are complemented by and have contributed to advances in our understanding of the interactions between cells and tissues during embryonic development.

In this volume, several growth factors are discussed individually in terms of their roles in development and their signal transduction mechanisms. Although some growth factors—fibroblast growth factors (FGFs), insulin-like growth factor-I (IGF-I), and platelet-derived growth factor (PDGF)—are similar in that they activate tyrosine kinase receptors, they are distinguished by other aspects of their signalling mechanisms which involve interactions of their receptors and the growth factors themselves with other macromolecules; these unique aspects include the intracellular signalling proteins with which the receptors interact, and the binding proteins and other macromolecules with which the growth factors interact. For some growth factors, such as interleukin 2 (IL-2), which do not activate a tyrosine kinase receptor, the intracellular signal is transmitted by a protein tyrosine kinase that associates with the receptor.

Animal development almost certainly relies on the interaction of complex combinations of growth factors and their receptors. The discovery that many growth factors have the option of more than one receptor for signal transduction and that these receptors can be expressed differently on different cell types and in the developing organism points to the potential for fine regulation of the individual cellular response to growth factors. Sometimes more than one related growth factor interacts with the same receptor. Two such growth factor families that are described in detail in this volume are the FGFs and the transforming growth factors beta (TGFβs).

Another important aspect of the action of growth factors is the influence of the cellular environment on their ability to initiate a signal through their receptors. The activities of individual growth factors are regulated differently. For example, TGFβ is secreted in a latent form in association with another protein and must be activated by cleavage of an N-terminal extension. The FGFs often require interaction with heparan-sulfate proteoglycans for proper receptor activation. Interaction of IGF-I with its receptor is regulated by the IGF binding proteins, of which there are several.

Another feature of growth factor signalling is the route taken by the growth factor to reach its target receptor. Most often, growth factors seem to act in paracrine or autocrine modes in the developing organism, where the growth factor influences the proliferation and differentiation of a subset of adjacent or nearby cells. However, some growth factors such as the IGFs are distributed systemically and can be considered as also playing an endocrine role. Others, such as some of the FGFs, do not possess a signal peptide and also have a nuclear localization domain. These FGFs have been proposed to regulate cellular function by an intracrine mode, never leaving the cell in which they are synthesized. Yet others, such as epidermal growth factor (EGF) and TGFα, are produced in the form of active transmembrane precursor proteins that can signal an adjacent cell to stimulate its proliferation in a restricted paracrine mode.

In this volume, discussion of the growth factor families and their roles in development is extended in chapters describing the signal transduction pathways that they regulate. In particular, we have focused on the most completely understood signal transduction pathway, which is initiated by tyrosine protein kinase receptors and involves the Ras protooncogene and several serine/threonine kinases. Protein phosphatases are also intimately involved in regulating the action of this signal transduction pathway in which the primary activating protein modification is phosphorylation.

The vigor of the current thrust towards understanding the mechanism of action of growth factors in development results from the fact that the actions of growth factors directly bear on mechanisms studied in various disciplines which include biochemistry, molecular genetics, oncovirology, and cellular and developmental biology. Contributions from these fields and from others have led to the realization that developmental processes in all animals have a common basis. This very important breakthrough has resulted in the productive exchange of information between individuals studying distinct developing systems, each of which provides different advantages for investigation. In this volume, the well-studied amphibian, nematode, and insect developing systems are discussed independently, and significant aspects of mouse development are included in chapters that focus on particular growth factors. Comparative analyses of the role of growth factors and their signal transduction mechanisms in these different developing systems can be found in most chapters that deal with specific growth factors and in the two chapters that discuss signal transduction pathways.

The rapid expansion of these fields and the identification of many new genes and gene families has created the difficulty that in volumes such as this one, many new acronyms are introduced which do not provide much information regarding the nature of the entity discussed. To compound the problem, many proteins and genes in the same organism have been given several names, which are still concurrently in use; and the equivalent proteins and genes in different organisms are often variously named. A glossary of genes and proteins that are referred to in this volume is presented to help the reader understand the relationships between genes and their protein products and the activities (when known) of the proteins.

The excitement in the study of growth factors in development is palpable at any congress of scientists interested in the subject. This volume provides a glimpse into the complexities and subtle differences that discriminate different signal transduction mechanisms and the means by which these pathways are used to control animal development. Much more will surely be uncovered over the ensuing years to evoke our continued admiration of the natural world.

Marit Nilsen-Hamilton

Acknowledgment

My thanks to the authors who contributed to this volume for their adherence to the deadlines so that this volume could be published in time to provide as current a view of growth factors as possible in a field that is moving so rapidly. Thanks also go to Lori Craig who helped in many aspects of the editorial process and to ensure that the volume came together in a timely manner and to Richard Hamilton for his advice, constant support, and encouragement.

Growth Factors and Signal Transduction in Development: 1–18

The Insulin Family of Growth Factors and Signal Transduction During Development

C.R. Ward, W.T. Garside, R.N. Dhir, and S. Heyner

I. INSULIN AND THE INSULIN-LIKE GROWTH FACTORS

The members of the insulin family of peptides include insulin, insulin-like growth factor-I (IGF-I) and insulin-like growth factor-II (IGF-II). They are single-chain polypeptide hormones that have a high degree of amino acid homology with each other and with proinsulin. This family of polypeptides has a number of overlapping activities, perhaps due in part to the retained similarity in the tertiary structure, known as the "insulin fold" (Blundell and Humbel, 1980). Due to these structural similarities, each of these polypeptides can bind not only to its cognate cell surface receptor, but also to heterologous receptors, although with reduced affinity to the latter. While the importance of the IGFs in the regulation of postnatal growth and development is firmly established, the role that these peptides play during embryonic and early fetal growth is only now beginning to be understood.

Insulin has been regarded classically as a vertebrate hormone, elaborated and stored by the beta cells of the mammalian pancreas. Studies in recent decades have revealed that insulin has extraordinary evolutionary conservation. Thus, an insulin-like molecule has been detected in phyla as diverse as fungi, invertebrates, poikilotherm vertebrates, and mammals. In many cases, insulin has been implicated in embryogenesis [reviewed by Mattson et al., 1989]. The role of insulin in glucoregulation and in the regulation of related metabolic pathways has been firmly established by studies in a wide variety of *in vivo* and *in vitro* systems. *In vivo,* insulin is a primary regulator of rapid anabolic responses. Its physiological actions include the regulation of glucose uptake in muscle and fat cells, the regulation of glycogen synthesis in the liver, fat synthesis in adipocytes, and the stimulation of amino acid and ion uptake in muscle, as well as other target cells [Rosen, 1987]. Metabolic effects of insulin were formerly believed to be mediated solely via the insulin receptor, while the IGF-I receptor regulated growth-promoting effects [Van Wyk et al., 1985]. promoting effects [Van Wyk et al., 1985]. However, hepatoma cells lacking the IGF-I receptor respond to growth-promoting effects mediated by the insulin receptor [Shimizu and Shimizu, 1986; Taub et al., 1987]. In addition, evidence is accumulating to show that gene transcription may possibly be regulated in part by translocation of insulin (or the IGFs) to the nucleus; studies supporting this hypothesis include our own observation of insulin accumulation in nuclei of blastocyst cells [Heyner et al., 1989a] and similar observations in other cell types by Peralta-Soler et al. [1989].

The classic source of the IGFs is the liver, although it is now recognized that both IGF-I and IGF-II are produced in extrahepatic tissues. The classic role of IGF-I as a primary growth regulator in numerous tissues during peri- and postnatal life has been reviewed by Lowe [1991]. Additionally, observations in the chick embryo show that expression of the IGF-I receptor predates that of the insulin receptor and can be detected on whole embryos by day 2 [Bassas et al., 1985]. In contrast to IGF-I, the physiological function of IGF-II is not clearly understood, although it has been suggested to stimulate growth in undifferentiated tissues [Bhaumick and Bala, 1987] as well as to play a critical role in very early embryogenesis [DeChiara et al., 1990, 1991].

II. RECEPTORS FOR THE INSULIN FAMILY OF GROWTH FACTORS

A. General

Insulin and the IGFs bind to cell surface receptors (Fig. 1), and their actions are receptor mediated. The nucleotide sequences of human insulin and IGF-I receptor mRNAs have been derived from overlapping of cDNA clones and have been shown to share a high degree of sequence identity [Ebina et al., 1985; Ullrich et al., 1985, 1986]. The insulin and IGF-I receptors are structurally very similar, (Fig. 1) both consisting of a tetrameric receptor made up of two alpha (M_r = 135,000) and two beta (M_r = 90,000) chains linked by disulfide bonds. The beta subunits traverse the cell membrane and possess a tyrosine kinase domain that is activated following ligand binding to the alpha subunit. The IGF-II receptor is quite different in structure; it has a relatively long extracellular domain, and a short cytoplasmic domain with no tyrosine kinase activity. The receptor has been cloned and sequenced and has been demonstrated to be 80% and 99% homologous, respectively, to the bovine and human mannose-6-phosphate (Man-6-P) receptors that bind Man-6-P with high affinity and are implicated in mediating lysosomal enzyme sorting [Morgan et al., 1987; Lobel et al., 1988; Oshima et al., 1988]. Further research has demonstrated that these receptors bind both IGF-II and Man-6-P, and are identical proteins; this receptor is now referred to as the IGF-II/Man-6-P receptor [Roth et al., 1987; Kiess et al., 1988]. However, Man-6-P and IGF-II bind to different domains of the receptor [Braulke et al., 1988], and the IGF-II/Man-6-P receptor mediates differential signals in response to IGF-II or Man-6-P binding [Okamoto et al., 1990; Muryama et al., 1990].

There is significant cross-reactivity within the family of receptors. Insulin binds with high affinity to the insulin receptor, and with lower affinity to the IGF-I receptor, while IGF-I binds with high affinity to its own receptor, and with lower affinity to the insulin receptor [Czech, 1989]. The IGF-II receptor binds IGF-II with high affinity, IGF-I with low affinity, and does not bind insulin [Ewton et al., 1987]. In addition, there is evidence that some mitogenic responses to IGF-II are mediated via the IGF-I receptor [Kiess et al., 1987].

B. Evidence for Insulin/IGF-I Hybrid Receptors

The insulin and IGF-I receptors are expressed as individual membrane-restricted heterotetrameric receptors; however, there is abundant evidence for the presence of a subpopulation of hybrid receptor molecules containing both the insulin receptor (α and β) and the IGF-I receptor (α and β) peptides in a beta-alpha-alpha-beta configuration [Werner et al., 1991]. Such naturally occurring hybrid receptors have been immunoprecipitated from solubilized extracts of human placenta and membranes from several cultured cell lines [Soos and Siddle, 1989; Soos et al., 1990]. These receptors may comprise two distinct types; one being generated from the association of an $\alpha\beta$ insulin half-receptor with an $\alpha\beta$ IGF-I, while the other type is formed from a wild-type insulin or IGF-I $\alpha\beta$ half-receptor and a mutant insulin $\alpha\beta$ half-receptor. The biological functions of such hybrid receptors are unknown at this time. However, even though definitive data are not available, it appears that the hybrid wild type/mutant receptors are essentially inactive kinases when the mutant

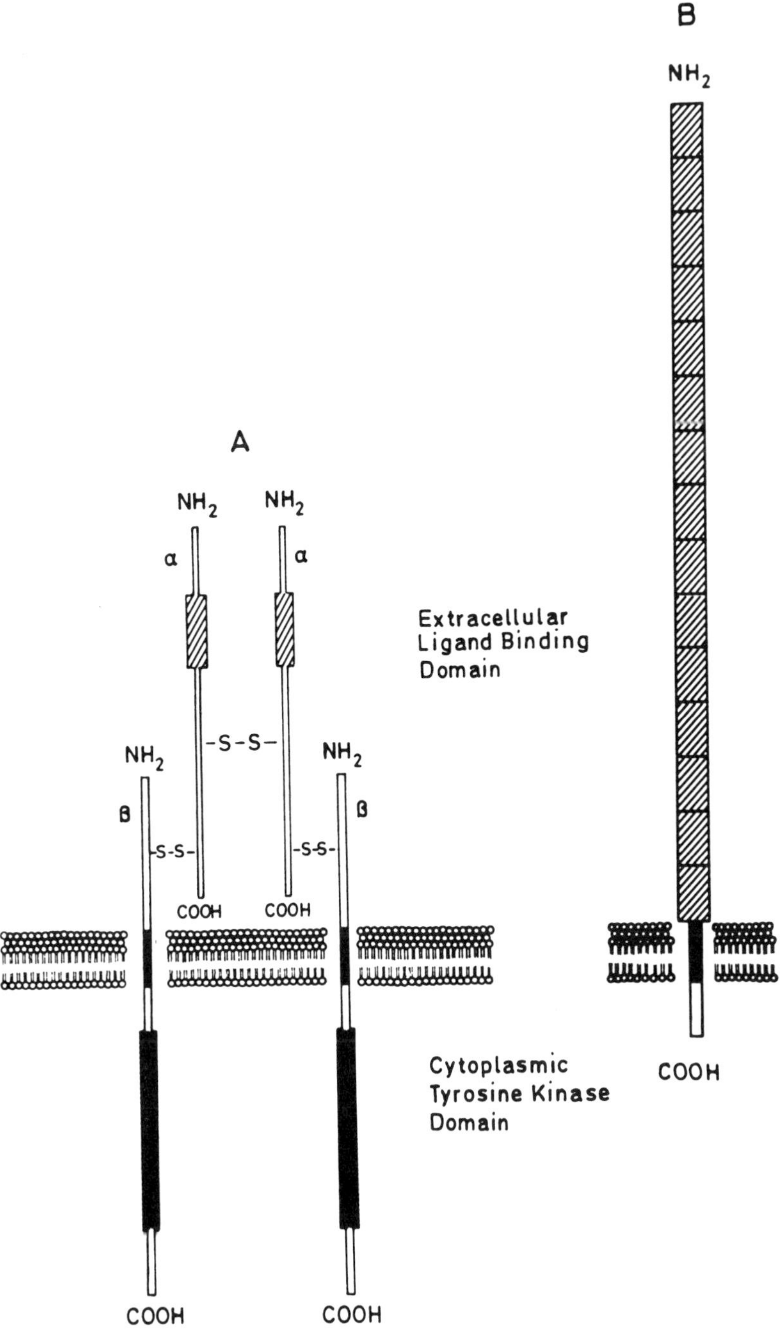

Fig. 1. *Schematic representation of the transmembrane insulin and IGF-I receptors (A), and the IGF-II/Man-6-P receptor (B). Insulin and IGF-I receptors are glycoproteins with an alpha, binding subunit, and a beta, cytoplasmic subunit that contains a tyrosine kinase domain. The IGF-II/Man-6-P receptor is a monomeric transmembrane glycoprotein with no tyrosine kinase domain.*

half-receptor is an active enzyme [reviewed in Frattali et al., 1992].

III. INSULIN-LIKE FACTOR BINDING PROTEINS

There is a well-established association of IGFs with specific binding proteins, which adds a further intricacy in elucidating the roles of the insulin family of peptides. At least six insulin-like growth factor binding proteins (IGFBPs) have been described, and the three well-characterized forms, (IGFBP-1, -2, and -3) have binding affinities for the ligand that are as high, or higher than that of the receptor. Although the precise role played by the IGFBPs is not understood, there is considerable evidence that a major function is the delivery of the IGF to a target tissue [reviewed in Clemmons, 1991]. With respect to their role in early development, there is a report of IGFBP production by porcine blastocysts [Corps et al., 1990], granulosa cells [Mondschein et al., 1990, 1991], and in the bovine endometrium [Geisert et al., 1991]. IGFBP-1 is a protein of molecular weight 28,000 to 32,000 that is found in highest concentrations in human amniotic fluid [Drop et al., 1984], decidual cells, and endometrial cells [Rutanen et al., 1985, 1986]. Thus, it may be the most significant binding protein to be functionally implicated in early mammalian development. However, in a recent high-resolution immunocytochemical electron microscopic study of IGF-I binding to early mouse embryos, binding of the IGF-I ligand did not appear to be modulated by IGFBP-1 in mouse blastocysts [Smith et al., 1993]. Although in the few studies available to date it is not clear how the IGFBPs function in very early mammalian development, a possibility is that IGFBP-1 may act to provide a store of IGF in the environment close to the embryo.

IV. SIGNAL TRANSDUCTION BY RECEPTORS OF THE INSULIN FAMILY OF PEPTIDES

In somatic cells, ligand binding to specific receptors initiates a series of receptor-coupled signal transduction processes that result in cellular activation. These direct signalling responses can include activation of GTP-binding proteins (G-proteins), stimulation of receptor- or non-receptor-associated tyrosine kinase activity, guanylyl cyclase activity, or ion channels. These initial signals can then be amplified to generate multiple second messengers that culminate in cellular responses. Such second messenger systems include activation of adenylyl cyclase, increases in cAMP and cGMP, phosphoinositide turnover, ion fluxes, serine kinase cascades, and changes in intracellular pH. Recent evidence (discussed below) indicates that some receptors containing tyrosine kinase activity may also activate G-protein-mediated signalling pathways. Furthermore, it has been clearly demonstrated that a single receptor can couple to more than one G-protein and that a single G-protein may interact with more than one effector [Ross, 1989; Birnbaumer, 1992]. Activation of certain second messengers such as phospholipase C may be mediated by both tyrosine kinases and G-proteins, although different isoenzymes may be activated by each signal transduction mechanism. These signalling pathways constitute a complicated and closely regulated system of intracellular communication, and it is easy to envision many opportunities for cross talk between different signal transduction systems.

A. Signalling Mechanisms of the Insulin and IGF-I Receptor

The insulin and IGF-I receptors are members of the family of growth factor receptors in which binding of the appropriate ligand activates an intrinsic protein kinase, which in turn activates cellular effector systems. These receptors consist of a large glycosylated, extracellular ligand binding domain; a single, hydrophobic transmembrane region; and a cytoplasmic domain containing a tyrosine kinase catalytic region [Yarden and Ullrich, 1988; Schlessinger, 1988; Williams, 1989]. As a consequence of this structure, the ligand binding domain and the tyrosine kinase activity are separated by the plasma membrane. Therefore, receptor activation by extracellular ligand binding must be translated across the membrane to result in activation of the cytoplasmic domain. In order to produce such an effect, contemporary studies suggest that ligand binding to insulin and IGF-I receptors results in conformational alterations of the extracellular domain. These conformational changes result in the interaction of both $\alpha\beta$ dimers so that α–α and β–β are associated [Boni-Schnetzler et al., 1988; O'Hare and Pilch, 1988] and this interaction is necessary for transmission of the signal and activation of the tyrosine kinase. These receptors, therefore, are capable of transmitting a signal initiated by a conformational change in the extracellular domain to the cytoplasmic domain, without altering the monomeric structures of the transmembrane domains (Fig. 2).

B. Mechanism of Action of the Insulin and IGF-I Receptors

The structure of the insulin receptor has been relatively well studied and structure–function relationships have begun to be elucidated. The IGF-I receptor has been studied in less detail and is frequently described by analogy to the insulin receptor. Extensive investigation concerning insulin receptor function has indicated that the intrinsic tyrosine kinase activity of the receptor β subunit is crucial to mediation of many of the cellular effects of insulin [Yarden and Ullrich, 1988]. Morgan and Roth [1987] demonstrated that injection of antibodies directed against the cytoplasmic domain of the insulin receptor into target cells diminished the effect of insulin on these cells. Subsequent investigations have involved the study of point mutations of specific sites on the receptors. For example, when cells that overexpress insulin receptors were mutated at the ATP-binding site in the tyrosine kinase domain, the kinase activity of the receptors was found to be completely abolished, and the mutated receptors failed to mediate the cellular action of insulin [Chou et al., 1987; Ebina et al., 1987; McClain et al., 1987]. Additional studies have demonstrated that some of the metabolic effects of insulin are blocked if one or more of the three major tyrosine phosphorylation sites are mutated [Ellis et al., 1986; Wilden et al., 1990; Murakami and Rosen, 1991]. In a similar line of research, cells expressing IGF-I receptors were mutated at the ATP-binding site of the tyrosine kinase domain. Subsequent treatment of these cells with the IGF-I ligand resulted in a decreased metabolic rate from that seen in cells expressing normal IGF-I receptors. These studies confirm the role of the tyrosine kinase domain of the insulin and IGF-I receptors in mediating the effects of insulin and IGF-I on the cell.

C. Targets of the Insulin Receptor Tyrosine Kinase

Although it is well established that insulin and IGF-I receptors mediate some or all of their effects by activation of an intrinsic protein tyrosine kinase, most of the cellular targets for phosphorylation remain to be identified. In addition to the receptor itself, insulin treatment of cells results in the tyrosine phosphorylation of numerous endogenous proteins of molecular weights ranging from 15,000 to 250,000 [Roth et al., 1988]. Recently, several of these substrates have been identified. They include a protein termed insulin receptor substrate 1 (IRS1; M_r = 175,000) that seems to be rapidly and directly phosphorylated on tyrosine in response to insulin and IGF-I [Sun et al., 1991; Keller et al., 1991; Rice et al., 1992]. Other identified substrates include the extracellular signal–regulated kinases 1 and 2 (ERK1 and ERK2). These proteins are serine/threonine kinases with molecular weights of 41,000 to 43,000 that are activated by phosphorylation of both their tyrosine and threonine residues [Cobb et al., 1991; Sturgill and Wu, 1991]. However, the activation of ERK1 and ERK2 does not appear to be the result of direct phosphorylation by the receptor [Cobb et al., 1991; Sturgill and Wu, 1991].

Insulin and IGF-I binding also promote dephosphorylation of many proteins, including the receptors themselves [Goldstein, 1992].

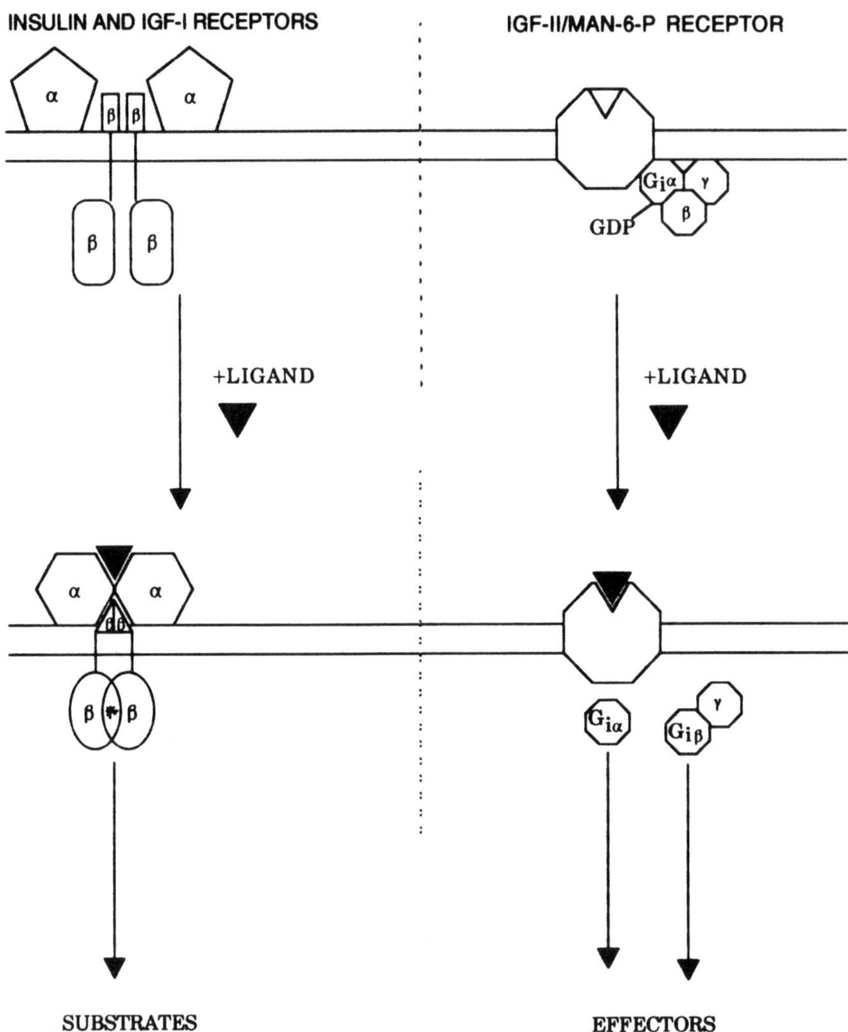

Fig. 2. *Schematic representation of the major signal transduction pathways mediated by activation of the insulin and IGF-I receptors and the IGF-II/Man-6-P receptor. The insulin and IGF-I receptors exist as heterotetrameric proteins. Upon ligand binding the receptor* α *and* β *subunits undergo conformational changes which result in activation of the intrinsic tyrosine kinase. This receptor is autophosphorylated and the tyrosine kinase also phosphorylates various cellular substrates. The IGF-II/Man-6-P receptor is directly coupled to a* G_i *protein. Subsequent to ligand binding, the* $G_{i\alpha}$ *subunit separates from the* $G_{i\beta\gamma}$ *subunits, which remain tightly associated. Each of these subunits activates effector systems resulting in cellular activation. * indicates activated tyrosine kinase.*

Phosphorylation/dephosphorylation cascades may mediate much of the action of insulin, although other pathways may involve phosphoinositide turnover and protein kinase C activation [Exton, 1991; Roth et al., 1992; Sung, 1992]. Currently, the precise mechanisms mediating cellular activation in response to insulin and IGF-I are unknown.

D. Evidence for G Protein Involvement in Insulin and IGF-I Signal Transduction

In addition to tyrosine kinases, another group of signal-transducing proteins that regulates intracellular second messenger systems has been identified. This is a superfamily of GTP-binding proteins that includes translational factors, tubulins, low molecular weight, monomeric G-proteins (including the protooncogenic *Ras* and *Ras*-related proteins, reviewed by Hamilton in Chapter 6, this volume) and heterotrimeric G-proteins [Gilman, 1987; Casey and Gilman, 1988; Kaziro et al., 1991].

Heterotrimeric G-proteins are classical signal transducers found in hormonally responsive somatic cells. They transmit signals derived from external domain ligand–receptor interaction through the plasma membrane to activate a number of second messenger systems. Heterotrimeric G-proteins consist of three subunits α (M_r = 39,000–52,000), β (M_r = 35,000–36,000), and γ (M_r = 7,000–10,000). The alpha subunit family is the most diverse, consisting of products of at least 16 different genes and 30 different cDNAs, including those that code for the $G_{s\alpha}$, $G_{i\alpha}$, $G_{q\alpha}$, and $G_{12\alpha}$ classes [Simon et al., 1991]. At the present time, four genes have been identified that encode beta subunits, and at least three encoding gamma subunits [Simon et al., 1991]. Classes of heterotrimeric G-proteins have been designated according to the different G_α subunits, and all the current evidence suggests that they may preferentially associate with distinct $G_{\beta\gamma}$ subunits [Strittmatter et al., 1990]. To date, three distinct $G_{i\alpha}$s have been identified, $G_{i\alpha1}$, $G_{i\alpha2}$, and $G_{i\alpha3}$. These G-proteins demonstrate 85% sequence homology and are products of distinct genes [Itoh et al., 1986; Itoh et al., 1988]. G-proteins regulate a host of enzymes including adenylyl cyclase, phospholipase C, phospholipase A_2, and retinal phosphodiesterase [Birnbaumer et al., 1990; Kaziro et al., 1991]. Furthermore, G_α and $G_{\beta\gamma}$ proteins have also been shown to couple directly to ion channels, including Ca^{2+} and K^+ channels [Brown and Birnbaumer, 1990].

G-proteins transmit signals across the membrane by alternating between both activation and deactivation cycles [Gilman, 1987] (see Fig. 2). Inactive G-proteins exist in the heterotrimeric, $\alpha\beta\gamma$ state in which G_α is GDP bound. G-proteins are believed to associate with membrane receptors by means of isoprenylation of G_γ [Yamane et al., 1990; Mumby et al., 1990; Fukada et al., 1990]; therefore, the G-protein in its heterotrimeric form is membrane bound. Activation of cell surface receptors causes the exchange of GTP for GDP bound to receptor-coupled G_α. The G_α-GTP then dissociates from the $G_{\beta\gamma}$, which remains tightly associated. G_α and $G_{\beta\gamma}$ can both subsequently activate intracellular effector systems, thus constituting early generation of multiple signal transducers for signal amplification. The intrinsic GTPase activity of G_α results in hydrolysis of GTP to GDP and causes release of the G_α-GDP from the effector. $G_{\beta\gamma}$ has a high affinity for G_α-GDP, so the subunits reassociate, thus restoring the GDP-$G_{\alpha\beta\gamma}$, inactive, holomeric form. Therefore, regulation of G-protein-mediated activation of cellular effector systems occurs by receptor activation of the G-protein and results in an increased rate of intrinsic G_α-GTPase activity.

Bacterial toxins have proven instrumental in the identification, localization, and function of G-proteins. Two toxins used to study heterotrimeric G-proteins are cholera toxin and pertussis toxin (PTX) [Ui, 1990]. These toxins consist of A and B subunits. The B subunit causes penetration of cellular membranes, and the A subunit catalyzes covalent transfer of an ADP-ribose group from intracellular NAD to G_α in a reaction termed ADP-ribosylation. By addition of [^{32}P]NAD, the modified G_α can be radiolabelled and identified. Pertussis toxin is specific for the G_i family of G-proteins (including G_0 and G_t) and uncouples G_i from its receptor, resulting in functional inactivation of G_i so that no second messengers are generated. G_i is sometimes referred to as a PTX substrate.

Although it appears that receptors with intrinsic tyrosine kinase activity cannot be directly coupled to heterotrimeric G-proteins, recent evidence suggests that some of the cellular effects of growth factors may be mediated by G-proteins as well. For example, a PTX-sensitive G-protein has been implicated in mediating the induction of Na^+ influx by colony-stimulating factor 1 [Imamura and Kufe, 1988], and in epidermal growth factor (EGF)-mediated phospholipid metabolism in hepatocytes [Yang et al., 1991, 1993]. Furthermore, Liang and Garrison [1991] demonstrated EGF receptor activation of a PTX-sensitive G-protein in hepatocytes, but not in cultured cells which overexpress the EGF receptor.

There is also an emerging body of evidence implicating G-proteins in mediating insulin action. Insulin binding to myocyte membranes was inhibited by high concentrations of the poorly hydrolyzable guanine nucleotide, guanosine 5′-*O*-(3-thiotriphosphate) ($GTP_\gamma S$) [Luttrell et al., 1990]. These investigators also demonstrated that incubation with insulin alters $GTP_\gamma S$ binding to membranes from myocytes. These effects were not seen in liver or adipocyte membranes, although $GTP_\gamma S$ did inhibit insulin-induced tyrosine phosphorylation of several proteins in these cells [Burdett et al., 1990; Davis and McDonald, 1990]. Russ et al. [1992] have recently demonstrated that treatment of partially purified insulin receptors from cardiomyocytes with [^{35}S]GTP increased receptor autophosphorylation; this suggests G-protein involvement in regulation of tyrosine kinase activity. Pertussis toxin has also been demonstrated to alter insulin binding and to inhibit some of insulin's effects in myocytes [Luttrell et al., 1988, 1990] and adipocytes [Rothenberg and Kahn, 1988; Ciaraldi and Maisel, 1989], thereby implicating a role for G_i in the regulation of a portion of the insulin receptor activity. Recent evidence has also been presented to indicate that treatment of adipocytes with insulin inhibits PTX-catalyzed ADP-ribosylation of a protein with molecular weight 41,000 [Jo et al., 1992], and may unmask the carboxy terminus of G_i [Record et al., 1993]. Furthermore, the results of several studies suggest that the insulin receptor can phosphorylate certain G_α subunits [Zick et al., 1988]; O'Brien et al., 1987; Krupinski et al., 1988], although these findings were not reproducible in whole cells [Rothenberg and Kahn, 1988]. These data suggest a direct interaction between insulin and IGF-I receptors, and heterotrimeric G-proteins, which would represent a novel and exciting signal transduction pathway. However, further detailed investigations, especially those demonstrating direct cellular G-protein activation in response to treatment with insulin and IGF-I, are required to link insulin and IGF-I receptors directly to the activation of G-proteins. Recent evidence also implicates *Ras* and *Ras*-related monomeric G-proteins in mediating the cellular action of insulin. It has been suggested that insulin receptor activation of the mitogen-activated and extracellular-regulated protein kinases (MAP/ERK kinases) may occur via the monomeric G-proteins. These pathways are reviewed in this volume by Hamilton (Chapter 6), by Terada et al. (Chapter 4) for interleukin-2 receptor signalling, and by Liebl and Hoffman (Chapter 8) in the context of Drosophila development.

E. Insulin-Like Growth Factor II

The polypeptide IGF-II is structurally related to insulin and IGF-I. It also exerts its cellular effects by means of a cell surface receptor that binds IGF-II with high affinity [Tally et al., 1987; Blanchard et al., 1988; Rogers and Hammerman, 1988]. Interestingly, this IGF-II/Man-6-P receptor appears to utilize a different mechanism for signal transduction from that described for the insulin or IGF-I receptors. The IGF-II/Man-6-P receptor does not contain intrinsic protein kinase activity and apparently does not undergo receptor oligomerization in response to treatment with IGF-II or Man-6-P. Nishimoto et al. [1987] demonstrated that the IGF-II/Man-6-P receptor interacts functionally with a G_i protein to activate a Ca^{2+}-permeable channel, thereby mediating Ca^{2+} influx in mouse 3T3 fibroblast cells. Further studies with reconstituted vesicles containing purified rat and human IGF-II/Man-6-P receptors demonstrated that these receptors directly couple to G_i [Okamoto et al., 1990; Murayama et al., 1990; Nishimoto et al., 1989]. These findings led to the identification of a 14-residue peptide of the human IGF II/Man-6-P receptor which directly activates G_i and selectively couples to G_{i2} over G_{i1} and G_{i3} [Okamoto et al., 1990].

V. ROLE OF THE INSULIN FAMILY OF GROWTH FACTORS IN DEVELOPMENT

Signals that govern the processes of cellular proliferation and differentiation in early embryos of vertebrates are poorly understood, although growth factors have received increasing attention during the past decade as putative effectors. While there is much information available regarding the action of these peptides during organogenesis [see recent reviews; Mercola and Stiles, 1988; Mummery and van den Eijnden-van Raaji, 1990; Cross and Dexter, 1991] there is little information regarding the earliest stages of development.

Insulin and the IGFs are among the most promising candidate growth factors for actions in early development, due to their widespread effects on cellular metabolism and phylogenetic distribution. Thus, insulin shows extraordinary evolutionary conservation, and insulin, or an insulin-like molecule has been detected in a number of invertebrate species, as well as in a number of lower vertebrates [see review by Mattson et al., 1989]. Studies in amphibia have revealed that both insulin and IGF-I stimulate maturation in *Xenopus* oocytes, apparently via stimulation of phosphodiesterase activity, with a concomitant inhibition of adenylate cyclase activity [Sadler and Maller, 1987]. Similarly, insulin has been shown to be required for newt limb regeneration, with a concomitant increase in the levels of cGMP over control levels, whereas cAMP levels remained unchanged [Kesik et al., 1986].

In chicken eggs, both the yolk and the white contain molecules immunologically related to insulin and IGF-I [reviewed in de Pablo, 1989]. In chick embryos, both insulin and IGF-I, acting through their cognate receptors, have been shown to stimulate metabolism, growth, and differentiation at stages 10–14 of Hamburger and Hamilton [Girbau et al., 1987]. In contrast, although the IGF-II ligand shows a binding profile that is remarkably parallel to that of [^{125}I]-IGF-I, to date, a receptor that possesses the peptide binding and cross-linking characteristics of the IGF-II receptor has not been detected in early avian embryos [de Pablo, 1989]. Other studies in the avian embryo have shown that IGF-I receptors are expressed earlier, and are significantly higher in concentration than insulin receptors at this stage [Bassas et al., 1985].

The earliest stages of mammalian embryogenesis eluded detailed investigation until a number of studies showed that these embryos can be cultured *in vitro* in simple media. The accepted paradigm for the study of mammalian preimplantation embryos *in vitro* has been the mouse. By means of superovulation techniques, mouse embryos can be recovered from the reproductive tract at defined stages (Fig. 3) and cultured *in vitro* in a medium consisting of a balanced salt solution, an energy source, and a source of nitrogen. The ability of mammalian preimplantation embryos to develop to the blastocyst stage in simple media suggests that they are autonomous. On the other hand, the well-documented observation that embryos *in vitro* lag behind their *in vivo* siblings, first documented by Bowman and McLaren [1970], suggests that the maternal tract also contributes a factor(s) enhancing successful preimplantation development, i.e., development that results in a viable pregnancy. Similarly, embryos from a variety of species cultured *in vitro* exhibit blocks to development. For example, in the mouse, it is difficult

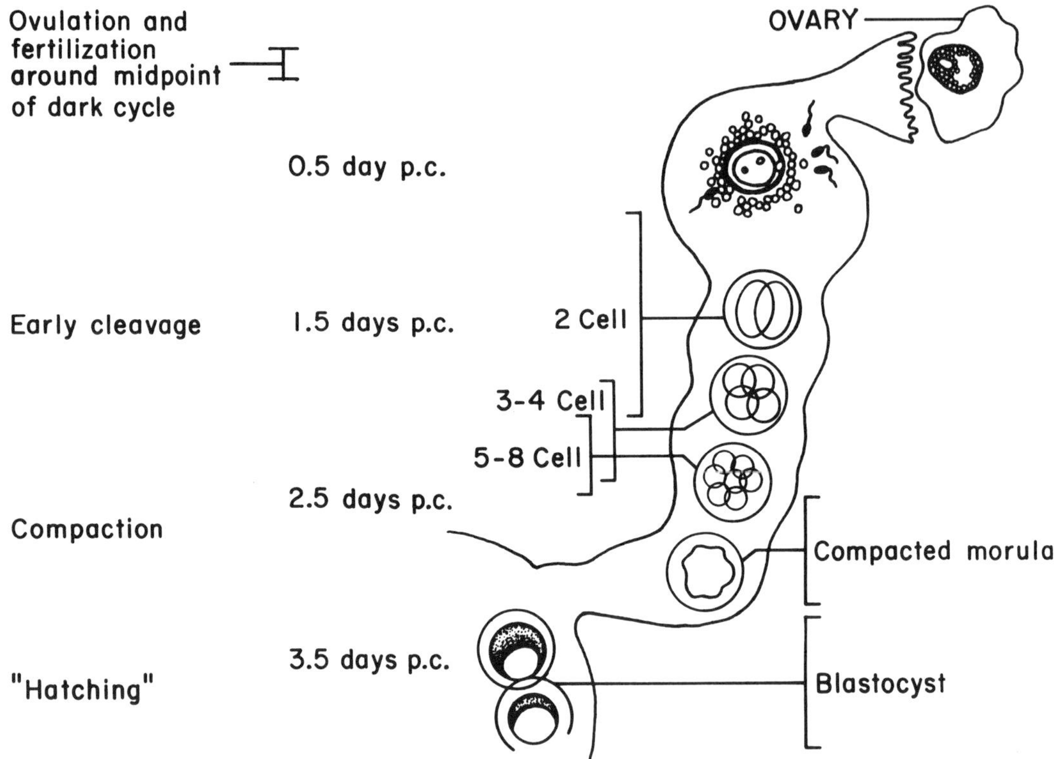

Fig. 3. *Schematic representation of preimplantation developmental stages in the mouse. Cleavage stages are shown in the oviduct, and the blastocyst is shown hatching from the zona pellucida in the uterus, prior to implantation.*

to culture zygotes from outbred strains past the 1-cell stage. This is referred to as the 2-cell block. In a like manner, cow and sheep embryos exhibit a block at the 8- to 16-cell stage. In general, these blocks correspond to the stage at which the embryonic genome takes over from the maternal messages in the embryo. While contributions from the maternal tract in the form of co-culture with cells, or addition of washings from the reproductive tract act to regulate embryonic development such that it approximates the *in vivo* situation [reviewed in Heyner et al., 1993], the precise factors involved remain poorly defined. However, evidence has been accumulating during the past decade that growth factors, with their well-documented mitogenic effects would fit the role of maternal mediators of embryonic development [Heyner et al., 1993; Schultz and Heyner, 1992], and some at least, may play crucial roles in the earliest stages of mammalian development. Among the first studies to support the concept of growth factor involvement in early mammalian development, Rosenblum et al. [1986] observed that insulin binds to the surface of morula and blastocyst stage mouse embryos in a stage-specific manner. Subsequently, Mattson et al. [1988] used autoradiography to demonstrate receptor-mediated binding of insulin to the cells of the morula, but not to earlier stages of the mouse preimplantation embryo. In addition, the expression of receptors binding ^{125}I-labelled IGF-I and IGF-II was examined on blastocyst outgrowths, with specific binding demonstrated. Therefore, the authors concluded that mouse preimplantation embryos express insulin receptors in a developmentally regulated manner. Furthermore, at the periimplantation stage, receptors for IGFs could also be visualized on the embryos. These results were confirmed at the mRNA level, using an mRNA phenotyping technique [Rappolee et al., 1988] to show that transcripts encoding the insulin and IGF-I receptor became detectable between the 8-cell and morula stage of development in the mouse, while the IGF-II receptor was detectable at the 2-cell stage [Heyner et al., 1989a; Rappolee et al., 1992].

The functional role of the insulin family of peptides in preimplantation mammalian embryos appears to parallel that of insulin and IGF-I in the chick. Thus, Harvey and Kaye [1988] showed that insulin stimulates protein synthesis in *in vitro* compacted mouse embryos at the 8-cell stage, while Heyner et al. [1989b] demonstrated significant increases in both DNA and RNA synthesis in mouse embryos exposed to physiological levels of insulin as compared to controls.

While the role of insulin concerning protein synthesis is clearly established, that of the IGFs is less well defined. In a comparison of the effects of IGFs and insulin on protein synthesis at the blastocyst stage, neither IGF-I nor IGF-II significantly increased nucleic acid synthesis or incorporation of radiolabelled methionine into the acid insoluble protein fraction of mouse embryos [Rao et al., 1990]. In contrast, Harvey and Kaye [1991] reported a significant increase in incorporation of labelled leucine into blastocysts cultured in the presence of IGF-I. Rappolee et al. [1992] also examined the metabolic effects of insulin and IGFs on preimplantation mouse embryos *in vitro,* and described significant effects of both insulin and IGF-II on radiolabelled methionine incorporation as compared with controls, or with embryos cultured in the presence of IGF-I. Thus, while data on the positive effect of insulin are consonant from a number of laboratories, there remain some differences between laboratories regarding the effects of IGF-I and IGF-II on embryonic metabolism.

A number of investigators have examined the role of insulin and the IGFs on cellular proliferation during preimplantation development. In an earlier study, Paria and Dey [1990] were unable to detect a proliferative effect of IGF-I on mouse preimplantation embryos cultured *in vitro*. Subsequently, both IGFs and insulin have been reported to increase cell numbers in the blastocyst [Harvey and Kaye, 1992; Rappolee et al., 1992; Smith et al., 1993]. However, while insulin acts to significantly increase total blastocyst numbers, the IGFs' effects are restricted to the inner cell mass [Smith et al., 1993], as shown in Table I.

TABLE I. Cell Numbers of Blastocytes Grown in the Presence or Absense of Peptides of the Insulin Family[†]

Culture condition	Embryos (n)	Mean total cell number	Mean ICM cell number
Control	32	67.75±2.26	22.78±0.83
Insulin	24	80.08±3.76*	29.83±1.99**
IGF-I	30	70.20±2.05	30.17±1.49**
IGF-II	33	72.58±2.29	28.52±1.19**

[†]Cell numbers in embryos grown in medium alone (control) or in medium containing 40 ng/ml of insulin IGF-I, and IGF-II, respectively, were evaluated as described by Smith et al. (1993). The data are expressed as means ± sem. Experiments were analyzed using one factor ANOVA.
*The total cell number in embryos grown in the presence of insulin was significantly different from controls and from IGF-I and IGF-II-treatments ($P<0.05$).
**indicates that the mean cell number in the inner cell mass (ICM) in embryos grown in the presence of insulin, IGF-I and IGF-II respectively was significantly different from controls ($P<0.05$).
Reproduced from Smith et al., 1993, Biology of Reproduction, Academic Press, New York; with permission.

Although when taken together, these data appear to be confusing at first glance, a recent study has revealed that, whereas earlier studies showed that IGF-I mRNA is not expressed in preimplantation mouse embryos, reverse transcription polymerase chain reaction (RT-PCR) analysis carried out under different technical conditions has shown that IGF-I mRNA is elaborated by the preimplantation mouse embryo [R.M. Schultz, personal communication]. Thus, because the embryo produces endogenous IGF-I, addition of exogenous growth factor is likely to have varying effects on cellular proliferation.

Preimplantation development in vitro is also influenced by IGF-II. Thus, addition of IGF-II increases cell numbers in mouse embryos cultured to the blastocyst stage [Rappolee et al., 1990; Smith et al., 1993]. Conversely, culture in the presence of antisense IGF-II oligonucleotides retards the rate of development to the blastocyst stage, with a concomitant reduction in cell number at the blastocyst stage as compared with controls [Rappolee et al., 1992].

Additional evidence for the role of insulin in the development of preimplantation mammalian embryos comes from studies of blastocysts derived from streptozocin-induced diabetic mice and rats. Blastocysts collected from diabetic mice contained significantly fewer cells, and displayed a reduced protein synthetic rate, as measured by the incorporation of [^{3}H]leucine, than controls [Beebe and Kaye, 1991]. In diabetic rats, embryos also lagged developmentally, and contained fewer cells [Pampfer, 1991], while an examination of cell populations revealed that blastocyst-stage rat embryos contained fewer cells in the inner cell mass than controls from non-diabetic dams [Vercheval et al., 1990]. There are few data regarding the role of growth factors of the insulin family in mammalian species other than the mouse or rat. Estrada et al. [1991] cultured preimplantation porcine embryos in the presence of exogenous IGF-I. These authors demonstrated that the uptake of [^{35}S]methionine did not differ significantly from controls; however, specific ligand-induced peptides could be detected after analysis by 2D gel electrophoresis. The significance of these data remains to be evaluated. Lewis et al. [1992] showed that increased rates of protein synthesis and rate of increase in the size of the porcine blastocyst could be demonstrated in response to exogenously added insulin. Similarly, in related studies in the bovine, Zhang et al. [1991] showed that exogenously added insulin enhanced maturation and development of *in vitro* fertilized oocytes.

VI. SUMMARY

Cellular proliferation and differentiation of mammalian preimplantation embryos seem to depend in part upon growth factors, whether they be endogenous or exogenous. In the case of endogenous factors, there is evidence for autocrine circuits, see, for example, transforming growth factor alpha (TGF-α) as documented by Dardik et al. [1992]. In this study, the authors noted that TGF-α in the embryo is localized to the inner cell mass and adjacent trophectoderm cells, whereas the epidermal growth factor receptors (which bind TGF-α) are juxtacoelic. The localization of ligand and receptor suggest strongly that the growth factor plays a role in regulating the proliferation and differentiation of cells of the inner cell mass and polar trophectoderm. The mechanism for effecting this regulation may be production of TGF-α by the cells of the inner cell mass and binding of the factor to the receptors on the basolateral surface of the adjacent polar trophectoderm cells with resultant paracrine stimulation [Dardik et al., 1992]. This is of particular significance in studies of mammalian embryos at the blastocyst stage. The trophectoderm provides an impermeable seal [Ducibella et al., 1975], and therefore forms a microenvironment for the internal, basolateral surface of the trophectoderm, as well as for the inner cell mass. Endogenously produced factors therefore have the ability to stimulate cellular proliferation and/or differentiation by means of autocrine or paracrine circuits.

In general, a functional role in early development has been ascribed to a growth factor if that factor has been shown to enhance development after addition to medium for the culture of embryos. Signalling by means of transmembrane receptors is the mechanism whereby the growth factors exert their effects, although their signalling mechanisms have not been fully defined. Thus, while the cascade of events triggered by insulin or IGF-I binding to the external domain is known to involve protein phosphorylation, and the signalling by IGF-II involves interactions with heterotrimeric G-proteins, precise details are unknown. An added complication is that there seem to be a number of gene products within several growth factor families that enhance preimplantation development when added to embryos *in vitro*. It is not clear whether these act in concert, or form a multiplicity of products with similar functions. These questions will not be answered until a more definitive description of signalling mechanisms in the early embryo can be obtained.

In summary, the early embryo expresses ligands and receptors of the insulin family of peptides. IGF-II is endogenously produced, and recent evidence suggests that IGF-I is also an endogenous factor. The receptors that bind these embryonic ligands are characteristically located in a basolateral position in the blastocyst. In contrast, insulin is present in the maternal tract, and has not been detected in preimplantation mammalian embryos. Receptors that bind insulin are located on the apical surface of the trophectoderm, a position in which they will be bathed by the oviductal fluid, which contains insulin. Regulatory pathways at autocrine, juxtacrine, and paracrine levels govern early development. Despite a significant amount of knowledge concerning particular growth factors and receptors, there is much to be discovered regarding their interactions and regulatory circuits.

ACKNOWLEDGMENTS

Studies from the laboratories of the authors have been supported by grants from the National Institutes of Health (HD 23511 to SH and HD 00972 to CRW) and from the Lalor Foundation (CRW). We are appreciative of postdoctoral support (RND) from the Rockefeller Foundation. The authors thank Robert M. Smith for critical comments on the manuscript.

REFERENCES

Bassas L, dePablo F, Lesniak MA, Roth J (1985): Ontogeny of receptors for insulin-like peptides in chick embryo tissues: early dominance of insulin-like growth factors over insulin receptors in brain. Endocrinology 117:2321–2329.

Beebe LF, Kaye PL (1991): Maternal diabetes and retarded preimplantation development of mice. Diabetes 40:457–461.

Bhaumick B, Bala RM (1987): Receptors for insulin-like growth factors I and II in developing embryonic mouse limb bud. Biochim Biophys Acta 927:117–128.

Birnbaumer L (1992): Receptor-to-effector signaling through G proteins: roles for βγ dimers as well as α subunits. Cell 71:1069–1072.

Birnbaumer L, Abramowitz J, Yatani A, Okabe K, Mattera R, Graf R, Sanford J, Codina J, Brown AM (1990): Roles of G proteins in coupling of receptors to ionic channels and other effector systems. Crit Rev Biochem Mol Biol 25:225–244.

Blanchard MM, Barenton B, Sullivan A, Foster B, Guyda HJ, Posner BI (1988): Characterization of the insulin-like growth factor (IGF) receptor in K562 erythroleukemia cells, evidence for a biological function for the type II IGF receptor. Mol Cell Endocrinol 56:235–244.

Blundell TL, Humbel RE (1980): Hormone families: pancreatic hormones and homologous growth factors. Nature 287:781–787.

Boni-Schnetzler M, Kaligian A, DelVecchio R, Pilch PF (1988): Ligand-dependent intersubunit communication within the insulin receptor complex activates its intrinsic kinase activity. J Biol Chem 263:6822–6828.

Bowman P, McLaren A (1970): Cleavage rate of mouse embryos in vivo and in vitro. J Embryol Exp Morphol 24:203–207.

Braulke T, Causin C, Waheed A, Junghans U, Hasilik A, Masy P, Humbel RE, von Figura K (1988): Mannose-6-phosphate/insulin-like growth factor II receptor: distinct binding sites for mannose-6-phosphate and insulin-like growth factor II. Biochem Biophys Res Commun 150:1287–1293.

Brown AM, Birnbaumer L (1990): Ionic channels and their regulation by G protein subunits. Annu Rev Physiol 52:197–213.

Burdett E, Mills GB, Klip A (1990): Effect of $GTP_{\gamma}S$ on insulin binding and tyrosine phosphorylation in liver membranes and L6 muscle cells. Am J Physiol 258:C99–C108.

Casey PJ, Gilman AG (1988): G protein involvement in receptor-effector coupling. J Biol Chem 263:2577–2580.

Chou CK, Dull TJ, Russell DS, Gherzi R, Lebwohl D, Ullrich A, Rosen OM (1987): Human insulin receptors mutated at the ATP-binding site lack protein tyrosine kinase activity and fail to mediate postreceptor effects of insulin. J Biol Chem 262:1842–1847.

Ciaraldi TP, Maisel A (1989): Role of guanine nucleotide regulatory proteins in insulin stimulation of glucose transport in rat adipocytes. Biochem J 264:389–396.

Clemmons DR (1991): Insulin-like growth factor binding proteins. In LeRoith D (ed): Insulin-like Growth Factors: Molecular and Cellular Aspects. Boca Raton: CRC Press, pp 151–180.

Cobb MH, Boulton TG, Robbins DJ (1991): Extracellular signal-related kinases: ERK in progress. Cell Regul 2:965–978.

Corps AN, Brigstock DR, Littlewood CJ, Brown KD (1990): Receptors for epidermal growth factor and insulin-like growth factor-I on preimplantation trophectoderm of the pig. Development 110:221–227.

Cross M, Dexter TM (1991): Growth factors in development, transformation and tumorigenesis. Cell 64: 271–280.

Czech MP (1989): Signal transduction by the insulin-like growth factors. Cell 59:235–238.

Dardik A, Smith RM, Schultz RM (1992): Colocalization of transforming growth factor-α and a functional epidermal growth factor receptor (EGFR) to the inner cell mass and preferential localization of the EGFR on the basolateral surface of the trophectoderm in the mouse blastocyst. Dev Biol 154:396–409.

Davis HW, McDonald JM (1990): Insulin receptor function is inhibited by guanosine 5′-[γ-thio]triphosphate (GTP[γS]). Biochem J 270:401–407.

DeChiara TM, Efstratiadis A, Robertson EJ (1990): A growth deficiency phenotype in mice carrying an insulin-like growth factor II gene disrupted by targeting. Nature 345:78–80.

DeChiara TM, Robertson EJ, Efstratiadis A (1991): Parental imprinting of the mouse insulin-like growth factor II gene. Cell 64:849–859.

dePablo F (1989): Insulin and insulin-like growth factors (IFGs) in avian development. In Rosenblum IY, Heyner S (eds): Growth Factors in Mammalian Development. Boca Raton: CRC Press, pp 71–90.

Drop SLS, Kortleve DJ, Guyda HJ (1984): Isolation of a somatomedin-binding protein from preterm amniotic fluid: development of a radioimmunoassay. J Clin Endocrinol Metab 59:899–907.

Ducibella T, Albertini DF, Anderson E, Biggers JD (1975): The preimplantation mouse embryo: characterization of intercellular junctions and their appearance during development. Dev Biol 45:231–250.

Ebina Y, Ellis L, Jarnagin K, Edery M, Graf L., Clauser E, Qu J, Masiarz F, Kan YW, Goldfine I, Rother RA, Rutter WJ (1985): The human insulin receptor cDNA: the structural basis for hormone-activated transmembrane signalling. Cell 40:747–758.

Ebina Y, Araki E, Taira M, Shimada F, Mori M, Craik CS, Siddle K, Pierce SB, Roth A, Rutter WJ (1987): Replacement of lysine residue 1030 in the putative ATP-binding region of the insulin receptor abolishes insulin- and antibody-stimulated glucose uptake and receptor kinase activity. Proc Natl Acad Sci USA 84:704–708.

Ellis L, Clauser E, Morgan DO, Edery M, Roth RA, Rutter WJ (1986): Replacement of insulin receptor tyrosine residues 1162 and 1163 compromises insulin-stimulated kinase activity and uptake of 2-deoxyglucose. Cell 45:721–732.

Estrada JL, Jones EE, Johnson BH, Petters RM (1991): Effect of insulin-like growth factor I on protein synthesis in porcine embryonic discs cultured in vitro. J Reprod Fertil 93:53–61.

Ewton DZ, Falen SL, Florini JR (1987): The type-II insulin like growth factor (IGF) receptor has a low affinity for IGF-I analogs: pleiotropic actions of IGF on myoblasts are apparently mediated by the type I receptor. Endocrinology 120:115–12.

Exton JH (1991): Perspectives in diabetes: some thoughts on the mechanism of action of insulin. Diabetes 40:521–526.

Frattali AL, Treadway JS, Pessin JF (1992): Insulin/IGF-I hybrid receptors: implications for the dominant-negative phenotype in syndromes of insulin resistance. J Cell Biochem 48:43–50.

Fukada Y, Takao, T, Ohguro H, Yoshizawa T, Akino T (1990): Farnesylated gamma subunit of the photoreceptor G protein is indispensable for GTP binding. Nature 346:658–660.

Geisert RD, Lee C-Y, Simmen FA, Zavy MY, Fliss AE, Bazer FW, Simmen RC (1991): Expression of messenger RNAs encoding insulin-like growth factor-I, -II and insulin-like growth factor binding protein-2 in bovine endometrium during the estrous cycle and early pregnancy. Biol Reprod 45:975–983.

Gilman AG (1987): G proteins: transducers of receptor-generated signals. Annu Rev Biochem 56:615–649.

Girbau M, Gomez JA, Lesniak MA, dePablo F (1987): Insulin and insulin-like growth factor I both stimulate metabolism, growth, and differentiation in the post-neurula chick embryo. Endocrinology 121:1477–1482.

Goldstein BJ (1992): Protein-tyrosine phosphatases and the regulation of insulin action. J Cell Biochem 48:33–42.

Harvey MB, Kaye PL (1988): Insulin stimulates protein synthesis in compacted mouse embryos. Endocrinology 122:1182–1184.

Harvey MB, Kaye PL (1991): Mouse blastocysts respond metabolically to short term stimulation by insulin and IGF-I through the insulin receptor. Mol Reprod Dev 29:253–258.

Harvey MB, Kaye PL (1992): Insulin like growth factor-I stimulates growth of mouse preimplantation embryos in vitro. Mol Reprod Dev 31:195–199.

Heyner S, Rao LV, Jarett L, Smith RM (1989a): Preimplantation mouse embryos internalize maternal insulin via receptor-mediated endocytosis: pattern of uptake and functional correlations. Dev Biol 134:48–58.

Heyner S, Smith RM, Schultz GA (1989b): Temporally regulated expression of insulin and insulin-like growth factors and their receptors in early mammalian development. Bioessays 11:171–178.

Heyner S, Shah N, Smith RM, Watson AJ, Schultz GA (1993): The role of growth factors in embryo production. Theriogenology 39:151–161.

Imamura K, Kufe D (1988): Colony-stimulating factor 1-induced Na^+ influx into human monocytes involves activation of a pertussis toxin-sensitive GTP-binding protein. J Biol Chem 263:14093–14098.

Itoh H, Kozasa T, Nagata S, Nakamura S, Katada T, Ui M, Iwai S, Ohtsuka E, Kawasaki H, Suzuki K (1986): Molecular cloning and sequence determination of cDNAs for alpha subunits of the guanine nucleotide binding proteins Gs, Gi, and G_o from rat brain. Proc Natl Acad Sci USA 83:3776–3780.

Itoh H, Toyama R, Kozasa T, Tsukamoto T, Matsuoka M, Kaziro Y (1988): Presence of 3 distinct molecular species of G_i protein alpha subunits. Structure of rat cDNAs and human genome DNAs. J Biol Chem 263:6656–6664.

Jo H, Cha BY, Davis HW, McDonald JM (1992): Identification, partial purification, and characterization of two guanosine triphosphate-binding proteins associated with insulin receptors. Endocrinology 131: 2855–2862.

Kaziro Y, Itoh H, Kozasa T, Nakafuku M, Satoh T (1991): Structure and function of signal-transducing GTP-binding proteins. Annu Rev Biochem 60:349–400.

Keller SR, Kitagawa K, Aekersold R, Leinhard GE, Garner CW (1991): Isolation and characterization of the 160,000Da phosphotyrosyl protein, a putative participant in insulin signaling. J Biol Chem 266:12817–12820.

Kesik A, Vethamany-Globus S, Globus M (1986): Effect of insulin on cyclic nucleotide levels and promotion of mitosis by insulin and ionophore A23187 in cultured newt blastema. In Vitro Cell Dev Biol 22:465–468.

Kiess W, Haskell JF, Lee L, Greenstein LA, Miller BE, Aarons AL, Rechler MM, Nissley SP (1987): An antibody that blocks insulin-like growth factor (IGF) binding to the type II receptor is neither an agonist nor an inhibitor of IGF-stimulated biologic responses in L6 myoblasts. J Biol Chem 262:12745–12751.

Kiess W, Blickenstaff GD, Sklar MM, Thomas CL, Nissley SP, Sahagian GG (1988): Biochemical evidence that the type II insulin-like growth factor receptor is identical to the cation-independent mannose-6-phosphate receptor. J Biol Chem 263:9339–9344.

Krupinski J, Rajaram R, Lakonishok M, Benovic JL, Cerione RA (1988): Insulin-dependent phosphorylation of GTP-binding proteins in phospholipid vesicles. J Biol Chem 263:12333–12341.

Lewis AM, Kaye PL, Lising R, Cameron RDA (1992): Stimulation of protein synthesis and expression of pig blastocyst during early conceptus development in the pig. Reprod Fertil Dev 4:119–123.

Liang M, Garrison JC (1991): The epidermal growth factor receptor is coupled to a pertussis toxin-sensitive guanine nucleotide regulatory protein in rat hepatocytes. J Biol Chem 266:13342–13349.

Lobel P, Dahms NM, Kornfeld S (1988): Cloning and sequence analysis of the cation-independent mannose-6-phosphate receptor. J Biol Chem 263:2563–2570.

Lowe WL (1991): Biological actions of the insulin-like growth factors. In LeRoith D (ed): Insulin-like Growth Factors: Molecular and Cellular Aspects. Boca Raton: CRC Press, pp 49–86.

Luttrell LM, Hewlett EL, Romero G, Rogol AD (1988): Pertussis toxin treatment attenuates some effects of insulin in BC_3H-1 murine myocytes. J Biol Chem 263:6134–6141.

Luttrell L, Kilgour E, Larner J, Romero G (1990): A pertussis toxin-sensitive G protein mediates some aspects of insulin action in BC_3H-1 murine myocytes. J Biol Chem 265:16873–16879.

Mattson BA, Rosenblum IY, Smith RM, Heyner S (1988): Autoradiographic evidence for insulin and insulin-like growth factor binding to early mouse embryos. Diabetes 37:585–589.

Mattson BA, Chambers SA, dePablo F (1989): Comparative aspects of insulin and the insulin receptor. In Rosenblum IY, Heyner S (eds): Growth Factors in Mammalian Development. Boca Raton: CRC Press, pp 47–70.

McClain DA, Maigawa H, Lee J, Dull TJ, Ullrich A, Olefsky JM (1987): A mutant insulin receptor with defective tyrosine kinase displays no biologic activity and does not undergo endocytosis. J Biol Chem 262:14663–14671.

Mercola M, Stiles CD (1988): Growth factor superfamilies and mammalian embryogenesis. Development 102:451–460.

Mondschein JS, Etherton TD, Hammond JM (1991): Characterization of insulin-like growth factor binding proteins of porcine ovarian follicular fluid. Biol Reprod 44:315–320.

Mondschein JS, Smith SA, Hammond JM (1990): Production of insulin-like growth factor binding proteins (IGFBPs) by porcine granulosa cells: identification of IGFBP-2 and -3 and regulation by hormones and growth factors. Endocrinology 127:2298–2306.

Morgan DO, Edman JC, Standring DN, Fried VA, Smith MC, Roth RA, Rutter WJ (1987): Insulin-like growth factor II receptor as a multifunctional binding protein. Nature 329:301–307.

Morgan DO, Roth RA (1987): Acute insulin action requires insulin receptor kinase activity: introduction of an inhibitory monoclonal antibody into mammalian cells blocks the rapid effects of insulin. Proc Natl Acad Sci USA 84:41–45.

Mumby SM, Casey PJ, Gilman AG, Gutowski S, Sternweis PC (1990): G protein gamma subunits contain a 20 carbon isoprenoid. Proc Natl Acad Sci USA 87:5873–5877.

Mummery CL, van den Eijenden-van Raaji AJM (1990): Growth factors and their receptors in differentiation and early murine development. Cell Differ Dev 30:1–18.

Murakami MS, Rosen OM (1991): The role in insulin receptor autophosphorylation in signal transduction. J Biol Chem 266:22653–22660.

Murayama Y, Okamoto T, Ogata E, Asano T, Iiri T, Katada T, Ui M, Grubb JH, Sly WS, Nishimoto I (1990): Distinctive regulation of the functional linkage between the human cation-independent mannose 6-phosphate receptor and GTP binding proteins by insulin-like growth factor II and mannose 6-phosphate. J Biol Chem 265:17456–17462.

Nishimoto I, Hata Y, Ogata E, Kojima I (1987): Insulin-like growth factor II stimulates calcium influx in competent BALB/c 3T3 cells primed with epidermal growth factor. J Biol Chem 262:12120–12126.

Nishimoto I, Murayama Y, Katada T, Ui M, Ogata E (1989): Possible direct linkage of insulin-like growth factor-II receptor with guanine nucleotide–binding proteins. J Biol Chem 264:14029–14038.

O'Brien RM, Houslay MD, Milligan G, Siddle K (1987): The insulin receptor tyrosine kinase phosphorylates holomeric forms of the guanine nucleotide regulatory proteins G_i and G_0. FEBS Lett 212:281–288.

O'Hare T, Pilch PF (1988): Separation and characterization of three insulin receptor species that differ in subunit composition. Biochemistry 27:5693–5700.

Okamoto T, Katada T, Murayama T, Ui M, Ogata E, Nishimoto I (1990): A simple structure encodes G protein-activating function of the IGF-II/mannose-6-phosphate receptor. Cell 62:709–717.

Oshima A, Nolan CM, Kyle JW, Grubb JH, Sly WS (1988): The human cation-independent mannose-6-phosphate receptor cloning and sequence of the full-length cDNA and expression of functional receptor in COS cells. J Biol Chem 263:2553–2562.

Pampfer S, De Hertogh R, Vanderheyden I, Michiels B, Vercheval M (1991): Decreased inner cell mass proportion in blastocyst from diabetic rats. Diabetes 39:471–476.

Paria BC, Dey SK (1990): Preimplantation embryo development in vitro: cooperative interactions among embryos and role of growth factors. Proc Natl Acad Sci USA 87:4756–4760.

Peralta Soler A, Thompson KA, Smith RM, Jarett L (1989): Immunological demonstration of the accumulation of insulin, but not insulin receptors, in nuclei of insulin-treated cells. Proc Natl Acad Sci USA 86:6640–6644.

Rao LV, Farber M, Smith RM, Heyner S (1990): The role of insulin in preimplantation mouse development. In Heyner S, Wiley L (eds): Early Embryo Development and Paracrine Relationships. New York: Wiley-Liss, pp 109–124.

Rappolee DA, Brenner CA, Schultz R, Mark D, Werb Z (1988): Developmental expression of PDGF, TGF-alpha, and TGF-beta genes in preimplantation mouse embryos. Science 241:1823–1825.

Rappolee DA, Sturm KS, Behrendtsen O, Schultz GA, Pederson RA, Werb Z (1992): Insulin-like growth factor II acts through endogenous growth pathway regulated by imprinting in early mouse embryos. Genes Dev 6:939–952.

Rappolee DA, Strum KS, Schultz GA, Pederson RA, Werb Z (1990): The expression of growth factor ligand and receptor in preimplantation embryos. In Heyner S, Wiley L (eds): Early Embryo Development and Paracrine Relationship. Wiley-Liss, pp 11–25.

Record RD, Smith RM, Jarett L (1993): Insulin induces an unmasking of the carboxyl terminus of G_i proteins in rat adipocytes. Exp Cell Res 206: (in press).

Rice KM, Lienhard GE, Garner CW (1992): Regulation of the expression of pp160, a putative insulin receptor signal protein, by insulin, dexamethasone, and 1-methyl-3-isobutylxanthine in 3T3-L1 adipocytes. J Biol Chem 267:10163–10167.

Rogers SA, Hammerman MR (1988): Insulin-like growth factor II stimulates production of inositol triphosphate in proximal tubular basolateral membranes from canine kidney. Proc Natl Acad Sci USA 85:4037–4041.

Rosen OM (1987): After insulin binds. Science 237:1452–1458.

Rosenblum IY, Mattson BA, Heyner S (1986): Stage-specific insulin binding in mouse preimplantation embryos. Dev Biol 116:261–262.

Ross EM (1989): Signal sorting and amplification through G protein–coupled receptors. Neuron 3:141–152.

Roth RA, Steele-Perkins G, Hari J, Stover C, Pierce S, Turner J, Edman JC, Rutter WJ (1988): Insulin and

insulin-like growth factor receptor and responses. Cold Spring Harbor Symp Quant Biol 53:537–543.
Roth RA, Stover C, Hari J, Morgan DO, Smith MC, Sara V, Fried VA (1987): Interactions of the receptor for insulin-like growth factor II with mannose-6-phosphate and antibodies to the mannose-6-phosphate receptor. Biochem Biophys Res Commun 149:600–606.
Roth RA, Zhang B, Chin JE, Kovacina K (1992): Substrates and signalling complexes: the tortured path to insulin action. J Cell Biochem 48:12–18.
Rothenberg PL, Kahn CR (1988): Insulin inhibits pertussis toxin-catalyzed ADP-ribosylation of G proteins. J Biol Chem 263:15546–15552.
Russ M, Reinauer H, Eckel J (1992): Regulation of cardiac insulin receptor function by guanosine nucleotides. FEBS Lett 314:72–76.
Rutanen EM, Koistinen R, Wahlstrom T, Bohn H, Ranta T, Seppala M (1985): Synthesis of placental protein 12 by human decidua. Endocrinology 116:1304–1309.
Rutanen EM, Koistinen R, Sjoberg J, Julkunen M, Wahlstrom T, Bohn H, Seppala M (1986): Synthesis of placental protein 12 by human endometrium. Endocrinology 118:1067–1071.
Sadler SE, Maller JL (1987): In vivo regulation of cyclic AMP phosphodiesterase in Xenopus oocytes. J Biol Chem 262:10644–10650.
Schlessinger J (1988): Signal transduction by allosteric receptor oligomerization. Trends Biochem Sci 13:443–447.
Schultz GA, Heyner S (1992): Gene expression in preimplantation mammalian embryos. Mutat Res 296:17–31.
Shimizu Y, Shimizu N (1986): Rat hepatoma cell variants resistant to insulin-diphtheria toxin A fragment conjugates. J Biol Chem 261:7342–7346.
Simon MI, Strathmann MP, Gautam N (1991): Diversity of G proteins in signal transduction. Science 252:802–808.
Smith RM, Garside WT, Aghayan M, Shi C-Z, Shah N, Jarett L, Heyner S (1993): Mouse preimplantation embryos exhibit receptor-mediated binding and transcytosis of maternal insulin-like growth factor I. Biol Reprod 49:1–12.
Soos MA, Siddle K (1989): Immunological relationships between receptors for insulin and insulin-like growth factor I. Biochem J 263:553–563.
Soos MA, Whittaker J, Lammers R, Ullrich A, Siddle K (1990): Receptors for insulin and insulin-like growth factor-I can form hybrid dimers. Characterization of hybrid receptors in transfected cells. Biochem J 270:383–390.
Strittmatter SM, Valenzuel D, Kennedy TE, Neer EJ, Fishman MC (1990): Go is a major growth cone protein subject to regulation by GAP-13. Nature 344:836–841.
Sturgill TW, Wu J (1991); Recent progress in characterization of protein kinase cascades for phosphorylation of ribosomal protein S6. Biochim Biophys Acta 1092:350–357.
Sun XJ, Rothenberg P, Kahn CR, Backer JM, Araki E, Wilden PA, Cahill DA, Goldstein BJ, White MF (1991): Structure of the insulin receptor substrate IRS-I defines a unique signal transduction protein. Nature 352:73–77.
Sung, CK (1992): Insulin receptor signaling through non-tyrosine kinase pathways: evidence from anti-receptor antibodies and insulin receptor mutants. J Cell Biochem 48:26–32.
Tally M, Li CH, Hall K (1987): IGF-2 stimulated growth mediated by somatomedin type 2 receptor. Biochem Biophys Res Commun 148:811–816.
Taub R, Roy A, Dieter R, Koontz J (1987): Insulin as a growth factor in rat hepatoma cells. J Biol Chem 262:10893–10897.
Ui M (1990): Pertussis toxin as a valuable probe for G protein involvement in signal transduction. In Moss J, Vaughan M (eds): ADP-ribosylating Toxins and G Proteins: Insights into Signal Transduction. Washington, DC: American Society for Microbiology, pp 45–78.
Ullrich A, Bell J, Chen E, Herrera R, Petruzzelli LM, Dull TJ, Gray A, Coussens L, Liao Y, Tsukobawa M, Mason A, Seeburg P, Grunfeld C, Rosen O, Ramachandran J (1985): Human insulin receptor and its relationship to the tyrosine kinase family of oncogenes. Nature 313:756–761.
Ullrich A, Gray A, Tam A, Yang-Feng T, Tsubokawa M, Collins C, Henzel W, LeBon T, Kathuria S, Chen E, Jacobs S, Francke U, Ramachandran J, Fujita-Yamaguchi Y (1986): Insulin-like growth factor I receptor primary structure: comparison with insulin receptor suggests structural determinants that define functional specificity. EMBO J 5:2503–2512.
Van Wyk JJ, Graves DC, Casella SJ, Jacobs S (1985): Evidence from monoclonal antibody studies that insulin stimulates deoxyribonucleic acid synthesis through the type I somatomedin receptor. J Clin Endocrinol Metab 61:639–643.
Vercheval M, De Hertogh R, Pampfer S, Vanderheyden I, Michiels B, De Bernardi P, De Meyer R (1990): Experimental diabetes impairs rat embryo development during the preimplantation period. Diabetologia 33:187–191.
Werner H, Woloschak M, Stannard B, Shen-Orr Z, Roberts CT, LeRoith D (1991): The insulin-like growth factor-I receptor: Molecular biology, heterogeneity and regulation. In LeRoith D (ed): Insulin-like Growth Factors: Molecular and Cellular Aspects. Boca Raton: CRC Press, pp 17–48.
Wilden PA, Backer JM, Kahn CR, Cahill DA, Schroeder GJ, White MF (1990): The insulin receptor with phenylalanine replacing tyrosine-1146 provides evidence

for separate signals regulating cellular metabolism and growth. Proc Natl Acad Sci USA 87:3358–3362.

Williams LT (1989): Signal transduction by the platelet-derived growth factor receptor. Science 243:1564–1570.

Yamane HK, Fransworth CC, Xie H, Howald W, Fung BK-K (1990): Brain G protein gamma subunits contain all trans-geranylgeranyl cysteine methyl esters at their carboxyl termini. Proc Natl Acad Sci USA 87:5868–5872.

Yang L, Baffy G, Rhee SG, Manning DM, Hansen CA, Williamson JR (1991): Pertussis toxin-sensitive G_i protein involvement in epidermal growth factor-induced activation of phospholipase C-γ in rat hepatocytes. J Biol Chem 266:22451–22458.

Yang L, Camoratto AM, Baffy G, Raj S, Manning DR, Williamson JR (1993): Epidermal growth factor-mediated signaling of G_i-protein to activation of phospholipase in rat-cultured hepatocytes. J Biol Chem 268:3739–3746.

Yarden Y, Ullrich A (1988): Growth factor receptor tyrosine kinases. Annu Rev Biochem 57:443–448.

Zhang L, Blakewood EG, Denniston RS, Godke RA (1991): The effect of insulin on maturation and development of in vitro fertilized oocytes. Theriogenology 35:301.

Zick Y, Sagi-Eisenberg R, Pines M, Gierschik P, Speigel AM (1986): Multisite phosphorylation of the α subunit of transducin by the insulin receptor. Proc Natl Acad Sci USA 83:9294–9297.

ABOUT THE AUTHORS

CYNTHIA R. WARD is a research associate and resident in small animal medicine at the University of Pennsylvania, in Philadelphia, Pennsylvania. After receiving her B.A. from Bryn Mawr College, she pursued veterinary medical training at the University of Pennsylvania School of Veterinary Medicine where she received her V.M.D. in 1987. Subsequently she began doctoral research at the University of Pennsylvania School of Medicine in the laboratories of Drs. Gregory S. Kopf and Bayard T. Storey studying the molecular interactions between sperm and egg immediately prior to fertilization. She received her Ph.D. in 1992. She then began her residency in veterinary internal medicine (small animal) at the Veterinary Hospital of the University of Pennsylvania and is concurrently continuing her research into the signal transduction mechanisms mediating sperm-egg interaction.

WILLIAM T. GARSIDE is a senior research investigator at the University of Pennsylvania Medical Center in Philadelphia, Pennsylvania. He also supervises the In Vitro Fertilization laboratory and instructs clinical fellows and technologists associated with that laboratory. Dr. Garside received a B.A. *magna cum laude* from Glassboro State College in New Jersey prior to joining Nina Hillman's laboratory at Temple University in Philadelphia. There he earned a Ph.D. for studies on the transmission ratio distortion and sterility of *t*-haplotype bearing male mice. After receiving the doctoral degree he remained in the Hillman laboratory continuing his research on the *t*-haplotypes. Since joining the University of Pennsylvania in 1991 he has been studying the effects of the insulin family of growth factors on early mammalian embryogenesis and been actively engaged in research involving assisted reproductive technologies.

RAVINDRA N. DHIR is a postdoctoral fellow at the University of Pennsylvania Medical Center in Philadelphia, Pennsylvania, and is supported by the Rockefeller Foundation. After receiving a B.Sc. in zoology from the University of Rajasthan, Jaipur, India, he pursued doctoral research at the Department of Zoology in the same University; here he earned a Ph.D. in 1986 for work on the effects and mode of action of certain plant extracts during preimplantation stages of pregnancy in the rat. After receiving the doctoral degree, Dr. Dhir joined the National Institute of Health and Family Welfare, New Delhi, India. Here he started work in the steroid biochemistry laboratory on protein purification and hormone assays. Since joining the University of Pennsylvania in 1992 he has been studying the effect of insulin and insulin-like growth factors on preimplantation mouse development.

SUSAN HEYNER is Professor of Obstetrics and Gynecology and Professor of Pathology and Laboratory Medicine at the University of Pennsylvania Medical Center, in Philadelphia, Pennsylvania. In addition to her research laboratory, she directs the In Vitro Fertilization and Andrology Laboratories, and supervises postdoctoral and clinical fellows. She received the B.Sc. in zoology with special honours from the University of Southampton, England, and completed the Ph.D. degree from London University in the laboratory of John D. Biggers. Her early work was concerned with the culture and development of cartilage, and this was later extended into an examination of the immunobiology of cartilage, with Rupert E. Billingham. Subsequently, Dr. Heyner's interests turned to preimplantation development, and she has contributed in this area for the past 15 years. She has served on a number of editorial boards and is currently an associate editor for *Molecular Reproduction and Development.* In 1978 she was awarded a Research Career Development Award from the National Institutes of Health. She has served on the Human Embryology and Development, and Reproductive Biology Study Sections of the NIH and is currently a member of the Physiological Sciences Study Section.

Growth Factors and Signal Transduction in Development: 19–49

Developmental Regulation and Signal Transduction Pathways of Fibroblast Growth Factors and Their Receptors

Keith Miller and Angie Rizzino

I. INTRODUCTION

The first member of the fibroblast growth factor (FGF) family was discovered in the mid 1970s. It has since become evident that the FGFs comprise a complex family of at least nine* members. Early work demonstrated that FGFs can stimulate cell proliferation, but it soon became apparent that FGFs influence a wide range of physiological processes. More recently, the temporal and spatial expression of several FGFs has been shown to be regulated tightly during mammalian development, which argues that FGFs play numerous roles during development. Although the evidence for this hypothesis in the case of mammals is circumstantial, several lines of evidence argue strongly that FGFs are essential for amphibian development (Section II.B, this chapter; Dawid, Chapter 10, this volume).

Within the past 2 years, it has become evident that the complexity and diversity of FGF receptors is even greater than that of the FGFs. At least five genes code for high-affinity FGF receptors and multiple forms of these receptors are generated by alternative splicing. In addition, it is now evident that the binding of FGFs at the cell surface involves heparan sulfate. Recognition of this complexity has raised a familiar question: Why are there so many forms of this growth factor and why is the diversity of its receptors so great? While one can only speculate at this time, we suggest that the answer to this question will emerge once the major FGF signal transduction pathways are identified and described in detail. In this regard, progress is now being made in understanding how FGF receptors are activated and how these receptors interact with intracellular targets.

Perhaps the most intriguing development in the FGF field is the mounting evidence that FGFs can localize to the nucleus. Currently, the functional roles of FGFs in the nucleus and the mechanisms by which intranuclear FGFs exert their putative effects are poorly understood. Nonetheless, examination of the possible role(s) of FGFs in the nucleus are under way in numerous laboratories and significant progress in this area is likely to occur during the next several years.

Because other reviews provide excellent coverage of the earlier work in the FGF field [Basilico and Moscatelli, 1992; Burgess and Maciag, 1989], this review will focus heavily on the newest findings, while providing sufficient basic background information for those new to the FGF field. For readers interested in working with FGFs and their receptors, we suggest several methodological reviews in the area of FGF isolation and purification [Shing et al., 1991], FGF biological assays [Rizzino, 1987], FGF radioreceptor assays [Rizzino and Kazakoff, 1991; Kan et al., 1991], and FGF localization to the nucleus [Gabriel et al., 1991].

*Note Added in Proof: Although we refer to only eight members of the FGF family in this review, a ninth member has been cloned recently [Miyamoto et al., 1993].

II. FGF FAMILY OF GROWTH FACTORS

A. Structure–Function Relationships

The FGF family consists of at least seven polypeptide factors: acidic FGF (aFGF), basic FGF (bFGF), int-2, k-FGF, FGF-5, FGF-6, and keratinocyte growth factor (KGF). The common names for individual members of the family reflect their particular attributes, the way they were discovered, or the order in which they were discovered. Recently, however, systematic nomenclature has been proposed for the FGF family, in which members of the family are designated either as heparin-binding growth factors (HBGFs) or simply as FGFs, and each member is assigned a number (1 through 7 for aFGF, bFGF, int-2, k-FGF, FGF-5, FGF-6, and KGF, respectively). Thus, aFGF is renamed either FGF-1 or HBGF-1, bFGF is renamed either FGF-2 or HBGF-2, etc. For the purposes of this review, we will use the designations FGF-1 through FGF-7. In addition to the seven known forms of FGF, an androgen-induced growth factor has been cloned recently from mouse mammary carcinoma cells, and this factor appears to be a good candidate for inclusion as the eighth member of the FGF family [Tanaka et al., 1992]. This factor exhibits 30%–40% amino acid sequence similarity with FGF-2. However, because this factor has not been characterized with regard to its developmental expression or its interaction with FGF receptors, it will not be discussed further in this review.

The FGFs are encoded by distinct genes that are believed to be related phylogenetically, and individual FGFs tend to be conserved strongly between species. The sequence identity between human and mouse FGF homologs, for example, is typically greater than 90% at the amino acid level. On the other hand, within the FGF family of a given species the sequence identity ranges anywhere from 30% to 70% overall. The sequence similarities tend to be greatest within a so-called "core" region of homology, which comprises nearly the full length of FGF-2. FGF-2 is considered to be a prototype FGF and has been used extensively to study the structure–function relationships of FGFs (Fig. 1).

The ability to bind heparin is apparently a feature shared by all FGFs, and cell surface heparin-like molecules appear to be necessary for some, if not all, of the biological activities of FGFs (see below). Several regions of the FGF-2 primary sequence have been implicated as potential heparin-binding domains. In this regard, synthetic peptides corresponding to residues 33–77, 115–124, and 130–155 (full length is 1–155) of FGF-2 have significant heparin-binding activity [Baird et al., 1988]. The peptide, FGF-2 (130–155), corresponds to the FGF-2 carboxyl(C)-terminus, and the importance of this region is supported by the finding that mutants of FGF-2 with deletions at the C-terminus have reduced affinity for heparin [Seno et al., 1990]. Furthermore, two of these regions (115–124 and 130–155) contain clusters of basic residues whose positive charge may contribute to the affinity of these peptides for heparin, and this is consistent with the three-dimensional crystallographic structure of FGF-1 and FGF-2 [Zhu et al., 1990]. Interestingly, FGF-2 contains at least one additional cluster of basic residues at FGF-2 (27–31). Point mutations that change the charge of this region have been shown to alter the mitogenic activity of recombinant FGF-2 [Heath et al., 1991]. Thus, regions that bind to heparin may be scattered throughout the FGF-2 molecule.

Because the FGFs are thought to exert many of their biological effects through interactions with their cell surface protein receptors, several investigators have attempted to define receptor-binding domains within the primary structure of FGF-2. Synthetic peptides corresponding to residues 33–77 and 115–124 compete effectively with FGF-2 for high-affinity cell surface binding, and exhibit partial agonist and antagonist activities in mitogenic assays [Baird et al., 1988]. Interestingly, it has been shown recently that threonine 121, within the second putative receptor binding domain, is a substrate for phosphorylation by protein kinase A, and that

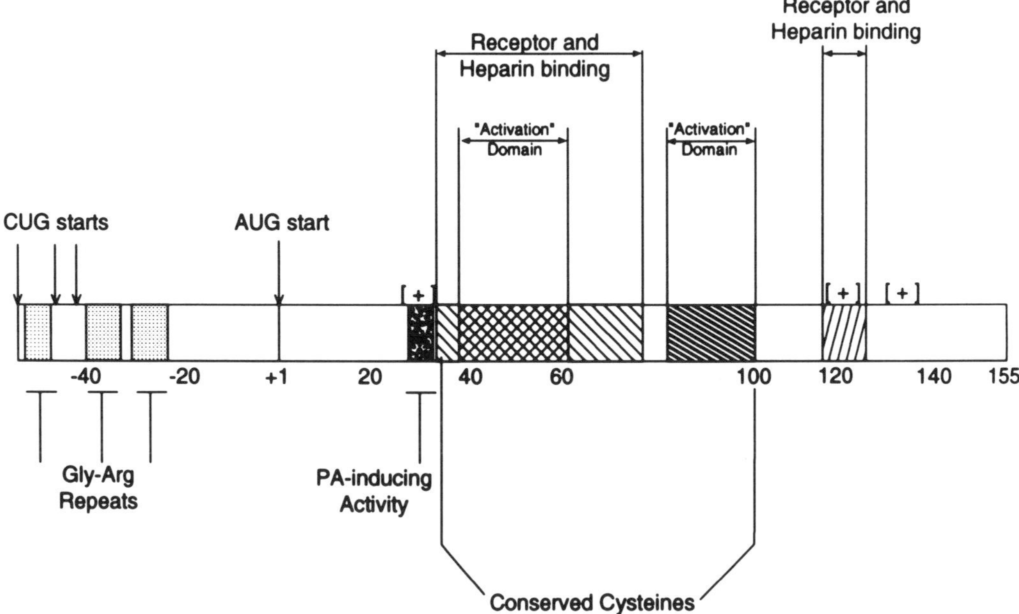

Fig. 1. *Schematic diagram of proposed structural and functional features of FGF-2. Numbers below the diagram indicate amino acid residues of full-length, AUG-initiated FGF-2. N-terminal sequences are the result of initiation of translation at alternative CUG start sites, as shown. [+], indicates clusters of basic residues that may participate in heparin binding. See text for further details.*

phosphorylation of this residue increases by approximately eight-fold the affinity of FGF-2 for its receptor [Feige and Baird, 1989]. This supports the hypothesized role for this region in receptor binding. However, as described above, these regions also appear to interact with heparin, and interactions with heparin-like molecules at the cell surface are likely to have an important role in high-affinity FGF binding (see below). In addition, neutralizing antibodies have been identified that are able to block the binding of FGF-2 to cell surface protein receptors without interfering with its binding to heparin [Kurokawa et al., 1989]. These results argue that the domains responsible for heparin binding and for receptor binding are distinct, and they suggest that it should be possible to create mutants of FGF-2 that exhibit normal affinity for heparin, but that are defective in high-affinity binding to the receptor.

Several other domains within FGF-2 have been associated with specific biological activities. A recent study has demonstrated that synthetic peptides corresponding to residues 38–61 and 82–101 each exhibit partial agonist and antagonist activities in FGF-2 mitogenic assays, but, surprisingly, they compete poorly with FGF-2 for receptor binding [Presta et al., 1991]. The authors of this study have argued that these regions represent "activation" domains that are required for mitogenic activity without being involved directly in binding to FGF receptors. Another study has suggested that there is an important function for a single lysine residue [Burgess et al., 1990]; mutation of lysine-132 of human FGF-2 resulted in a reduced affinity for heparin and a dramatic reduction in mitogenic activity. However, this mutant FGF exhibited normal receptor-binding activity, and induced the same pattern of tyrosine phosphorylation and *c-fos* induction as the wild-type factor. One other recent report has shown that an FGF-2 deletion mutant lacking amino acids 27–32 has normal receptor-binding and mitogenic activities, but is at least 100 times less efficient at inducing plasminogen activator activity in cultured endothelial cells [Isacchi et al., 1991]. Together, these studies suggest that (a) different biological activities of FGF-2 can be dissociated at the structural level, (b) activation of the tyrosine kinase activity of FGF receptors may not be sufficient to transmit a complete mitogenic signal, and (c) FGFs are likely to interact directly with multiple cellular targets.

The presence of four cysteine residues within the primary amino acid sequence of FGF-2 and the fact that two of the four cysteines are conserved throughout the FGF family has led to speculation that native FGF-2 and possibly other FGFs contain one or more disulfide bonds. However, there is no clear consensus regarding the importance of the conserved cysteine residues for the biological activity of FGFs, or of their participation in disulfide linkages. The role of the conserved cysteine residues in the function of FGF-1 (aFGF) has been reviewed in detail elsewhere [Thomas et al., 1991]. Several lines of evidence suggest that the conserved cysteine residues in FGF-1 are not required for its biological activity. Furthermore, oxidation of FGF-1 to force disulfide bond formation yields an inactive molecule. However, other studies suggest that the conserved cysteines may influence the ability of heparin to stabilize FGF-1.

Excluding the two prototypes, FGF-1 and FGF-2, all of the FGFs contain at their N-termini stretches of hydrophobic residues that are likely to function as signal peptides for secretion. In contrast, FGF-1 and FGF-2 lack signal sequences and, in fact, there is little evidence that these two factors are secreted. However, several recent reports have provided indirect evidence that FGF-2 may be exported from cells that produce it [Sakaguchi et al., 1988; Sato and Rifkin, 1988; Mignatti et al., 1991]. While the mechanism of extracellular release of FGF-2 remains to be determined, there is some evidence that its release may be achieved by exocytosis that is independent of the endoplasmic reticulum–Golgi (ER–Golgi) pathway [Mignatti et al., 1992], or by export in association with cell surface proteoglycans [Bashkin et al., 1992].

Although these findings indicate that all of the FGFs have the potential to be exported from their producing cells, several convergent lines of evidence suggest that some, if not all, of the FGFs have an altogether different intracellular fate. Several members of the FGF family have now been detected within the nuclei of a variety of cell types, both *in vitro* and *in vivo*. Furthermore, functional nuclear localization signals have been described for two members of the FGF family, and with the possible exception of FGF-5, all of the members contain amino acid sequences similar to known nuclear localization signals. This provocative area of study will be discussed in greater detail in a later section of this review.

B. Biological Actions of FGFs

Expression studies have provided many useful insights into the potential roles of FGFs during mammalian embryonic development (developmental regulation of FGFs is discussed below). Further progress in this field has been limited severely because of technical difficulties in working with mammalian embryos. However, significant advances have been made in our understanding of FGF function during mesoderm formation in *Xenopus laevis*. This early patterning event is believed to depend on inductive signals passed from the presumptive endoderm of the blastula vegetal pole to the ectodermal blastomeres of the animal cap, and several factors, including the FGFs, have been implicated in this process [reviewed in Smith, 1989; Dawid et al., 1992; Dawid, Chapter 10, this volume]. Indeed, the importance of FGFs in mesoderm formation was demonstrated clearly in a recent study, which employed a mutant FGF receptor that blocks the function of wild-type FGF receptors when they are coexpressed [Amaya et al., 1991]. Overexpression of this dominant negative mutant FGF receptor in *Xenopus* embryos resulted in major deficiencies in mesodermal development. While embryos expressing the mutant receptor initiated gastrulation normally, subsequent development of mesodermal derivatives was abnormal, with especially prominent defects in trunk and posterior development. Overexpression of the wild-type receptor suppressed the abnormal phenotype and allowed normal development.

While these results testify to the importance of FGFs in amphibian development, the specific member(s) of the FGF family that are important for this activity are unknown. FGF-2 is an obvious candidate for the mesoderm-inducing activity in *Xenopus,* because it is expressed at the appropriate time [Kimelman et al., 1988]. However, a recent study has cast doubt on the importance of FGF-2 as a primary mesoderm-inducing factor [Slack, 1991]. In this study, the properties of soluble mesoderm-inducing factors (MIFs) were studied by placing vegetal explants (the inducing tissue) and ectoderm (the responding tissue) on either side of a permeable membrane. In control experiments, the vegetal explants were able to induce the ectodermal blastomeres to form mesodermal tissues, indicating that the inductive signal is a secreted factor that is able to penetrate the permeable membrane. Interestingly, while heparin was able to inhibit formation of mesodermal tissues in the transfilter induction studies, antibodies specific for FGF-2 had no effect. This suggests that FGF-2 is not a major component of the signal emitted by the vegetal cells of the *Xenopus* blastula. In addition, recent immunolocalization studies have shown that both FGF-1 and FGF-2 are localized throughout the mesoderm- and endoderm-forming regions of the *Xenopus* embryo prior to gastrulation [Shiurba et al., 1991]. The staining for both factors was almost entirely intracellular, and underwent a shift from predominantly cytoplasmic to mostly nuclear at the midblastula stage.

The detection of FGF-1 and FGF-2 within cells in the mesoderm-forming region suggests that these cells produce one or both factors. If cells of the mesoderm-forming region produce their own FGF-2, it is unclear why they would need to receive a secreted FGF-2 signal from the vegetal pole. One possibility is that production of endogenous FGF-2 and/or FGF-1 by these cells regulates their competence to respond to additional signals emitted from the presumptive endoderm [Godsave and Shiurba, 1992]. Recently, a novel member of the FGF family has been cloned and sequenced from *Xenopus* embryo cDNA [Isaacs et al., 1992]. This FGF is most closely related to mammalian FGF-4 and FGF-6, and like these factors it contains a signal peptide for secretion. In addition, its expression in *Xenopus* development is consistent with a role in mesoderm induction, and the recombinant factor has mesoderm-inducing activity. Thus, it is the best candidate yet identified for the endogenous FGF-related mesoderm-inducing factor in *Xenopus.* It will be interesting to determine whether neutralizing antibodies specific for this factor inhibit the formation of mesodermal tissues in transfilter induction studies.

Members of the FGF family may also regulate the differentiation of other cell types during development. Several lines of evidence support a potential role for FGFs during muscle and limb development [Suzuki et al., 1992; Niswander and Martin, 1992]. At least two FGFs (FGF-4 and FGF-5) and at least one FGF receptor (FGFR-4) are expressed in embryonic muscle precursors of the mouse [Niswander and Martin, 1992; Haub and Goldfarb, 1991; Stark et al., 1991], and FGFs have been shown to inhibit differentiation of myoblasts *in vitro* [Lathrop et al., 1985; Clegg et al., 1987], and to regulate the transcription of other important myogenic regulatory genes [Vaidya et al., 1989; Brunetti and Goldfine, 1990]. FGFs have also been shown to influence the differentiation and survival of neurons *in vitro,* and several members of the FGF family are expressed widely in the developing nervous system [Kalcheim and Neufeld, 1990]. Furthermore, FGF-1, FGF-2, and FGF-5 expression persists in the adult central nervous system [Alterio et al., 1988; Haub et al., 1990; Shimasaki et al., 1988], and FGFs produced by glial cells may have a trophic effect on surrounding neurons in the adult [Hatten et al., 1988].

Members of several growth factor families, including the FGFs, have also been implicated in angiogenesis and wound healing. Angiogenesis is thought to involve the mobilization of vascular endothelial cells from preexisting blood vessels to invade surrounding tissues and organize into new blood vessels. *In vitro* studies have shown that FGFs elicit a variety of responses in cultured vascular endothelial cells that are consistent with their proposed function in angiogenesis [reviewed in Burgess and Maciag, 1989; Basilico and Moscatelli, 1992]. These responses include mitogenesis, chemotaxis, production of proteolytic enzymes, and the ability to penetrate basement membranes. In addition, several *in vivo* models of angiogenesis have demonstrated directly that FGFs promote formation of granulation tissue and new blood vessels *in vivo*. However, other factors have also been shown to promote angiogenesis [reviewed in Hom and Maisel, 1992], and it is unclear which factor(s) are most important *in vivo*. Nonetheless, members of the FGF family remain strong candidates as angiogenic factors in wound healing, and it is possible that FGFs participate in forming the vascular system during embryonic development.

Early studies demonstrated that FGF-1 and FGF-2 are able to induce the soft-agar growth of several nontransformed cell lines [Rizzino and Ruff, 1986], and three of the FGFs (FGF-3, FGF-4, and FGF-5) were identified originally as oncogenes. Indeed, recent reports have demonstrated clearly that several of the FGF genes can behave as dominant transforming oncogenes *in vitro*. Therefore, members of the FGF family may be involved in tumorigenesis. These results have been reviewed in detail elsewhere [Basilico and Moscatelli, 1992].

C. Developmental Regulation of FGFs

Efforts to understand the roles of FGFs during mammalian development have focused heavily on determining the patterns of expression of FGFs and of their receptors during development. The developmental expression of FGF receptors has been examined recently, and this topic is discussed in a later section. Thus far, at least six of the FGFs have been detected in developing embryos at the RNA level, and it is apparent that their expression is regulated both temporally and spatially. These results suggest that expression of the FGFs is regulated independently and, consequently, each is likely to have distinct roles in development.

FGF-4 appears to be the member of the FGF family expressed earliest in mouse development. Previous studies have shown that the FGF-4 gene is expressed as early as the 4-cell stage of mouse development and continues to be expressed until at least the blastocyst stage [Rappolee et al., 1990]. However, a recent report suggests that FGF-4 expression persists well beyond the blastocyst stage [Niswander and Martin, 1992]. In these studies, developing mouse embryos from fertilization to the 14th day of gestation were studied using *in situ* hybridization to detect FGF-4 transcripts. FGF-4 expression was detected initially at the late blastocyst stage, where it was expressed uniformly in cells of the inner cell mass. This suggests that FGF-4 may influence preimplantation development. By the early gastrula stage, FGF-4 expression is restricted to cells near or within the primitive streak where mesoderm formation occurs. Furthermore, tissue-specific FGF-4 expression was shown to persist during later stages of development, suggesting that FGF-4 exerts diverse effects on growth and differentiation during mammalian development.

The developmental expression of two other members of the FGF family (FGF-3 and FGF-5) has been studied in detail [Haub and Goldfarb, 1991; Hébert et al., 1991; Wilkinson et al., 1989; Wilkinson et al., 1988]. In contrast to FGF-4, the expression of FGF-3 and FGF-5 prior to gastrulation appears to be restricted to parietal endoderm and visceral endoderm, respectively. Subsequently, FGF-3, FGF-4, and FGF-5 messenger RNAs (mRNAs) are all detectable during mesoderm formation. At this stage, the expressions of transcripts for these three FGFs are spatially distinct, suggesting that each may participate in mesoderm formation, but exert different effects. Like FGF-4, FGF-3 and FGF-5 continue to be expressed in highly restricted regions following the major patterning events, suggesting that these factors also play important roles in the development of specific tissues.

Although the developmental expression of FGF-1, FGF-2, and FGF-6 has not been described in detail, Northern analysis indicates that FGF-1 and FGF-2 are expressed at least as early as day 10.5 of gestation in mouse embryos, and their expression is detectable in a variety of tissues at day 13.5 [Hébert et al., 1990]. Likewise, FGF-6 transcripts are expressed during middle and late gestation with a peak at day 15 [de Lapeyriere et al., 1990].

D. Mechanisms That Regulate FGF Expression

Given the functional similarities among the FGFs in many biological assays, it is likely that differential regulation of FGF gene expression is an important mechanism that enables each member of the FGF family to subserve distinct functions during development. Significant advances have been made recently in our understanding of the mechanisms that regulate FGF expression. Many of these studies have made use of embryonal carcinoma (EC) cells as a model system. Embryonal carcinoma cells are multipotent tumor stem cells that are similar phenotypically to the totipotent cells of the early embryo. In 1988, it was determined that differentiation of EC cells represses the expression of an FGF-like activity [Rizzino et al., 1988a]. Subsequently, it was shown that EC cells express FGF-4 and that differentiation of EC cells results in a large reduction in the steady-state levels of FGF-4 mRNA [Yoshida et al., 1988; Peters et al., 1989; Velcich et al., 1989; Tiesman and Rizzino, 1989a].

More recently, several groups have employed promoter-reporter gene constructs to map the cis-regulatory elements that control transcription of the human FGF-4 gene [Curatola and Basilico, 1990] and the mouse FGF-4 gene [Ma et al., 1992] in EC cells and their differentiated cells. Work with the human FGF-4 5′-flanking region suggests that a strong enhancer exists in the untranslated region of the third exon of the human FGF-4 gene. This conclusion is supported by work with the mouse FGF-4 gene, and several lines of evidence suggest that this region contains more than one important cis-regulatory element, including a putative octamer-binding site, that control the expression of the FGF-4 gene in EC cells [Ma et al., 1992]. Furthermore, both positive and negative cis-regulatory elements have been identified in the upstream region of the mouse FGF-4 gene.

Embryonal carcinoma cells are also being used to study the mechanisms that regulate FGF-3 (Int-2) expression. Previous studies have shown that differentiation of mouse and human EC cells into endoderm-like cells leads to an increase in the level of FGF-3 mRNA [Jakobovits et al., 1986; Smith et al., 1988; Brookes et al., 1989], and that this increase appears to be due to an increase in transcription of the FGF-3 gene [Grinberg et al., 1991]. Furthermore, multiple size classes of FGF-3 mRNA have been detected in differentiated mouse EC cells, and it appears that the complex expression of FGF-3 is the result of initiation of transcription from three possible promoter regions and by the use of two alternative polyadenylation signals [Mansour and Martin, 1988; Smith et al., 1988]. Transcripts resulting from initiation at the distal promoter contain an additional exon that is not included in transcripts that are initiated at downstream sites. However, the presence of the alternative upstream exon does not affect the predicted coding sequence of the FGF-3 transcript [Mansour and Martin, 1988; Smith et al., 1988], or the stability of the transcript [Grinberg et al., 1991]. Curiously, only two size classes of FGF-3 mRNA have been detected in human EC cells treated with retinoic acid, and human FGF-3 transcripts lack heterogeneity at their 5′ ends, which suggests that regulation of the expression of human FGF-3 may not be as complex as expression of mouse FGF-3 [Brookes et al., 1989]. The biological significance of multiple promoters in the mouse FGF-3 gene remains unclear. One possibility is that the different promoters may regulate complex patterns of tissue-specific gene expression during development. Further studies are needed to determine the specific elements that regulate the expression of the FGF-3 gene, and to determine the functional differences, if any, among the mouse FGF-3 promoter regions.

The FGF-2 gene promoter has also been characterized partially. Although multiple size classes of FGF-2 mRNA have been detected, primer extension analysis indicates a single site of transcription initiation [Shibata et al., 1991]. The sequence upstream of the putative initiation site lacks consensus TATA and CAAT boxes [Abraham et al., 1986a,b]. However, five potential SP-1 binding sites, as well as one potential activator protein-1 (AP-1) binding site, have been identified in this region. Recently, promoter-reporter gene constructs have been used to identify the regions of the FGF-2 promoter that are important for its transcriptional activity [Shibata et al., 1991]. These results indicate that the region containing the five putative SP-1 binding sites constitutes the functional core of the promoter. In addition, it seems that the promoter contains at least two regions that have a negative regulatory function: one residing 3′ of the cluster of GC boxes, and one residing between approximately 500 and 850 nucleotides upstream of the transcription start site. Additional studies are needed to characterize these regions further. In addition, further work is needed to determine the mechanisms by which expression of other members of the FGF gene family is regulated.

III. FGF RECEPTORS

A. The Family of FGF Receptor Genes

During the past several years, major advances have been made in the characterization and cloning of genes that code for FGF cell surface protein (CSP) receptors. In their mature forms, these receptors have been reported to have affinities as strong as 4 pM, and typically they have affinities in the range of 20–100 pM [Neufeld and Gospodarowicz, 1985; Moscatelli, 1988; Dionne et al., 1990; Mansukhani et al., 1990]. However, as will be discussed (Section III.F), high affinity FGF receptors may not be single proteins but complexes that involve other cell surface molecules, including heparan sulfate proteoglycans (HSPG). To date, at least five different FGF CSP receptor genes have been cloned and sequenced [Ruta et al., 1988; Kornbluth et al., 1988; Dionne et al., 1990; Hattori et al., 1990; Johnson et al., 1990; Keegan et al., 1991; Mansukhani et al., 1990; Partanen et al., 1991; Pasquale and Singer, 1989; Ruta et al., 1989; Reid et al., 1990; Safran et al., 1990], and it is apparent that RNA splicing can generate multiple forms of FGF receptor from at least two of the FGF receptor genes [Hou et al., 1991; Eisemann et al., 1991; Bernard et al., 1991; Johnson et al., 1991; Miki et al., 1992; Werner et al., 1992; Dell and Williams, 1992; Duan et al., 1992]. As in the case of the FGFs, FGF receptors have been given different names by different groups. For clarity, we will refer to them as FGFR-1, FGFR-2, FGFR-3, and FGFR-4. FGFR-1 is also known as Flg and Cek-1. FGFR-2 is also known as Bek, K-SAM, TK-14, and Cek-3; and splice variants of FGFR-2 that bind KGF are also known as KGFR. FGFR-3 is also referred to as Cek-2. Lastly, it should be noted that a fifth receptor referred to as Flg-2 is very closely related to FGFR-3. Although it has not been formally proven, it is likely that it is encoded by a distinct gene (discussed in section IIIB). In this review, we will refer to this receptor as Flg-2.

B. General Structure of FGF Receptors

Examination of predicted amino acid sequences of the five FGF CSP receptors reveals a number of common features, including an extracellular domain (approximately 360 amino acids), a short transmembrane domain (21 amino acids), and a cytoplasmic domain (approximately 420 amino acids). In addition, like other growth factor receptors, FGF CSP receptors are heavily glycosylated, with the carbohydrate moiety contributing roughly 25%–35% of the receptor mass. Typically, the extracellular domain contains, from its amino terminus to its carboxyl terminus, a signal peptide (18–24 amino acids) and three extracellular immunoglobulin-like loops, which are referred to as loops I, II and III, respectively, indicating their positions relative to the amino terminus. There is an aspartate, serine-rich acidic box located between loops I and II, and a putative heparan sulfate binding domain just before the beginning of loop II. The transmembrane domain is followed by a cytoplasmic domain that contains a juxtamembrane domain (74 to 81 amino acids), two tyrosine kinase domains separated by a short kinase insert (14 amino acids), and a C-terminal tail (59 to 69 amino acids) (Fig. 2). Comparison of the amino acid sequences of FGFR-1 through FGFR-4 indicates an overall sequence similarity that ranges from a high of 71% between FGFR-1 and FGFR-2 to a low of 56% between FGFR-1 and FGFR-4. Closer examination indicates that there is a high degree of similarity between these four FGF receptors in several regions, including loop II (63%–81%), loop III (74%–82%), and the kinase domains (75%–92%). In contrast, there is a much lower degree of similarity in other regions, including loop I (21%–38%), the juxtamembrane domain (31%–66%), and the kinase insert domain (14%–50%) (Partanen et al., 1991). Comparison of the amino acid sequences of FGFR-3 and Flg-2 indicates that they are closely related and exhibit an overall sequence similarity of 92%. However, two regions of FGFR-3 and Flg-2 (loop I and the kinase insert) exhibit lower amino sequence similarities (78% and 79%, respectively). Consequently, it appears that Flg-2 is encoded by a fifth FGF receptor gene. Detailed structural comparisons between the five FGF CSP receptors can be found in reports by Partanen et al. [1991] and Avivi et al. [1991].

C. Receptor Specificity

A major question raised by the diversity of the five major FGF receptors is the extent to which they differ in their ligand-binding properties, their signal transduction pathways, and their patterns of expression. Thus far, the specificity of the five receptors has not been characterized completely. Most of the information available is for FGF-1 and FGF-2, with less reported for FGF-4, FGF-5, and FGF-7. However, even the information reported cannot be considered complete because alternative splicing can generate multiple isoforms for at least FGFR-1 and FGFR-2, and these isoforms exhibit different affinities for the FGFs that have been examined (discussed in Section III.D). The information available indicates that one isoform of FGFR-1, which contains all three immunoglobulin-like loops, binds FGF-1 and FGF-2 with slightly different

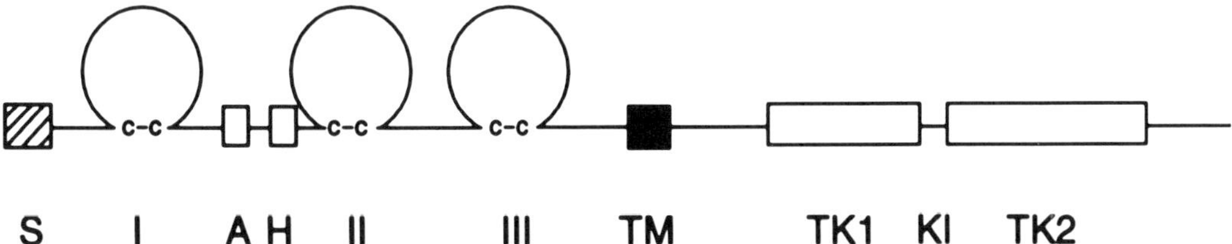

Fig. 2. *Schematic diagram of the three-loop structure of FGF CSP receptors. From the N-terminus at the left, the following domains are typically found: S, signal peptide; I, Ig-like loop I; A, acidic box; H, putative heparan-sulfate proteoglycan binding domain; II, Ig-like loop II; III, Ig-like loop III; TM, transmembrane domain; TK1, tyrosine kinase domain 1; KI, kinase insert; and TK2, tyrosine kinase domain 2.*

affinities [Dionne et al., 1990]. Another FGFR-1 isoform, which lacks the immunoglobulin-like loop I, binds FGF-4 with an affinity approximately 15 times lower than its affinity for FGF-2 [Mansukhani et al., 1990], and this isoform of FGFR-1 does not bind FGF-5 [Mansukhani et al., 1992]. However, a third isoform of FGFR-1, which is produced by alternative splicing and which contains a different amino acid sequence in immunoglobulin-like loop III, binds FGF-1 with an affinity roughly 50 times higher than its affinity for FGF-2 [Werner et al., 1992]. Moreover, there are secreted isoforms of FGFR-1 that bind FGF-2 with a 10-fold higher affinity than for FGF-1 [Duan et al., 1992].

In the case of FGFR-2, one isoform binds FGF-1 and FGF-2 with high affinity, although it exhibits a higher affinity for FGF-1 [Dionne et al., 1990]. In addition, this isoform binds FGF-4 [Dionne et al., 1990; Mansukhani et al., 1992], but it does not bind FGF-7 or FGF-5 [Miki et al., 1992; Mansukhani et al., 1992]. Interestingly, a splice variant of FGFR-2, which contains a different amino acid sequence in immunoglobulin-like loop III (see Section III.E), binds FGF-1 and FGF-7 with approximately the same affinity; whereas it binds FGF-2 very weakly if at all [Miki et al., 1992; Dell and Williams, 1992]. FGFR-3 has been reported to bind FGF-1, FGF-2, FGF-4, and FGF-5 [Keegan et al., 1991; Ornitz and Leder, 1992); whereas FGFR-4 binds FGF-1 with high affinity, and FGF-4 and FGF-2 with 10-fold lower affinities than that for FGF-1, but does not bind FGF-5 [Partanen et al., 1991; Ornitz and Leder, 1992]. In addition, a recent report claims that FGFR-4 binds FGF-6 [Vainikka et al., 1992]. Thus far, the receptors that bind FGF-3 have not been identified, and the ligand preferences and affinities for Flg-2 have not been described.

D. Receptor Specificity and FGF-Binding Domains

As indicated above, several lines of evidence demonstrate that receptor specificity is determined at least in part by generating multiple isoforms from alternative splicing. This is illustrated clearly by the dramatic effect of alternative splicing on the binding characteristics of the human FGFR-2 gene product [Miki et al., 1992]. This splice variant differs from FGFR-2 in a region of 52 amino acids that begins in loop III and extends 22 amino acids beyond this loop. Interestingly, the mouse homologue of the FGFR-2 splice variant lacks loop I (due to alternative splicing) and the omission of loop I does not appear to alter significantly the affinity of this receptor for FGF-1, FGF-7, or FGF-2 [Miki et al., 1992]. Similarly, at least one group has reported that FGF-2 and FGF-4 bind equally well to FGFR-1, with or without loop I [Mansukhani et al., 1990]. However, more recent studies suggest that FGF-1 binds with as much as a 10-fold higher affinity to FGFR-1 containing only loops II and III [Shi et al., 1991; W. McKeehan, personal communication]. Thus, loop I may not form the ligand-binding domain, but it may have a significant influence on ligand affinity.

Thus far, FGF binding determinants have not been identified on any of the FGF receptors; however, the studies discussed above argue that loops II and III of the receptor, which comprise approximately 225 amino acids, make up at least part of the FGF binding site, and that loop III plays an important role in determining receptor specificity. This conclusion is supported strongly by the results of a recent study in which genetically modified FGFR-1 was employed [Hou et al., 1992]. In this study, several key cysteine residues, which are believed to form disulfide bridges and which are believed to be important in forming the secondary structure of loops II and III, were replaced with either glycine or serine. The results from FGF-1 binding assays performed on COS cells transfected with the genetically modified receptor cDNAs argue that disruption of either loop II or loop III abolishes FGF binding. Furthermore, recombinant proteins that contain loop I but lack either loop II or loop III do not bind FGF. Lastly, a recent study used epitope-specific antibodies to probe and help predict the secondary and tertiary structure of loops I, II, and III [Xu et al., 1992]. Readers interested in specific details are directed to this thought-provoking study.

E. Multiple Forms of FGF Receptors and Alternative Splicing

Alternative splicing is capable of generating a surprisingly high level of diversity and complexity in the case of FGFR-1 [Hou et al., 1991; Eisemann et al., 1991; Werner et al., 1992; Duan et al., 1992], and similar levels of complexity appear to exist for FGFR-2 [Miki et al., 1992; Dell and Williams, 1992] and possibly FGFR-3 [Pasquale, 1990]. However, for FGFR-4 at least one of the important regions (loop III) does not appear to be subject to alternative splicing [Vainikka et al., 1992].

In the case of FGFR-1, reverse transcription polymerase chain reaction and cDNA cloning were used to identify three different extracellular motifs, two different acidic boxes (these are only present in two of the extracellular motifs), two juxtamembrane motifs, and two tyrosine kinase 2 motifs (Fig. 3). In addition, two different exons that code for the region containing loop III have been identified [Werner et al., 1992]. Together these motifs could generate as many as 24 different FGFR-1 isoforms, at least 6 of which have been identified at the mRNA level in HepG-2 cells [Hou et al., 1991]. Moreover, the expression of multiple FGFR-1 isoforms *in vivo* has been verified recently in rat prostate mesenchymal cells with the aid of form-specific FGFR-1 antibodies [Xu et al., 1992]. Thus far, over a dozen splice variants have been reported for FGFR-1 and FGFR-2 [Hou et al., 1991; Dionne et al., 1990; Keegan et al., 1991; Partanen et al., 1991; Miki et al., 1992; Duan et al., 1992], and it is anticipated that this number will increase as more studies are performed.

Although functional differences between the different FGFR-1 isoforms have not been established firmly, several hypotheses can be proposed based on an examination of the different FGFR-1 motifs. In the case of the extracellular motifs, one has been shown to contain all three immunoglobulin-like loops, whereas the other two extracellular motifs contain only loops II and III [Hou et al., 1991]. As discussed above, loop I may affect ligand specificity. However, the significance of the structural differences between the other two extracellular motifs is far from clear. One of them lacks a signal peptide and the acidic box that is located between loops I and II. Thus, this isoform could be an intracellular FGF receptor and this intriguing possibility is supported by the failure to detect this isoform on the surface of cells transfected with the appropriate cDNA expression plasmid [Xu et al., 1992].

The differences between the two juxtamembrane motifs and between the two kinase motifs of FGFR-1 also lead to speculation. In this regard, one of the juxtamembrane motifs contains an 8-amino-acid insert containing threonine. Although it remains to be determined whether this threonine can be phosphorylated, juxtamembrane phosphorylation sites are candidates for altering ligand affinity, kinase activity, and/or downregulating other growth factor receptors [Ullrich and Schlessinger, 1990]. In the case of the two kinase motifs, one contains a functional kinase, whereas the other is truncated in the C-terminal kinase domain [Hou et al., 1991]. The functions of isoforms containing the latter motif remain to be determined, but these isoforms can form heterodimers with other isoforms that contain functional tyrosine kinase domains. Thus, these isoforms could serve as dominant negative repressors (discussed below) and block activation of FGF receptor isoforms that contain functional tyrosine kinases. Finally, cDNAs for FGFR-1 and FGFR-4 isoforms that lack a transmembrane domain and an intracellular domain have been reported and these could code for secreted receptors [Dionne et al., 1990; Johnson et al., 1991; Duan et al., 1992; Horlick et al., 1992]. Currently, the roles of secreted receptors are unclear, but they could serve to influence the stability and the availability of specific secreted FGFs,

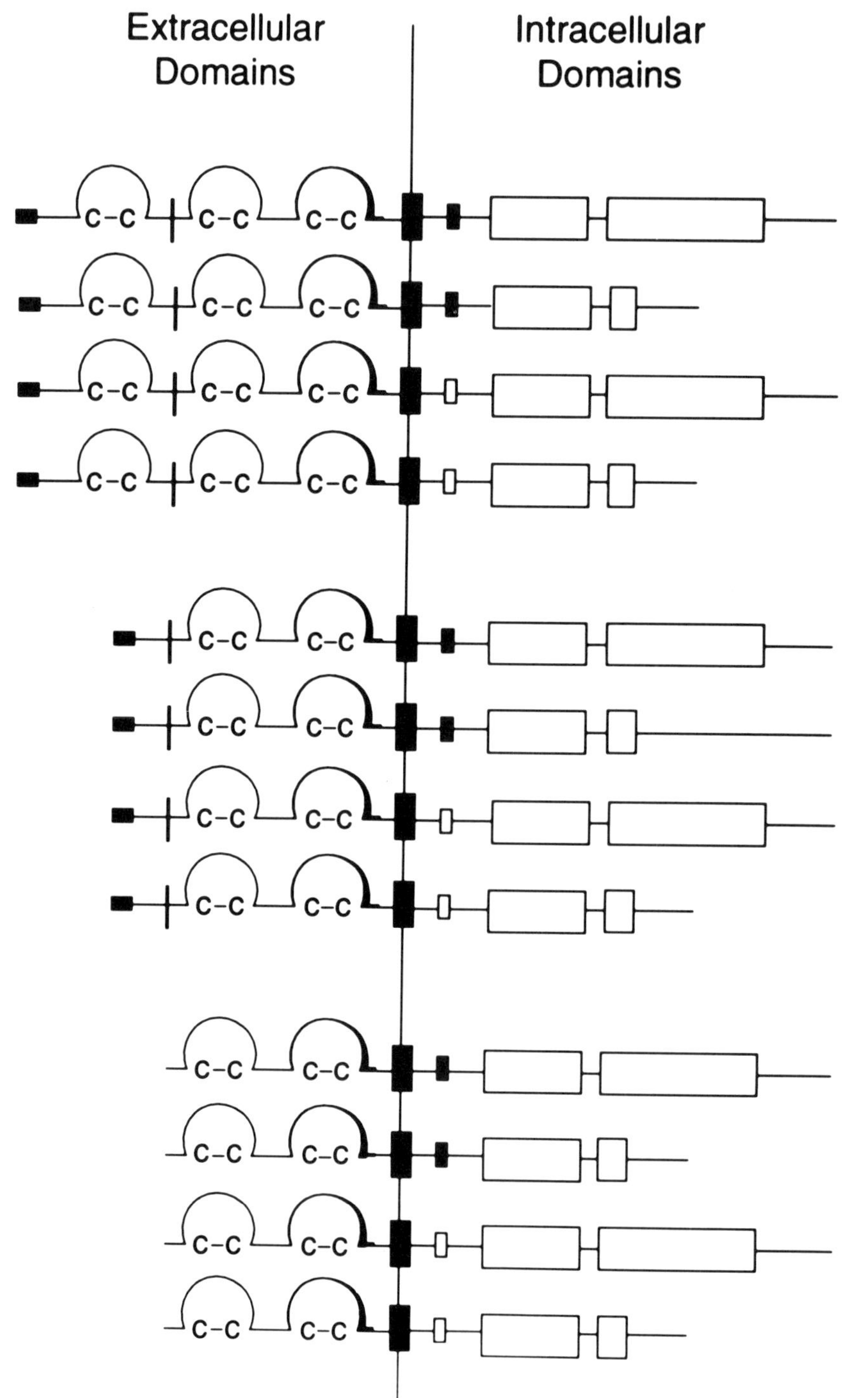

Fig. 3. *Predicted FGFR-1 isoforms. The top four isoforms contain: the signal peptide; the acidic box; Ig loops I, II, III; the transmembrane domain followed by two different juxtamembrane domains (indicated by open and closed rectangles); tyrosine kinase domain 1; the kinase insert; and either a complete or a truncated tyrosine kinase domain 2. The middle four isoforms lack Ig loop I, and the bottom four isoforms lack the signal peptide, the acidic domain and Ig loop I. As many as an additional 12 isoforms could exist due to alternative splicing of part of the Ig loop III domain, which is shown as a darkened line. In addition, secreted isoforms have been observed that lack the transmembrane domain and the intracellular portion of FGFR-1. This figure was adapted from Hou et al (1991).*

because these receptors can bind FGFs with different affinities [Duan et al., 1992].

F. Low-Affinity FGF Receptors

Cells also express low-affinity FGF-binding sites that exhibit apparent dissociation constants in the range of 2 nM (typically 20- to 100-fold lower than the high-affinity receptors). These low-affinity receptors are present in much higher numbers than high-affinity FGF receptors [Moscatelli, 1987]. Several lines of evidence argue that low-affinity FGF receptors are composed, at least in part, of HSPG located on the cell surface [Moscatelli, 1988] and in the extracellular matrix [Vlodavsky et al., 1987]. Interestingly, a recent kinetic analysis of FGF-1 binding to Balb/c3T3 cells indicates that the "on rates" (k_{on}) for HSPGs and FGF CSP receptors are very similar. In contrast, the "off rate" (k_{off}) for HSPG is over 20 times greater than that for FGF CSP receptors [Nugent and Edelman, 1992a]. Furthermore, the k_{off} for isolated FGF protein receptors is approximately 20 times higher than that for cell-associated FGF protein receptors. Based on the latter finding, it appears that FGF CSP receptors may be more complex structures than single polypeptides.

Several studies argue that HSPGs are essential for the binding of FGF to cells. One of the first studies to support this argument convincingly was performed with Chinese hamster ovary (CHO) cells that are defective in the production of HSPGs and which exhibit few high-affinity FGF receptors. These cells produce only about 5% of the HSPGs produced by wild-type CHO cells and the number of low-affinity FGF receptors exhibited is about 10 times smaller than their parental clone. When these cells are transfected with an expression plasmid containing an FGFR-1 cDNA, the cells bind FGF-1 with high affinity only when heparin (but not chondroitin sulfate) is added to the binding medium [Yayon et al., 1991].

The conclusion that HSPGs are essential for the high-affinity binding of FGFs to the cell surface is supported by two other lines of evidence. Treatment of Swiss 3T3 cells with either sodium chlorate (which blocks sulfation) or heparinase greatly reduces the binding of FGF-1 to both the low-affinity receptors and the FGF CSP receptors [66%–80%, Rapraeger et al., 1991]. Similarly, MM14 myoblasts, which can be prevented from differentiating by culturing them in the presence of FGFs, fail to respond to FGF-1, FGF-2, or FGF-4 and differentiate when treated with chlorate. While additional studies remain to be conducted, the work reported for MM14 myoblasts argues that HSPGs are required for the binding of at least these three FGFs to high-affinity receptors.

G. HSPGs and the Formation of High-Affinity FGF Receptors

The findings discussed above raise a number of important mechanistic questions concerning the binding of FGFs. First, are FGFs bound initially by HSPGs and subsequently "presented," possibly in an altered conformation, to the high-affinity protein receptors? Alternatively, do FGFs bind simultaneously to HSPGs and their protein receptors? Lastly, do HSPGs interact directly with cell surface protein receptors and form a high-affinity coreceptor for FGF? Complete answers to these questions remain to be determined, but a recent report provides strong evidence that HSPGs interact with FGF CSP receptors and that this interaction is required for the formation of high-affinity FGF binding sites.

Initial evidence for the possibility that FGF CSP receptors interact with HSPGs was provided by the finding that heparin can protect specific regions of recombinant FGFR-1 from proteolytic digestion [Kan et al., 1993]. Isolation and sequence analysis of the heparin-binding, trypsin-resistant fragments ultimately helped identify an 18-amino-acid region of FGFR-1 (KMEKKLHAVPAAKTVKFK) that binds heparin. This sequence is located just upstream of immunoglobulin-like loop II, and is similar to amino acid sequences found in the heparin-binding domains of other heparin-binding proteins [Cardin and Weintraub, 1989]. Interestingly, this sequence is highly conserved at the amino acid level in each of the five FGF CSP receptors, with minor conservative amino acid substitutions observed in each case. Other studies indicate that an antibody to this 18-amino-acid polypeptide, and the synthetic polypeptide itself, both block the binding of FGF to intact cells and to a recombinant FGFR-1/heparin complex [Kan et al., 1993]. Moreover, these reagents block the induction of cell proliferation by FGF. Finally, the role of this putative heparin-binding domain was tested by modifying FGFR-1 genetically. Using polymerase chain reaction (PCR)-mediated mutagenesis, the first three lysine residues in this sequence were modified with the result that the binding of FGF-1 and FGF-2 to cells expressing the modified FGFR-1 was abolished [Kan et al., 1993].

Collectively, the studies discussed in this section provide considerable evidence that HSPGs are essential for FGF to bind to FGF CSP receptors. However, there is ample evidence to suggest that HSPGs are likely to influence FGF utilization in at least three other ways. First, HSPGs may increase the stability of FGFs by limiting proteolytic degradation [Gospodarowicz and Cheng, 1986; Saksela et al., 1987]. Second, HSPGs may serve as a reservoir for FGF by providing extracellular sites for long-term storage [Flaumenhaft et al., 1989]. Third, FGFs complexed to heparan sulfate diffuse more rapidly in an *in vitro* model system, which suggests that HSPGs cleaved from the cell surface could increase the distribution of FGFs *in vivo* [Flaumenhaft et al., 1990].

H. Candidate HSPGs and FGF-Binding Proteins

The studies discussed in the previous section suggest that HSPGs participate in the formation of high-affinity FGF receptors; however, the specific HSPGs involved have not been identified. The family of HSPGs known as syndecans appears to be a good candidate. Thus far, four different members of the syndecan family have been cloned. Each codes for a protein core that consists of a signal peptide, a large extracellular domain (which is poorly conserved and to which heparin sulfate is attached), a conserved putative protease cleavage site (which is adjacent to the transmembrane domain) and a short, highly conserved, intracellular domain. Readers interested in the structure, developmental expression, and tissue distribution of the syndecan family are directed to an excellent review by Bernfield et al. [1992]. Additional information regarding HSPGs can be found in reviews by Ruoslahti [1991] and Jackson et al. [1991].

The possible involvement of syndecans in the binding and utilization of FGFs is supported by a number of different studies. Purified and cell-associated syndecans are known to bind FGFs, whereas their core proteins do not [Bernfield et al., 1992]. Expression of hamster syndecan-1 by a human lymphoblastoid cell line is known to induce the formation of low-affinity FGF-2 binding sites [Kiefer et al., 1990, 1991]. Furthermore, the expression of syndecan-1 during amphibian development parallels the induction of axial mesoderm formation. Given these suggestive findings, the next important step should be to determine whether the selective inactivation or removal of specific syndecans can alter the binding of different FGFs to their respective FGF CSP receptors.

There are two other molecules that are known to bind FGFs. Several years ago a unique FGF-binding protein (150 kD) was identified. Recently the cDNA for this glycoprotein was cloned and sequenced [Burrus et al., 1992]. This protein is composed of a large cysteine-rich extracellular domain, a transmembrane domain, and a short (13-amino-acid) intracellular domain. A small amount of this protein localizes to the cell surface, but most of it appears to be associated with the Golgi apparatus [B. Olwin, personal communication]. Thus far, the FGF-binding determinants of this protein have not been determined nor has the function of this protein been determined. However, it has been shown to bind FGF-1, FGF-2, and FGF-4.

FGF-2 has also been shown recently to bind betaglycan, which was first identified as a transforming growth factor-β (TGF-β) cell surface–binding protein. However, unlike TGF-β, which binds to the betaglycan protein core, FGF-2 binds via the heparan sulfate side chains of betaglycan [Andres et al., 1992]. Interestingly, treatment of osteoblast-enriched primary cultures of fetal rat calvaria with FGF-2, but not TGF-β, reduced the heparan sulfate content of betaglycan. As in the case of the FGF-binding protein discussed above, the physiological role of betaglycan, if any, in the actions of FGFs remains to be determined. Lastly, two recent reports demonstrate that TGF-β1 increases both the production of HSPGs and the binding of FGF-2 [Nugent and Edelman, 1992b; Jiang et al., 1992]. Although its mechanism remains to be determined, this phenomenon could help explain the ability of TGF-β1 to potentiate the biological effects of FGF-2 on several cell types, including corneal endothelial cells [Plouët and Gospodarowicz, 1989], osteoblasts [Globus et al., 1988], and 3T3 cells lacking the epidermal growth factor (EGF) receptor [Rizzino and Ruff, 1986]. Readers interested in the effects of TGF-β on the extracellular matrix are referred to several detailed review articles [Rizzino, 1988; Massagué, 1990; Akhurst, this volume].

I. Developmental Regulation of FGF Receptors

Examinations of the expression of FGF receptors during mouse embryonic development have begun recently. As is the case for FGFs, the expression of FGF CSP receptor genes is highly regulated temporally and spatially. Although not reviewed here, similar findings have been made recently for the expression of FGF CSP receptors during chicken development [Patstone et al., 1993]. Consequently, it is very likely that different FGF receptors mediate distinct developmental events. FGFR-1 and FGFR-2 expression has been demonstrated both prior to and during gastrulation in early mouse embryos. The earliest detectable FGF receptor expression occurs in the blastocyst stage embryo, where transcripts for FGFR-1 have been demonstrated using the polymerase chain reaction to amplify reverse transcribed mRNA [Campbell et al., 1992]. In the early egg-cylinder stage embryo, *in situ* hybridization studies have been used to show that FGFR-1 and FGFR-2 are expressed diffusely throughout the primitive ectoderm prior to gastrulation (Orr-Urtreger et al., 1991), but are not expressed in the primitive endoderm. With the onset of gastrulation, expression of the two receptors becomes distinct, with persistent expression of FGFR-2 in the primitive ectoderm, and FGFR-1 transcripts appearing in the nascent mesodermal layer. FGFR-1 expression at this stage appears to follow the anterolateral migration of the gastrulating mesoderm as it leaves the primitive streak, and, by the late primitive streak stage, FGFR-1 is evident throughout the mesoderm and ectoderm [Yamaguchi et al., 1992]. In contrast, FGFR-2 expression remains restricted to the ectodermal layer. This distinct pattern of FGF receptor expression is reminiscent of the expression of the ligands, FGF-3, FGF-4, and FGF-5, in the ectoderm and mesoderm during gastrulation, and it suggests a complex role for FGFs during mesoderm formation.

FGFR expression continues in a wide variety of tissues during organogenesis, indicating that FGFs influence the development of many tissues. Despite this apparent complexity, several patterns have emerged. In general, FGFR-1 expression is most prominent in mesenchyme, whereas FGFR-2 expression is strongest in surface ectoderm and in the epithelia of several developing organs [Peters et al., 1992a; Orr-Urtreger et al., 1991]. In contrast, transcripts for FGFR-4 are most conspicuous in muscle precursors and in differentiated skeletal muscle, as well as in definitive endoderm (Stark et al., 1991). In several instances where organ formation is likely to involve reciprocal epithelial-mesenchymal interactions, FGFR-1 and FGFR-2 exhibit a complementary pattern of expression. A striking example of this occurs in the skin, where

FGFR-2 transcripts are localized within a discrete band corresponding to the proliferating basal layer of epidermis and within the ectodermal components of developing hair follicles. FGFR-1 expression, on the other hand, is restricted to the mesoderm-derived dermal layer immediately beneath the basal skin layer and the mesodermal components of developing follicles. A similar complementary pattern of expression was observed in other developing organs, including the limb bud, lung, gut, and kidney. These findings support the proposal that FGFs are important in mediating inductive interactions between cells of different developmental lineages.

The expression of FGFR-4 transcripts suggests that this receptor has a prominent role in muscle development. During the differentiation of somites, the myotome forms as the precursor of the skeletal muscle of the trunk and limbs. FGFR-4 transcripts are expressed at high levels in the myotome [Stark et al., 1991], whereas transcripts for FGFR-1 and FGFR-2 are absent. Furthermore, FGFR-4 expression appears to persist in all skeletal muscle during subsequent development, but is absent from cardiac muscle. FGFR-4 is also expressed strongly in definitive gut endoderm and other endoderm-derived tissues, such as the liver and pancreas, but its expression is negligible in the central and peripheral nervous system. Interestingly, FGFR-3 appears to be expressed in the central nervous system and several other tissues, including intestine, lung, and bone [Partanen et al., 1991]. The pattern of FGFR-3 expression during development has not been reported, but a study describing the developmental expression of FGFR-3 apparently will be published in the near future to show that the pattern of FGFR-3 expression during development differs from that of FGFR-1 and FGFR-2 [Ornitz and Leder, 1992]. Lastly, although Flg-2 mRNA has been shown to be expressed in skin, brain, and lung [Avivi et al., 1991], its expression during development has not been reported.

The expression of FGF receptors during early *Xenopus* development has also been described [Musci et al., 1990]. FGF receptor transcripts are expressed at relatively constant levels from the mature oocyte to the neurula stages, with no apparent regional differences in transcript levels. Whereas isolated animal caps are able to respond to exogenous FGFs, it seems that the oocyte does not express functional FGF receptors. This suggests that post-transcriptional regulation may play an important role in FGF receptor expression during amphibian development. This point is worth bearing in mind when interpreting the results of *in situ* hybridization studies of FGF receptor expression in mammalian embryos. Furthermore, it is important to note that *in situ* hybridization studies have not distinguished among the various isoforms of FGF receptor transcripts generated by alternative splicing. This may prove to be important because the affinities of different FGF receptor isoforms for specific FGFs can vary by three orders of magnitude. Lastly, it should be kept in mind that FGF receptors, at least *in vitro,* can be influenced by cell density. In this case, as cell density increases, FGF binding decreases [Rizzino et al., 1988b; Veomett et al., 1989].

Studies of the expression of FGF receptors during embryonic development argue that the regulation of FGF receptors, in addition to the regulation of their ligands, is a critical mechanism that ensures the specificity of cell–cell signalling events in development. Consequently, it will be important to study the mechanisms that regulate the expression of FGF receptors. The recent cloning of promoter regions of the FGFR-1 and FGFR-2 genes [Avivi et al., 1992; Saito et al., 1992] should enable these studies to progress rapidly. In addition, several *in vitro* systems have been described that could serve as suitable models for understanding FGF receptor regulation. In this regard, mouse EC cells have been used extensively to study the regulation of the ligands of the FGF family, because the patterns of FGF expression change dramatically upon differentiation [Rizzino et al., 1988a; Ma et al., 1992]. These cells also regulate expression of their FGF receptors; differentiation of mouse F9 EC cells leads to a 7- to 10-fold increase in high-affinity FGF binding [Rizzino et al., 1988a]. Only the differentiated cells exhibit a mitogenic response to exogenous FGFs. Other studies indicate that EC cells express transcripts for FGFR-1 and FGFR-2 [Campbell et al., 1992], whereas EC cells do not express transcripts for FGFR-4 until they are induced to differentiate into cultures containing myoblasts [Stark et al., 1991]. Another fruitful model system for the study of FGF receptor regulation is the *in vitro* differentiation of myoblasts. Previous studies have shown that cultured myoblasts express transcripts for FGFR-1 [Moore et al., 1991] and that FGFs stimulate the growth and inhibit the differentiation of these cells [Linkhart et al., 1980; Lathrop et al., 1985; Spizz et al., 1986; Clegg et al., 1987]. Furthermore, terminal differentiation of cultured myoblasts is associated with a loss of FGF receptors.

IV. FGF RECEPTOR ACTIVATION AND SIGNAL TRANSDUCTION PATHWAYS

A. Activation of Tyrosine Kinase Activity

Work during the past several years suggests strongly that the activation of FGF receptors follows a pattern of events very similar to that observed during the activation of several other growth factor receptors, including receptors for epidermal growth factor and platelet-derived growth factor [Schlessinger and Ullrich, 1992; Tiesman and Rizzino, 1989b]. After FGF binds to its receptor, ligand-bound receptors undergo dimerization, which elevates receptor tyrosine kinase activity and, in turn, induces receptor autophosphorylation by an intermolecular mechanism (Fig. 4). Surprisingly, although FGF bound to its receptor is internalized quickly [Moscatelli, 1988; Moenner et al., 1989], it is not degraded rapidly as is the case for epidermal growth factor (EGF) and platelet-derived growth factor (PDGF). Thus far, most of the evidence for this receptor activation model has been derived by examining FGFR-1 and FGFR-2, but it is very likely that this model also applies to FGFR-3, FGFR-2, and flg-2.

Support for this model derives primarily from work with FGF receptors that lack tyrosine kinase activity. For some of these studies, recombinant tyrosine kinase–deficient FGF receptors were produced by substituting alanine for lysine at the ATP-binding site (positions 514 and 517 in FGFR-1 and FGFR-2, respectively). Tyrosine kinase–deficient FGF receptors are able to bind FGF with high affinity and undergo dimerization [Bellot et al., 1991]. In addition, it appears that tyrosine kinase–deficient FGF receptors can be phosphorylated when complexed with the wild-type receptor in a heterodimer. This argues that FGF receptor tyrosine autophosphorylation occurs by an intermolecular mechanism [Bellot et al., 1991], as appears to be the case for several other growth factor receptors [Ullrich and Schlessinger, 1990].

If it is generally true that tyrosine kinase–deficient receptors in a heterodimer can be phosphorylated, then it is likely that activation of FGF receptor signal transduction pathways requires that both receptors in the complex undergo autophosphorylation on specific tyrosine residues. This conclusion is based on the finding that overexpression of tyrosine kinase–deficient FGF receptors in amphibian oocytes can suppress activation of the wild-type receptor [Amaya et al., 1991; Ueno et al., 1992]. Although it is widely believed that the dominant-negative effect of tyrosine kinase–deficient FGF receptors is due to the formation of inactive heterodimers, this important conclusion should be confirmed by more direct studies. As the reagents needed for these studies are available [Xu et al., 1992; Hou et al., 1992; Peters et al., 1992b; Mohammadi et al., 1992], this issue could be resolved quickly.

The formation of heterodimers and the apparent need for intermolecular phosphorylation to activate FGF receptors raises the possibility that FGF receptor isoforms that lack functional tyrosine kinase domains may limit the response of cells to FGFs and generate a biphasic response of cells to FGFs. In one scenario, cells may express two isoforms of FGFR-1, one that binds FGF-1 with high affin-

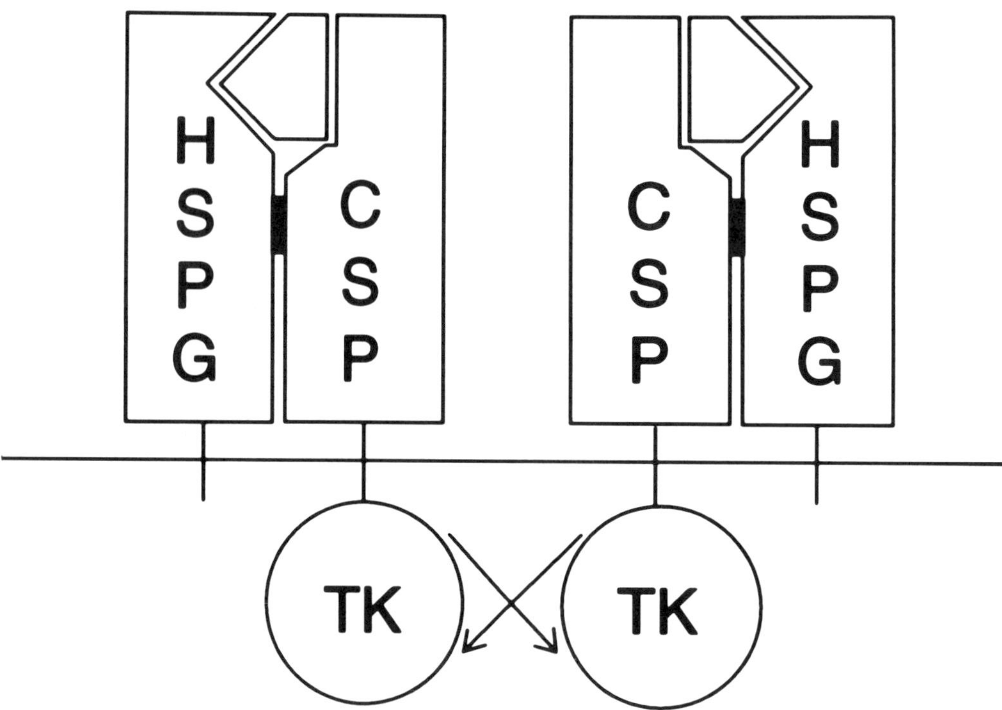

Fig. 4. *Model for activation of FGF receptors. Ternary complexes, which form upon the binding of an FGF (pentagon) to a heparan-sulfate proteoglycan (HSPG) and an FGF CSP receptor (CSP), can dimerize to produce homodimers (shown here) or heterodimers. Upon dimerization, the intracellular tyrosine kinase of each ternary complex is activated, which results in intermolecular (and possibly intramolecular) autophosphorylation of specific tyrosine residues. After autophosphorylation occurs, activated FGF receptors phosphorylate various intracellular targets, which in turn activate an intracellular cascade of signal transduction pathways.*

ity and contains a functional tyrosine kinase, and one that binds FGF-1 with lower affinity and lacks a functional tyrosine kinase domain. In this case, active homodimers would form at low concentrations of FGF-1 and stimulate cell growth. However, as the concentration of FGF-1 increases, it would bind to both isoforms and allow inactive heterodimers to form, and this in turn could reduce the growth response of the cells to FGF-1. This model has been proposed to explain the biphasic growth response of the human hepatoblastoma cell line, HepG2, to FGF-1 [Kan et al., 1988; W. MeKeehan, personal communication]. Recent studies demonstrate that these cells not only express the two receptor isoforms in question but, as expected, these cells do not appear to express the receptor isoform that binds FGF-1 with high affinity and lacks a functional tyrosine kinase domain [W. MeKeehan, personal communication].

B. Autophosphorylation Sites

Thus far, two tyrosine residues have been identified that undergo autophosphorylation after FGF receptor dimerization. One is tyrosine 766 (located approximately 20 amino acids downstream of the second tyrosine kinase domain) [Peters et al., 1992b; Mohammadi et al., 1992] and the other is tyrosine 653 (located within the second tyrosine kinase domain) [Hou et al., 1993]. Although these studies were performed with FGFR-1, both tyrosine residues are conserved in all five FGF CSP receptors. Furthermore, it seems that tyrosine 766 [Peters et al., 1992b] Mohammadi et al., 1992] and tyrosine 653 [W. McKeehan, personal communication] are both involved in the FGF signal transduction pathways that stimulate phosphatidylinositol hydrolysis via the activation of phospholipase C-γl (PLC)-γl. In the case of tyrosine 766, recombinant FGF receptors containing phenylalanine at this position are unable to activate phosphatidylinositol hydrolysis. However, expression of these recombinant receptors in L6 myoblasts, which lack FGF receptors, enables these cells to respond to stimulation with FGF by an increase in DNA synthesis. Thus, it appears that increased hydrolysis of phosphatidylinositol and presumed activation of protein kinase C are not always required for FGFs to generate a mitogenic signal.

Other intracellular regions of the FGF receptor are undoubtedly important in activation of FGF receptor signal transduction pathways, but little information regarding their location is available. A recent study has focused attention on the kinase insert domain [Wennström et al., 1992]. In this study, the kinase insert domain (17 amino acids) of the FGF receptor was replaced by the kinase insert (100 amino acids) of the PDGF-β receptor to produce a chimeric receptor. Upon transfection into porcine aortic endothelial cells, the chimeric receptor was observed to exhibit PDGF receptor–specific signalling properties when the cells were treated with FGF. While this study does not implicate the FGF receptor kinase insert in FGF signalling, it illustrates the utility of using chimeric receptors to help identify regions of the receptor involved in different FGF signal transduction pathways. Moreover, as the signalling domains of the receptor are identified, they can in turn be used to screen expression libraries for proteins that bind to specific domains of FGF receptors, as was done recently for EGF receptors [Skolnik et al., 1991; Margolis et al., 1992].

C. Substrates for Activated FGF Receptors

Thus far, only a few substrates for activated FGF receptors have been identified. As mentioned above, activated FGF receptors are believed to phosphorylate PLC-γ1. Once activated by phosphorylation, PLC-γ1 hydrolyzes phosphotidylinositol to produce two important second messengers: inositol 1,4,5-triphosphate, which can induce the release of intracellular calcium, and diacylglycerol, which can activate protein kinase C [Pandiella and Meldolesi, 1992]. Activated FGF receptors have also been shown to interact with the phosphatidyl inositol (PI)-3-kinase subunit, p85 [Raffioni and Bradshaw, 1992]. PI-3-kinase is thought to be a heterodimer composed of an 84-kD regulatory subunit and a 110-kD catalytic subunit, which, when activated, can phosphorylate the D-3 position of inositol to produce a novel phospholipid, phosphatidylinositol-3-phosphate [Whitman et al., 1988]. Although the roles of these second messengers are far from clear, the proteins involved in their metabolism share an interesting property. Both PLC-γ1 and p85 contain regions that exhibit significant sequence similarity to two regions in $pp60^{src}$. These regions are referred to as Src-homology (SH) domains: SH2 (approximately 100 amino acids) and SH3 (approximately 50 amino acids). PLC-γ1 and p85 each contain two SH2 domains and a single SH3 domain [reviewed by Pawson and Gish, 1992]. Although further study is needed to fully understand how these proteins interact with FGF receptors, PLC-γ1 is believed to interact with FGF receptors via one or both of its SH2 domains [Mohammadi et al., 1991]. Similarly, it is anticipated that the same is true for p85; however, based on work with PDGF receptors, it is anticipated that p85 and PLC-γ1 interact with different regions of the FGF receptor.

In addition to increases in the hydrolysis of phosphoinositides, FGFs are also known to induce a number of other early events, but in most cases the mechanisms involved are not understood in detail. These early events include increases in intracellular calcium and intracellular pH [Tsuda et al., 1985; Halperin and Lobb, 1987]; increases in the phosphorylation of cellular proteins, such as Raf-1 [Morrison et al., 1988]; and increases in transcription of a few genes, such as c-*fos* [Stumpo and Blackshear, 1986]. However, unlike receptors for PDGF, FGFR-1 does not seem to interact with the GTPase activating protein, GAP [W. McKeehan, personal communication]; whereas, FGF receptors may be coupled to a membrane-associated heterotrimeric G-protein [Jarvis et al., 1992]. A major challenge for the future will be to understand the roles of these early events and to understand how they occur mechanistically.

V. INTRANUCLEAR LOCALIZATION OF FGFS

A. Nuclear Localization of FGFs

As described above, FGF-1 and FGF-2 lack recognizable signal sequences for vectorial translation into the endoplasmic reticulum and eventual secretion. Indeed, FGF-1 and FGF-2 seem to be secreted inefficiently, although potential mechanisms for their export from the cell have been reported [Mignatti et al., 1992; Bashkin et al., 1992]. However, a growing body of data suggests that FGF-1, FGF-2, and FGF-3, and possibly other members of the FGF family, may exert some of their effects directly in the cell nucleus. Thus, secretion may not be a requirement for all biological activities of the FGFs. Thus far, FGF-2 has been detected within the nucleus of a variety of cell lines and tissues, including vascular endothelial cells [Speir et al., 1991; Bouche et al., 1987], smooth muscle cells [Speir et al., 1991], cardiac myocytes [Kardami and Fandrich, 1989], hepatoma cells [Renko et al., 1990], male germ cells [Suzuki et al., 1991], early embryonic blastomeres of *Xenopus* [Shiurba et al., 1991], and various cell types transfected with FGF-2 expression plasmids.

Several lines of evidence suggest that nuclear translocation of FGF-2 is mediated by specific mechanisms, and that its presence in the nucleus is likely to serve important functions. In synchronized cultures of adult bovine aortic endothelial cells, the accumulation of FGF-2 in the cytoplasm is continuous throughout the cell cycle, whereas translocation to the nucleus occurs specifically during the late G_1 phase of the cell cycle [Baldin et al., 1990]. Similarly, a recent study with serum-starved coronary venular endothelial cells demonstrated that FGF-2 accumulates in the nucleus more rapidly than it accumulates in the cytoplasm [Hawker and Granger, 1992]. These results suggest that cytoplasmic and nuclear uptake are kinetically distinct events that are independently regulated. Interestingly, full-length FGF-2 persists in the nucleus for up to 24 hours, with little evidence of proteolytic degradation [Hawker and Granger, 1992]. In this regard, FGF-2 that has been recovered and purified from nuclear fractions retains its biological activity [Dell'Era et al., 1991]. In contrast, cytoplasmic FGF-2 is quickly degraded, presumably by the proteolytic action of lysosomal enzymes [Hawker and Granger, 1992]. Thus, nuclear FGF-2 may serve a continuing function within the nucleus well after the onset of DNA synthesis and cell proliferation.

Further evidence that FGF-2 is transported to the nucleus by a specific mechanism is provided by the finding that different forms of FGF-2 are distributed differently between nuclear and cytoplasmic fractions. Previous studies have shown that some cell types coexpress different species of FGF-2 that range in molecular weight from approximately 18.5 kD to approximately 24 kD (see Section V.B). Recent reports have demonstrated that the higher molecular weight forms of FGF-2 localize almost exclusively within the nucleus, whereas the 18.5-kD species is found in both the cytosol and in the nucleus [Renko et al., 1990; Bugler et al., 1991].

B. Nuclear Localization Signals in FGFs

Although the mechanism by which FGF-2 is transported into the nucleus is unknown, it appears that the high molecular weight species of FGF-2 contain a functional nuclear localization sequence (NLS) that is not present in the 18.5-kD form. In this regard, the nucleotide sequences of the human and bovine FGF-2 cDNAs both predict a single translation product of 155 amino acid residues resulting from initiation of translation at an ATG start codon [Abraham et al., 1986a,b]. However, it is now well established that the higher molecular weight forms of FGF-2 are generated from the same FGF-2 mRNA by initiation at one of three upstream alternative CTG codons [Prats et al., 1989; Florkiewicz and Sommer, 1989]. Thus, the translation products initiated at upstream CTG codons are identical to the ATG-initiated (18.5 kD) form, except for an N-terminal extension of 41, 46, or 55 amino acid residues. It appears that each of the alternative codons is used *in vivo,* because some cell lines coexpress FGF-2 species of the predicted sizes [Florkiewicz and Sommer, 1989].

As described above, the higher molecular weight forms of FGF-2 localize preferentially to the nuclei of cells that produce endogenous FGF-2. Recently, it has been demonstrated that the N-terminal extension, which is unique to the larger forms of FGF-2, contains a functional NLS. In this regard, recombinant fusion products consisting of the N-terminal extension of FGF-2 and a reporter gene, whose product is normally cytoplasmic, are directed instead to the nucleus of transiently transfected cells [Bugler et al., 1991; Quarto et al., 1991]. These results demonstrate that the N-terminal sequences of the larger FGF-2s contain a functional NLS that is able to direct heterologous gene products into the nucleus. The N-terminal sequence of FGF-2 is particularly rich in arginine residues, and it has been shown that some of these arginine residues are methylated [Burgess et al., 1991]. This is noteworthy because the presence of methylarginine residues has been demonstrated in proteins that associate with nucleic acids [Lapeyre et al., 1986].

Like FGF-2, FGF-1 has been detected within the nucleus of a variety of cell lines and tissues, including endothelial cells [Speir et al., 1991], smooth muscle cells [Speir et al., 1991], neurons and neurological cells [Tourbah et al., 1991], and cardiac myocytes [Speir et al., 1992]. Recent efforts to characterize a NLS within FGF-1 have focused on a region near the N-terminus of the truncated forms of FGF-1. This sequence (Asn-Tyr-Lys-Lys-Pro-Lys-Leu) lies three residues N-terminal to the first cysteine residue that is conserved throughout the FGF family. Moreover, the four residues, Lys-Lys-Pro-Lys, fit the consensus sequence recently defined for NLSs [Chelsky et al., 1989]. The function of this region has been studied recently by generating a recombinant form of FGF-1 in which the putative NLS has been deleted [Imamura et al., 1990]. The recombinant mutant form of FGF-1 exhibited normal heparin-binding affinity and was able to induce tyrosine phosphorylation and *c-fos* transcription in serum-starved NIH-3T3 cells. However, the mutant was unable to induce DNA synthesis and cell proliferation when added to murine lung capillary endothelial (LEII) cells. Moreover, addition of the yeast histone 2B NLS to the N-terminus of the truncated FGF-1 mutant restored the wild-type activity. Subsequent studies have shown that the putative NLS is required for translocation of FGF-1 to the nucleus of LEII cells transfected with FGF-1 expression plasmids [Imamura et al., 1992]. However, because these results were obtained when FGF-1 was produced from a recombinant vector, further work is needed to determine whether exogenous FGF-1 is also translocated to the cell nucleus upon addition to cell cultures. In addition, it will be important to show that the putative NLS of FGF-1 in heterologous gene constructs can drive these gene products to the nucleus.

Recent results indicate that nuclear localization is not unique to the two FGFs that lack a signal peptide. As was described above for FGF-2, it is now known that at least two forms of FGF-3 are generated by initiation of translation at one of two possible start codons [Dickson and Acland, 1990]. Studies with COS-1 cells transfected with recombinant FGF-3 expression vectors show that the N-terminally extended form is localized almost exclusively in the nucleus, even though both of the FGF-3 translation products contain a signal peptide [Dickson and Acland, 1990; Acland et al., 1990]. However, it is unclear how, in the presence of an apparently functional signal peptide, the larger FGF-3 translation product is directed into the nucleus. The N-terminal extension may contain a NLS that suppresses or masks the signal peptide. Alternatively, there may be a cryptic NLS within the interior of the FGF-3 sequence that is able to direct the product to the nucleus only when the presence of extra N-terminal residues interferes with the activity of the signal peptide. Further studies are needed to resolve these possibilities. However, the finding that endogenous FGF-3 is able to be translocated into the nucleus indicates that any of the FGFs, even those containing signal peptides, may exert some of their biological effects by directly influencing events in the nucleus.

VI. CONCLUDING REMARKS

It is evident from the work discussed that major advances have been made in the FGF field during the past year. Nonetheless, there are many more questions than answers. For example, the temporal and spatial expression of several FGFs have been examined during mammalian development, but the developmental roles of the FGFs are still entirely speculative. While there are several ways of attempting to address this important issue, inactivation of each FGF gene by gene targeting [reviewed by Wilder and Rizzino, 1993] offers the greatest opportunity for fully understanding the roles of the different FGFs during development. Equally important, and an area that has received far too little attention, are the mechanisms that regulate the temporal and spatial expression of the different FGFs and their receptors. For this work, and the gene targeting studies suggested above, it will be important to isolate mouse genomic clones for each of the FGFs and their receptors.

An area of research that has not been addressed, and one that is critical to fully understanding the roles of the different FGFs, is the determination of which FGFs bind to which FGF receptor *in vivo*. At present, there does not seem to be any simple method for answering this question. However, detailed comparisons of the temporal and spatial expression of the different FGFs and the different FGF receptor isoforms, coupled with inactivation of different FGF receptor genes, should help provide important clues for answering this question.

FGF signal transduction is another area that warrants intensive investigation. Thus far, few FGF signal transduction pathways have been identified and, as in the case of other growth factors, we know virtually nothing of the signal transduction pathways utilized to induce DNA synthesis and cell division. Here again, simple strategies do not exist for answering the questions at hand. However, as receptor domains required for triggering signal transduction pathways are identified, they can be used to screen expression libraries and help identify the initial intracellular targets involved in the signaling pathways.

Equally challenging questions remain regarding the targeting of FGFs to the nucleus. For example, it is unclear how exogenous FGFs are transported from the extracellular compartment into the nucleus. The findings for FGF-1, described above, suggest that NLSs within the primary sequence may play a role in this process. In addition, it is unclear whether exogenous and endogenous FGFs are transported to the nucleus by similar mechanisms; however, a recent study suggests that similar mechanisms may be involved [Zhan et al., 1992]. Perhaps the most important question is, if FGFs exert some of their effects in the nucleus, what are the mechanisms involved? An intriguing possibility has been suggested by the finding that FGF-2 can regulate gene transcription directly in a cell-free transcription system [Nakanishi et al., 1992]. Whether this effect is mediated by direct interaction of FGF-2 with DNA or by indirect mechanisms is unclear. Consequently, an essential goal of this work should be the identification of potential nuclear receptors for FGFs.

We are a long way from knowing why there are so many FGFs and why there are so many splice variants of the different FGF receptor genes. The most likely explanation is that this level of complexity makes it possible for FGFs to activate different combinations of signal transduction pathways in each cell type that responds to FGFs. In this regard, our current knowledge suggests that different FGFs affect each receptor isoform somewhat differently by virtue of different binding affinities. Similarly, it is very likely that most of the receptor isoforms have different signalling domains. If this is the case, it suggests that each FGF receptor isoform could activate a different combination of signal transduction pathways. At present there is relatively little evidence for this, but recent studies indicate that cells expressing FGFR-1 and FGFR-4 contain different phosphorylated cellular proteins [Vainikka et al., 1992]. Moreover, it appears that FGFR-1 is more likely to phosphorylate PLC-γ1 than FGFR-4. Lastly, additional diversity could be generated by regulating the intracellular machinery that makes up the signal transduction pathways. This could occur in cell differentiation with fine modulation of the intracellular levels of factors involved in signal transduction. In addition, it is equally likely that most FGF-responding cells only express the components necessary for a small fraction of the different FGF signal transduction pathways. In this regard, some cytoplasmic proteins, which are believed to be involved in signal transduction, are now known to be expressed in only some cell lineages [Katzac et al., 1989; Adams et al., 1992; Margolis et al., 1992].

Finally, one should not be too surprised by unexpected developments in the FGF field. For example, it would not be surprising to find that FGFs exert some of their effects on inactive transcription factors that reside in the cytoplasm until phosphorylated directly by activated FGF receptors. Just such a case has been documented recently for the interferon-α receptor [Schindler et al., 1992; David and Larner, 1992]. Another interesting possibility is that FGF receptors may exert other enzymatic activities. In this regard, a new class of proteins, including the insulin receptor [Baltensperger et al., 1992; R. Lewis personal communication], has been shown to exhibit both tyrosine kinase activity and serine-threonine kinase activity [Lindberg et al., 1992].

ACKNOWLEDGMENTS

Wally McKeehan, Bradley Olwin, and Rob Lewis are thanked for discussing their unpublished data and for providing copies of unpublished manuscripts. The authors also wish to thank Tony Hollingsworth and Heather Rizzino for reading this manuscript and for making helpful suggestions. Work conducted in the laboratory of Angie Rizzino is supported by grants from the National Institute of Child Health and Human Development (HD 19837) and the National Cancer Institute (Laboratory Cancer Research Center grant CA 36727). Keith Miller was supported by a fellowship from the Nebraska Research Initiative in Biotechnology.

REFERENCES

Abraham JA, Mergia A, Whang JL, Tumolo A, Friedman J, Hjerrild KA, Gospodarowicz D, Fiddes JC (1986a): Nucleotide sequence of a bovine clone encoding the angiogenic protein, basic fibroblast growth factor. Science 233:545–548.

Abraham JA, Whang JL, Tumolo A, Mergia A, Friedman J, Gospodarowicz D, Fiddes JC (1986b): Human basic fibroblast growth factor: nucleotide sequence and genomic organization. EMBO J 5:2523–2528.

Acland P, Dixon M, Peters G, Dickson C (1990: Subcellular fate of the Int-2 oncoprotein is determined by choice of initiation codon. Nature 343:662–665.

Adams JM, Houston H, Allen J, Lints, T, Harvey R (1992): The hemapoietic expressed *vav* protooncogene shares homology with the *dbl* GDP-GTP exchange factor, *bcr* gene and a yeast gene CDC24 involved in cytoskeletal organization. Oncogene 7:611–618.

Alterio J, Halley C, Brou C, Soussi T, Courtois Y, Laurent M (1988): Characterization of a bovine acidic

FGF cDNA clone and its expression in brain and retina. FEBS Lett 242:41–46.

Amaya E, Musci TJ, Kirschner MW (1991): Expression of a dominant negative mutant of the FGF receptor disrupts mesoderm formation in *Xenopus* embryos. Cell 66:257–270.

Andres JL, DeFalcis D, Noda M, Massagué J (1992): Binding of two growth factor families to separate domains of the proteoglycan betaglycan. J Biol Chem 267:5927–5930.

Avivi A, Zimmer Y, Yayon A, Yarden Y, Givol D (1991): Flg-2, a new member of the family of fibroblast growth factor receptors. Oncogene 6:1089–1092.

Avivi A, Skorecki A, Yayon A, Givol D (1992): Promoter region of the murine fibroblast growth factor receptor 2 (*bek*/KGFR) gene. Oncogene 7:1957–1962.

Baird A, Schubert D, Ling N, Guillemin R (1988): Receptor- and heparin-binding domains of basic fibroblast growth factor. Proc Natl Acad Sci USA 85:2324–2328.

Baldin V, Roman A-M, Bosc-Bierne I, Amalric F, Bouche G (1990): Translocation of bFGF to the nucleus is G_1 phase cell cycle specific in bovine aortic endothelial cells. EMBO J 9:1511–1517.

Baltensperger K, Lewis RE, Woon W-C, Vissavajjhala P, Ross AH, Czech MP (1992): Catalysis of serine and tyrosine autophosphorylation by the human insulin receptor. Proc Natl Acad Sci USA 89:7885–7889.

Bashkin P, Neufeld G, Gitay-Goren H, Vlodavsky I (1992): Release of cell surface-associated basic fibroblast growth factor by glycosylphosphatidylinositol-specific phospholipase C. J Cell Phys 151:126–137.

Basilico C, Moscatelli D (1992): The FGF family of growth factors and oncogenes. Adv Cancer Res 59:115–165.

Bellot F, Crumley G, Kaplow JM, Schlessinger J, Jaye M, Dionne CA (1991): Ligand-induced transphosphorylation between different FGF receptors. EMBO J 10:2849–2854.

Bernard O, Li M, Reid HH (1991): Expression of two different forms of fibroblast growth factor receptor 1 in different mouse tissues and cell lines. Proc Natl Acad Sci USA 88:7625–7629.

Bernfield M, Kokenyesi R, Kato M, Hinkes MT, Spring J, Gallo RL, Lose EJ (1992): Biology of the syndecans: a family of transmembrane heparan sulfate proteoglycans. Annu Rev Cell Biol 8:365–393.

Bouche G, Gas N, Prats H, Baldin V, Tauber J-P, Teissié J, Amalric F (1987): Basic fibroblast growth factor enters the nucleolus and stimulates the transcription of ribosomal genes in ABAE cells undergoing $G_0 \rightarrow G_1$ transition. Proc Natl Acad Sci USA 84:6770–6774.

Brookes S, Smith R, Casey G, Dickson C, Peters G (1989): Sequence organization of the human *int*-2 gene and its expression in teratocarcinoma cells. Oncogene 4:429–436.

Brunetti A, Goldfine ID (1990): Role of myogenin in myoblast differentiation and its regulation by fibroblast growth factor. J Biol Chem 265:5960–5963.

Bugler B, Amalric F, Prats H (1991): Alternative initiation of translation determines cytoplasmic or nuclear localization of basic fibroblast growth factor. Mol Cell Biol 11:573–577.

Burgess WH, Shaheen AM, Ravera M, Jaye M, Donohue PJ, Winkles JA (1990): Possible dissociation of the heparin-binding and mitogenic activities of heparin-binding (acidic fibroblast) growth factor-1 from its receptor-binding activities by site-directed mutagenesis of a single lysine residue. J Cell Biol 111:2129–2138.

Burgess WH, Maciag T (1989): The heparin-binding (fibroblast) growth factor family of proteins. Annu Rev Biochem 58:575–606.

Burgess WH, Bizik J, Mehlman T, Quarto N, Rifkin DB (1991): Direct evidence for methylation of arginine residues in high molecular weight forms of basic fibroblast growth factor. Cell Regul 2:87–93.

Burrus LW, Zuber ME, Lueddecke BA, Olwin BB (1992): Identification of a cysteine-rich receptor for fibroblast growth factors. Mol Cell Biol 12:5600–5609.

Campbell WJ, Miller KA, Anderson TM, Shull JD, Rizzino A (1992): Expression of fibroblast growth factor receptors by embryonal carcinoma cells and early mouse embryos. In Vitro Cell Dev Biol 28A:61–66.

Cardin AD, Weintraub HJR (1989): Molecular modeling of protein-glycosaminoglycan interactions. Arteriosclerosis 9:21–32.

Chelsky D, Ralph R, Jonak G (1989): Sequence requirements for synthetic peptide–mediated translocation to the nucleus. Mol Cell Biol 9:2487–2492.

Clegg CH, Linkhart TA, Olwin BB, Hauschka SD (1987): Growth factor control of skeletal muscle differentiation: commitment to terminal differentiation occurs in G_1 phase and is repressed by fibroblast growth factor. J Cell Biol 105:949–956.

Curatola AM, Basilico C (1990): Expression of the K-*fgf* proto-oncogene is controlled by 3′ regulatory elements which are specific for embryonal carcinoma cells. Mol Cell Biol 10:2475–2484.

David M, Larner AC (1992): Activation of trancription factors by interferon-alpha in a cell-free system. Science 257:813–815.

Dawid IB, Taira M, Good PJ, Rebagliati MR (1992): The role of growth factors in embryonic induction in *Xenopus laevis*. Mol Reprod Dev 32:136–144.

de Lapeyriere O, Rosnet O, Benharroch D, Raybaud F, Marchetto S, Planche J, Galland F, Mattei M, Copeland NG, Jenkins NA, Coulier F, Birnbaum D (1990):

Structure, chromosome mapping and expression of the murine *Fgf-6* gene. Oncogene 5:823–831.

Dell KR, Williams LT (1992): A novel form of fibroblast growth factor receptor 2. J Biol Chem 267:21225–21229.

Dell'Era P, Presta M, Ragnotti G (1991): Nuclear localization of endogenous basic fibroblast growth factor in cultured endothelial cells. Exp Cell Res 192:505–510.

Dickson C, Acland P (1990): Int-2: a member of the fibroblast growth factor family has different subcellular fates depending on the choice of initiation codon. Enzyme 44:225–234.

Dionne CA, Crumley G, Bellot F, Kaplow JM, Searfoss G, Ruta M, Burgess WH, Jaye M, Schlessinger J (1990): Cloning and expression of two distinct high-affinity receptors cross-reacting with acidic and basic fibroblast growth factors. EMBO J 9:2685–2692.

Duan D-SR, Werner S, Williams LT (1992): A naturally occurring secreted form of fibroblast growth factor (FGF) receptor 1 binds basic FGF in preference over acidic FGF. J Biol Chem 267:16076–16080.

Eisemann A, Ahn JA, Graziani G, Tronick SR, Ron D (1991): Alternative splicing generates at least five different isoforms of the human basic-FGF receptor. Oncogene 6:1195–1202.

Feige J-J, Baird A (1989): Basic fibroblast growth factor is a substrate for protein phosphorylation and is phosphorylated by capillary endothelial cells in culture. Proc Natl Acad Sci USA 86:3174–3178.

Flaumenhaft R, Moscatelli D, Saksela O, Rifkin DB (1989): The role of extracellular matrix in the action of basic fibroblast growth factor: matrix as a source of growth factor for long-term stimulation of plasminogen activator production and DNA synthesis. J Cell Physiol 140:75–81.

Flaumenhaft R, Moscatelli D, Rifkin DB (1990): Heparin and heparan sulfate increase the radius of diffusion and action of bFGF. J Cell Biol 111:1651–1659.

Florkiewicz RZ, Sommer A (1989): Human basic fibroblast growth factor gene encodes four polypeptides: three initiate translation from non-AUG codons. Proc Natl Acad Sci USA 86:3978–3981.

Gabriel B, Baldin V, Roman AM, Bosc-Bierne I, Noaillac-Depeyre N, Prats H, Teissié J, Bouche G, Amalric F (1991): Localization of peptide growth factors in the nucleus. In Barnes D, Mather JP, Sato GH (eds): Methods in Enzymology, Peptide Growth Factors, Part C, Vol 198. San Diego: Academic Press, pp 480–501.

Globus RK, Patterson-Buckendahl P, Gospodarowicz D (1988): Regulation of bovine bone cell proliferation by fibroblast growth factor and transforming growth factor β. Endocrinology 123:98–105.

Godsave SF, Shiurba RA (1992): *Xenopus* blastulae show regional differences in competence for mesoderm induction: correlation with endogenous basic fibroblast growth factor levels. Dev Biol 151:506–515.

Gospodarowicz D, Cheng J (1986): Heparin protects basic and acidic FGF from inactivation. J Cell Physiol 128:475–484.

Grinberg D, Thurlo J, Watson R, Smith R, Peters G, Dickson C (1991): Transcriptional regulation of the *int*-2 gene in embryonal carcinoma cells. Cell Growth Differ 2:137–143.

Halperin JA, Lobb RR (1987): Effect of heparin-binding growth factors on monovalent cation transport in Balb/c 3T3 cells. Biochem Biophys Res Commun 144:115–122.

Hatten ME, Lynch M, Rydel RE, Sanchez J, Joseph-Silverstein J, Moscatelli D, Rifkin DB (1988): *In vitro* neurite extension by granule neurons is dependent upon astroglial-derived fibroblast growth factor. Dev Biol 125:280–289.

Hattori Y, Odagiri H, Nakatani H, Miyagawa K, Naito K, Sakamoto H, Katoh O, Yoshida T, Sugimura T, Terada M (1990): K-*sam*, an amplified gene in stomach cancer, is a member of the heparin-binding growth factor receptor genes. Proc Natl Acad Sci USA 87:5983–5987.

Haub O, Drucker B, Goldfarb M (1990): Expression of the murine fibroblast growth factor 5 gene in the adult central nervous system. Proc Natl Acad Sci USA 87:8022–8026.

Haub O, Goldfarb M (1991): Expression of the fibroblast growth factor-5 gene in the mouse embryo. Development 112:397–406.

Hawker JR Jr, Granger HJ (1992): Internalized basic fibroblast growth factor translocates to nuclei of venular endothelial cells. Am Physiol 262:H1525–H1537.

Heath WF, Cantrell AS, Mayne NG, Jaskunas SR (1991): Mutations in the heparin-binding domains of human basic fibroblast growth factor alter its biological activity. Biochemistry 30:5608–5615.

Hébert JM, Basilico C, Goldfarb M, Haub O, Martin GR (1990): Isolation of cDNAs encoding four mouse FGF family members and characterization of their expression patterns during embryogenesis. Dev Biol 138:454–463.

Hébert JM, Boyle M, Martin GR (1991): mRNA localization studies suggest that murine FGF-5 plays a role in gastrulation. Development 112:407–415.

Hom DB, Maisel RH (1992): Angiogenic growth factors: their effects and potential in soft tissue wound healing. Ann Otol Rhinol Laryngol 101:349–354.

Horlick RA, Stack SL, Cooke GM (1992): Cloning, expression and tissue distribution of the gene encoding rat fibroblast growth factor receptor subtype 4. Gene 120:291–295.

Hou J, Kan M, Wang F, Xu J-M, Nakahara M, McBride G, McKeehan K, McKeehan WL (1992): Substitution of putative half-cystine residues in heparin-binding fibroblast growth factor receptors. J Biol Chem 267:17804–17808.

Hou J, Kan M, McKeehan K, McBride G, Adams P, McKeehan WL (1991): Fibroblast growth factor receptors from liver vary in three structural domains. Science 251:665–668.

Hou J, McKeehan K, Kan M, Carr SA, Huddleston MJ, Crabb JW, McKeehan WL (1993): Identification of tyrosines 154 and 307 in the extracellular domain and 653 and 766 in the intracellular domain as phosphorylation sites in the heparin-binding fibroblast growth factor receptor tyrosine kinase (flg), Protein Science 2:86–92.

Imamura T, Engleka K, Zhan X, Tokita Y, Forough R, Roeder D, Jackson A, Maier JAM, Hla T, Maclag T (1990): Recovery of mitogenic activity of a growth factor mutant with a nuclear translocation sequence. Science 249:1567–1570.

Imamura T, Tokita Y, Mitsui Y (1992): Identification of a heparin-binding growth factor-1 nuclear translocation sequence by deletion mutation analysis. J Biol Chem 267:5676–5679.

Isaacs HV, Tannahill D, Slack JMW (1992): Expression of a novel FGF in the *Xenopus* embryo. A new candidate inducing factor for mesoderm formation and anteroposterior specification. Development 114:711–720.

Isacchi A, Statuto M, Chiesa R, Bergonzoni L, Rusnati M, Sarmientos P, Ragnotti G, Presta M (1991): A six-amino-acid deletion in basic fibroblast growth factor dissociates its mitogenic activity from its plasminogen activator-inducing capacity. Proc Natl Acad Sci USA 88:2628–2632.

Jackson RL, Busch SJ, Cardin AD (1991): Glycosaminoglycans: molecular properties, protein interactions, and role in physiological processes. Physiol Rev 71:481–539.

Jakobovits A, Shackleford G, Varmus H, Martin G (1986): Two proto-oncogenes implicated in mammary carcinogenesis, int-1 and int-2, are independently regulated during mouse development. Proc Natl Acad Sci USA 83:7806–7810.

Jarvis M, Gessner GW, Martin GE, Michael J, Ravera MW, Dionne GA (1992): Characterization of [^{125}I]acidic fibroblast growth factor binding to the cloned human growth factor receptor, FGF-flg, on NIH 3T3 cell membranes: inhibitory effects of heparin, pertussin toxin and guanine nucleotides. J Pharm Exp Therap 263:253–263.

Jiang Z, Savona C, Chambaz EM, Feige J-J (1992): Transforming growth factor β_1 and adrenocorticotropin differentially regulate the synthesis of adrenocortical cell heparan sulfate proteoglycans and their binding of basic fibroblast growth factor. J Cell Physiol 153:266–276.

Johnson DE, Lu J, Chen H, Werner S, Williams LT (1991): The human fibroblast growth factor receptor genes: a common structural arrangement underlies the mechanisms for generating receptor forms that differ in their third immunoglobulin domain. Mol Cell Biol 11:4627–4634.

Johnson DE, Lee PL, Lu J, Williams LT (1990): Diverse forms of a receptor for acidic and basic fibroblast growth factors. Mol Cell Biol 10:4728–4736.

Kalcheim C, Neufeld G (1990): Expression of basic fibroblast growth factor in the nervous system of early avian embryos. Development 109:203–215.

Kan M, DiSorbo D, Hou J, Hoshi H, Mansson P-E, McKeehan WL (1988): High and low affinity binding of heparin-binding growth factor to a 130-kDa receptor correlates with stimulation and inhibition of growth of a differentiated human hepatoma cell. J Biol Chem 263:11306–11313.

Kan M, Wang F, Xu J, Crabb JW, Hou J, McKeehan WL (1993): An essential heparin-binding domain in the fibroblast growth factor receptor kinase. Science 259:1918–1921.

Kan M, Shi E-G, McKeehan WL (1991): Identification and assay of fibroblast growth factor receptors. In Barnes D, Mather JP, Sato GH (eds): Methods in Enzymology, Peptide Growth Factors, Part C, Vol 198, pp 158–171.

Kardami E, Fandrich RR (1989): Basic fibroblast growth factor in atria and ventricles of the vertebrate heart J Cell Biol 109:1865–1875.

Katzac S, Martin-Zanca D, Barbacid M (1989): *vav,* a novel human oncogene derived from a locus ubiquitously expressed in hematopoietic cells. EMBO J 8:2283–2290.

Keegan K, Johnson DE, Williams LT, Hayman MG (1991): Isolation of an additional member of the fibroblast growth factor receptor family. FGFR-3. Proc Natl Acad Sci USA 88:1095–1099.

Kiefer MC, Ishihara M, Swiedler SJ, Crawford K, Stephans JC, Barr PJ (1991): The molecular biology of heparan sulfate fibroblast growth factor receptors. In Baird A, Klagsbrun M (eds): The Fibroblast Growth Factor Family. Annals NY Acad Sci 638:167–176.

Kiefer MC, Stephans JC, Crawford K, Okino K, Barr PJ (1990): Ligand-affinity cloning and structure of a cell surface heparan sulfate proteoglycan that binds basic fibroblast growth factor. Proc Natl Acad Sci USA 87:6985–6989.

Kimelman D, Abraham JA, Haaparanta T, Palisi TM, Kirschner MW (1988): The presence of fibroblast growth factor in the frog egg: its role as a natural mesoderm inducer. Science 242:1053–1056.

Kornbluth S, Paulson E, Hanafusa H (1988): Novel tyrosine kinase identified by phosphotyrosine antibody screening of cDNA libraries. Mol Cell Biol 8:5541–5544.

Kurokawa M, Doctrow SR, Klagsbrun M (1989): Neutralizing antibodies inhibit the binding of basic fibroblast growth factor to its receptor but not to heparin. J Biol Chem 264:7686–7691.

Lapeyre B, Amalric F, Ghaffari SH, Venkatarama Rao SV, Dumbar TS, Olson MOJ (1986): Protein and cDNA sequence of a glycine-rich, dimethylarginine containing region located near the carboxy-terminal end of nucleolin (C23 and 100 kDa). J Biol Chem 261:9167–9173.

Lathrop B, Olson E, Glaser L (1985): Control by fibroblast growth factor of differentiation in the BC3H1 muscle cell line. J Cell Biol 100:1540–1547.

Lindberg RA, Quinn AM, Hunter T (1992): Dual-specificity protein kinases: Will any hydroxyl do? Trends Biochem Sci 17:114–119.

Linkhart TA, Clegg CH, Hauschka SD (1980): Control of mouse myoblast commitment to terminal differentiation by mitogens. J Supramolec Struct 14:483–498.

Ma Y-G, Rosfjord E, Huebert C, Wilder P, Tiesman J, Kelly D, Rizzino A (1992). Transcriptional regulation of the murine k-FGF gene in embryonic cell lines. Dev Biol 154:45–54.

Mansour SL, Martin GR (1988): Four classes of mRNA are expressed from the mouse *int*-2 gene, a member of the FGF family. EMBO J 7:2035–2041.

Mansukhani A, Dell'Era, P, Moscatelli D, Kornbluth S, Hanafusa H, Basilico C (1992): Characterization of the murine BEK fibroblast growth factor (FGF) receptor: activation by three members of the FGF family and requirement for heparin. Proc Natl Acad Sci USA 89:3305–3309.

Mansukhani A, Moscatelli D, Talarico D, Levytska V, Basilico C (1990): A murine fibroblast growth factor (FGF) receptor expressed in CHO cells is activated by basic FGF and Kaposi FGF. Proc Natl Acad Sci USA 87:4378–4382.

Margolis B, Silvennoinen O, Comoglio F, Roonprapunt C, Skolnik E, Ullrich A, Schlessinger J (1992): High-efficiency expression/cloning of epidermal growth factor-receptor-binding proteins with Src homology 2 domains. Proc Natl Acad Sci USA 89:8894–8898.

Massagué J (1990): The transforming growth factor-β family. Annu Rev Cell Biol 6:597–641.

Mignatti P, Morimoto T, Rifkin DB (1991): Basic fibroblast growth factor released by single, isolated cells stimulates their migration in an autocrine manner. Proc Natl Acad Sci USA 88:11007–11011.

Mignatti P, Morimoto T, Rifkin DB (1992): Basic fibroblast growth factor, a protein devoid of secretory signal sequence, is released by cells via a pathway independent of the endoplasmic reticulum–Golgi complex. J Cell Physiol 151:81–93.

Miki T, Bottaro DP, Fleming TP, Smith CL, Burgess WH, Chan AM-L, Aaronson SA (1992): Determination of ligand-binding specificity by alternative splicing: two distinct growth factor receptors encoded by a single gene. Proc Natl Acad Sci USA 89:246–250.

Miyamoto M, Naruo K-I, Seko C, Matsumoto S, Kondo T and Kurokawa T (1993): Molecular cloning of a novel cytokine cDNA encoding the ninth member of the fibroblast growth factor family, which has a unique secretion property. Mol Cell Biol 13:4251–4259.

Moenner M, Bannoun-Zaki L, Badet J, Barritault D (1989): Internalization and limited processing of basic fibroblast growth factor on Chinese hamster lung fibroblasts. Growth Factors 1:115–123.

Mohammadi M, Dionne CA, Li W, Li N, Spivak T, Honegger AM, Jaye M, Schlessinger J (1992): Point mutation in FGF receptor eliminates phosphatidylinositol hydrolysis without affecting mitogenesis. Nature 358:681–684.

Mohammadi M, Honegger AM, Rotin D, Fischer R, Bellot F, Li W, Dionne CA, Jaye M, Rubinstein M, Sclessinger J (1991): A tyrosine-phosphorylated carboxy-terminal peptide of the fibroblast growth factor receptor (Flg) is a binding site for the SH2 domain of phospholipase C-γ1. Mol Cell Biol 11:5068–5078.

Moore JW, Dionne C, Jaye M, Swain JL (1991): The mRNAs encoding acidic FGF, basic FGF and FGF receptor are coordinately downregulated during myogenic differentiation. Development 111:741–748.

Morrison DK, Kaplan DR, Rapp U, Roberts TM (1988): Signal transduction from membrane to cytoplasm: growth factors and membrane-bound oncogene products increase Raf-1 phosphorylation and associated protein kinase activity. Proc Natl Acad Sci USA 85:8855–8859.

Moscatelli D (1988): Metabolism of receptor-bound and matrix-bound basic fibroblast growth factor by bovine capillary endothelial cells. J Cell Biol 107:753–759.

Moscatelli D (1987): High and low affinity binding sites for basic fibroblast growth factor on cultured cells: absence of a role for low affinity binding in the stimulation of plasminogen activator production by bovine capillary endothelial cells. J Cell Physiol 131:123–130.

Musci TJ, Amaya E, Kirschner MW (1990): Regulation of the fibroblast growth factor receptor in early *Xenopus* embryos. Proc Natl Acad Sci USA 87:8365–8369.

Nakanishi Y, Kihara K, Mizuno K, Masamune Y, Yoshitake Y, Nishikawa K (1992): Direct effect of basic fibroblast growth factor on gene transcription in a cell-free system. Proc Natl Acad Sci USA 89:5216–5220.

Neufeld G, Gospodarowicz D (1985): The identification and partial characterization of the fibroblast growth factor receptor of baby hamster kidney cells. J Biol Chem 260:13860–13868.

Niswander L, Martin GR (1992): *Fgf-4* expression during gastrulation, myogenesis, limb and tooth development in the mouse. Development 114:755–768.

Nugent MA, Edelman ER (1992a): Kinetics of basic fibroblast growth factor binding to its receptor and heparan sulfate proteoglycan: a mechanism for cooperativity. Biochemistry 31:8876–8883.

Nugent MA, Edelman ER (1992b): Transforming growth factor β1 stimulates the production of basic fibroblast growth factor binding proteoglycans in Balb/c3T3 cells. J Biol Chem 267:21256–21264.

Ornitz DM, Leder P (1992): Ligand specificity and heparin dependence of fibroblast growth factor receptors 1 and 3. J Biol Chem 267:16305–16311.

Orr-Urtreger A, Givol D, Yayon A, Yarden Y, Lonai P (1991): Developmental expression of two murine fibroblast growth factor receptors, *flg* and *bek*. Development 113:1419–1434.

Pandiella A, Meldolesi J (1992): Phosphoinositide hydrolysis and ensuing calcium and potassium fluxes: role in the action of EGF and other growth factors. Cell Physiol Biochem 2:196–212.

Partanen J, Mäkelä TP, Eerola E, Korhonen J, Hirvonen H, Claesson-Welsh L, Alitalo K (1991): FGFR-4, a novel acidic fibroblast growth factor receptor with a distinct expression pattern. EMBO J 10:1347–1354.

Pasquale EB, Singer SJ (1989): Identification of a developmentally regulated protein-tyrosine kinase by using anti-phosphotyrosine antibodies to screen a cDNA expression library. Proc Natl Acad Sci USA 86:5449–5453.

Pasquale EB (1990): A distinctive family of embryonic protein-tyrosine kinase receptors. Proc Natl Acad Sci USA 87:2812–2818.

Pastone G, Pasquale E, Maher P (1993): Different members of the fibroblast growth factor receptor family are specific to distinct cell types in the developing chicken embryo. Dev Biol 155:107–123.

Pawson T, Gish GD (1992): SH2 and SH3 domains: from structure to function. Cell 71:359–362.

Peters G, Brookes S, Smith R, Placzek M, Dickson C (1989): The mouse homolog of the *hst/k-FGF* gene is adjacent to *int*-2 and is activated by proviral insertion in some virally induced mammary tumors. Proc Natl Acad Sci USA 86:5678–5682.

Peters KG, Werner S, Chen G, Williams LT (1992a): Two FGF receptor genes are differentially expressed in epithelial mesenchymal tissues during limb formation and organogenesis in the mouse. Development 114:233–243.

Peters KG, Marie J, Wilson E, Ives HE, Escobedo J, Del Rosario M, Mirda D, Williams LT (1992b): Point mutation of an FGF receptor abolishes phosphatidylinositol turnover and Ca^{2+} flux but not mitogenesis. Nature 358:678–681.

Plouët J, Gospodarowicz (1989): Transforming growth factor β-1 positively modulates the bioactivity of fibroblast growth factor on corneal endothelial cells. J Cell Physiol 141:392–399.

Prats H, Kaghad M, Prats AC, Klagsbrun M, Lélias JM, Liauzun P, Chalon P, Tauber JP, Amalric F, Smith JA, Caput D (1989): High molecular mass forms of basic fibroblast growth factor are initiated by alternative CUG codons. Proc Natl Acad Sci USA 86:1836–1840.

Presta M, Rusnati M, Urbinati C, Sommer A, Ragnotti G (1991): Biologically active synthetic fragments of human basic fibroblast growth factor (bFGF): identification of two Asp-Gly-Arg-containing domains involved in the mitogenic activity of bFGF in endothelial cells. J Cell Phys 149:512–524.

Quarto N, Finger FP, Rifkin DB (1991): The NH_2-terminal extension of high molecular weight bFGF is a nuclear targeting signal. J Cell Physiol 147:311–318.

Raffioni S, Bradshaw RA (1992): Activation of phosphatidylinositol 3-kinase by epidermal growth factor, basic fibroblast growth factor, and nerve growth factor in PC12 pheochromocytoma cells. Proc Natl Acad Sci USA 89:9121–9125.

Rappolee DA, Sturm KS, Schultz GA, Pedersen RA, Werb Z (1990): The expression of growth factor ligands and receptors in preimplantation mouse embryos. In Heyner S, Wiley LM (eds): Early Embryo Development and Paracrine Relationships. UCLA Symposia on Molecular and Cellular Biology New Series, Vol 117. New York: Wiley-Liss, pp 11–25.

Rapraeger AC, Krufka A, Olwin BB (1991): Requirement of heparan sulfate for bFGF-mediated fibroblast growth and myoblast differentiation. Science 252: 1705–1708.

Reid HH, Wilks AF, Bernard O (1990): Two forms of the basic fibroblast growth factor receptor-like mRNA are expressed in the developing mouse brain. Proc Natl Acad Sci USA 87:1596–1600.

Renko M, Quarto N, Morimoto T, Rifkin DB (1990): Nuclear and cytoplasmic localization of different basic fibroblast growth factor species. J Cell Physiol 144:108–114.

Rizzino A (1987): Soft agar growth assays for transforming growth factors and mitogenic peptides. In Barnes D, Sirbasku DA (eds): Methods in Enzymology Peptide Growth Factors, Part A, Vol 146. Orlando: Academic Press, pp 341–352.

Rizzino A (1988): Review. Transforming growth factor-β: multiple effects on cell differentiation and extracellular matrices. Dev Biol 130:411–422.

Rizzino A, Kuszynski C, Ruff E, Tiesman J (1988a): Production and utilization of growth factors related to fibroblast growth factor by embryonal carcinoma cells and their differentiated cells. Dev Biol 129:61–71.

Rizzino A, Kazakoff P, Ruff E, Kuszynski C, Nebelsick J (1988b): Regulatory effects of cell density on the binding of transforming growth factor β, epidermal growth factor, platelet-derived growth factor, and fibroblast growth factor. Cancer Res 48:4266–4271.

Rizzino A, Kazakoff P (1991): Iodination of peptide

growth factors: platelet-derived growth factor and fibroblast growth factor. In Barnes D, Mather JP, Sato GH (eds): Methods in Enzymology, Peptide Growth Factors, Part C, Vol 198. San Diego: Academic Press, pp 467–479.

Rizzino A, Ruff E (1986): Fibroblast growth factor induces the soft agar growth of two nontransformed cell lines. In Vitro Cell Dev Biol 22:749–755.

Ruoslahti E, Yamaguchi Y (1991): Proteoglycans as modulators of growth factor activities. Cell 64:867–869.

Ruta M, Howk R, Ricca G, Drohan W, Zabelshansky M, Laureys G, Barton DE, Francke U, Schlessinger J, Givol D (1988): A novel protein tyrosine kinase gene whose expression is modulated during endothelial cell differentiation. Oncogene 3:9–15.

Ruta M, Burgess W, Givol D, Epstein J, Neiger N, Kaplow J, Crumley G, Dionne C, Jaye M, Schlessinger J (1989): Receptor for acidic fibroblast growth factor is related to the tyrosine kinase encoded by the *fms*-like gene (FLG). Proc Natl Acad Sci USA 86:8722–8726.

Safran A, Avivi A, Orr-Urtereger A, Neufeld G, Lonai P, Givol D, Yarden Y (1990): The murine *flg* gene encodes a receptor for fibroblast growth factor. Oncogene 5:635–643.

Saito H, Kouhara H, Kasayama S, Kishimoto T, Sato B (1992): Characterization of the promoter region of the murine fibroblast growth factor receptor 1 gene. Biochem Biophys Res Commun 183:688–693.

Sakaguchi M, Kajio T, Kawahare K, Kato K (1988): Antibodies against basic fibroblast growth factor inhibit the autocrine growth of pulmonary artery endothelial cells. FEBS Lett 233:163–166.

Saksela O, Moscatelli D, Rifkin DB (1987): The opposing effects of basic fibroblast growth factor and transforming growth factor beta on the regulation of plasminogen activator activity in capillary endothelial cells. J Cell Biol 105:957–964.

Sato Y, Rifkin DB (1988): Autocrine activities of basic fibroblast growth factor: regulation of endothelial cell movement, plasminogen activator synthesis, and DNA synthesis. J Cell Biol 107:1199–1205.

Schindler C, Shuai K, Prezioso VR, Darnell JE Jr (1992): Interferon-dependent tyrosine phosphorylation of a latent cytoplasmic transcription factor. Science 257: 809–813.

Schlessinger J, Ullrich A (1992): Growth factor signaling by receptor tyrosine kinases. Neuron 9:383–391.

Seno M, Sasada R, Kurokawa T, Igarashi K (1990): Carboxyl-terminal structure of basic fibroblast growth factor significantly contributes to its affinity for heparin. Eur J Biochem 188:239–245.

Shi E, Kan M, Hou J, Xu J, McBride G, McKeehan WL (1991): Receptor mediated biphasic growth response: protein expression and function of the receptor (flg) variants of heparin binding (fibroblast) growth factors (HBGFs) in human hepatoma cells. J Cell Biol 115:417a.

Shibata F, Baird A, Florkiewicz RZ (1991): Functional characterization of the human basic fibroblast growth factor gene promoter. Growth Factors 4:277–287.

Shimasaki S, Emoto N, Koba A, Mercado M, Shibata F, Cooksey K, Baird A, Ling N (1988): Complementary DNA cloning and sequencing of rat ovarian basic fibroblast growth factor and tissue distribution study of its mRNA. Biochem Biophys Res Commun 157:256–263.

Shing Y (1991): Biaffinity chromatography of fibroblast growth factors. In Barnes D, Mather JP, Sato GH (eds): Methods in Enzymology, Peptide Growth Factors, Part C, Vol 198. San Diego: Academic Press, pp 91–95.

Shiurba RA, Jing N, Sakahura T, Godsave SF (1991): Nuclear translocation of fibroblast growth factor during *Xenopus* mesoderm induction. Development 113:487–493.

Skolnik EY, Margolis B, Mohammadi M, Lowenstein E, Fischer R, Drepps A, Ullrich A, Schlessinger J (1991): Cloning of P13 kinase-associated p85 utilizing a novel method for expression/cloning of target proteins for receptor tyrosine kinases. Cell 65:83–90.

Slack JMW (1991): The nature of the mesoderm-inducing signal in *Xenopus:* a transfilter induction study. Development 113:661–669.

Smith R, Peters G, Dickson C (1988): Multiple RNAs expressed from the *int*-2 gene in mouse embryonal carcinoma cell lines encode a protein with homology to fibroblast growth factors. EMBO J 7:1013–1022.

Smith JC (1989): Mesoderm induction and mesoderm-inducing factors in early amphibian development. Development 105:665–677.

Speir E, Tanner V, Gonzalez AM, Farris J, Baird A, Casscells W (1992): Acidic and basic fibroblast growth factors in adult rat heart myocytes. Localization, regulation in culture, and effects on DNA synthesis. Circ Res 71:251–259.

Speir E, Sasse J, Shrivastav S, Casscells W (1991): Culture-induced increase in acidic and basic fibroblast growth factor activities and their association with the nuclei of vascular endothelial and smooth muscle cells. J Cell Physiol 147:362–373.

Spizz G, Roman D, Strauss A, Olson EN (1986): Serum and fibroblast growth factor inhibit myogenic differentiation through a mechanism dependent on protein synthesis and independent of cell proliferation. J Biol Chem 261:9483–9488.

Stark KL, McMahon JA, McMahon AP (1991): FGFR-4, a new member of the fibroblast growth factor receptor family, expressed in the definitive endoderm and skeletal muscle lineages of the mouse. Development 113:641–651.

Stumpo DJ, Blackshear PJ (1986): Insulin and growth factor effects on *c-fos* expression in normal and protein kinase C-deficient 3T3-L1 fibroblast and adipocytes. Proc Natl Acad Sci USA 83:9453–9457.

Suzuki K, Kamei T, Hakamata Y, Kikukawa K, Shiota K, Takahashi M (1991): Basic fibroblast growth factor-like substance in nuclei of male germ cells undergoing meiosis. Soc Exp Biol Med 198:728–731.

Suzuki HR, Sakamoto H, Yoshida T, Sugimura T, Terada M, Solursh M (1992): Localization of *HstI* transcripts to the apical ectodermal ridge in the mouse embryo. Dev Biol 150:219–222.

Tanaka A, Miyamoto K, Minamino N, Takeda M, Sato B, Matsuo H, Matsumoto K (1992): Cloning and characterization of an androgen-induced growth factor essential for the androgen-dependent growth of mouse mammary carcinoma cells. Proc Natl Acad Sci USA 89:8928–8932.

Thomas KA, Ortega S, Soderman D, Schaeffer M, DiSalvo J, Gimenez-Gallego G, Linemeyer D, Kelly L, Menke J (1991): Structural modifications of acidic fibroblast growth factor alter activity, stability, and heparin dependence. In Baird A, Klagsbrun M (eds): The Fibroblast Growth Factor Family. Ann NY Acad Sci 638:9–17.

Tiesman J, Rizzino A (1989a): Expression and developmental regulation of the k-FGF oncogene in human and murine embryonal carcinoma cells. In Vitro Cell Dev Biol 25:1193–1198.

Tiesman J, Rizzino A (1989b): Recent developments in the structure, function and regulation of platelet-derived growth factor and its receptors. Cytotechnology 2:333–350.

Tourbah A, Oliver L, Jeanny JC, Gumpel M (1991): Acidic fibroblast growth factor (aFGF) is expressed in the neuronal and glial spinal cord cells of adult mice. J Neurosci Res 29:560–568.

Tsuda T, Kaibuchi K, West B, Takai Y (1985): Involvement of Ca^{2+} in platelet-derived growth factor-induced expression of *c-myc* oncogene in Swiss 3T3 fibroblasts. FEBS Lett 187:43–46.

Ueno H, Gunn M, Dell K, Tseng A Jr, Williams L (1992): A truncated form of fibroblast growth factor receptor 1 inhibits signal transduction by multiple types of growth factor receptor. J Biol Chem 267:1470–1476.

Ullrich A, Schlessinger J (1990): Signal transduction by receptors with tyrosine kinase activity. Cell 61:203–212.

Vaidya TB, Rhodes SJ, Taparowsky EJ, Konieczny SF (1989): Fibroblast growth factor and transforming growth factor (beta) repress transcription of the myogenic regulatory gene MyoD1. Mol Cell Biol 9:3576–3579.

Vainikka S, Partanen J, Bellosta P, Coulier F, Basilico C, Jaye M, Alitalo K (1992): Fibroblast growth factor receptor-4 shows novel features in genomic structure, ligand binding and signal transduction. EMBO J 11:4273–4280.

Velcich A, Delli-Bovi P, Mansukhani A, Ziff EB, Basilico C (1989): Expression of the K-*fgf* protooncogene is repressed during differentiation of F9 cells. Oncogene Res 5:31–37.

Veomett G, Kuszynski C, Kazakoff P, Rizzino A (1989): Cell density regulates the number of cell surface receptors for fibroblast growth factor. Biochem Biophys Res Commun 159:694–700.

Vlodavsky I, Folkman J, Sullivan R, Fridman R, Ishai-Michaeli R, Sasse J, Klagsbrun (1987): Endothelial cell-derived basic fibroblast growth factor: synthesis and deposition into subendothelial extracellular matrix. Proc Natl Acad Sci USA 84:2292–2296.

Wennström S, Landgren E, Blume-Jensen P, Claesson-Welsh L (1992): The platelet-derived growth factor β-receptor kinase insert confers specific signaling properties to a chimeric fibroblast growth factor receptor. J Biol Chem 267:13749–13756.

Werner S, Duan D-SR, de Vries C, Peters KG, Johnson DE, Williams LT (1992): Differential splicing in the extracellular region of fibroblast growth factor 1 generates receptor variants with different ligand-binding specificies. Mol Cell Biol 12:82–88.

Whitman M, Downes CP, Keeler M, Keller T, Cantley L (1988): Type I phosphatidylinositol kinase makes a novel inositol phospholipid, phosphatidylinositol-3-phosphate. Nature 332:644–646.

Wilder PJ, Rizzino A (1993): Mouse genetics in the 21st century: using gene targeting to create a cornucopia of mouse mutants possessing precise genetic modifications. Cytotechnology 11:79–99.

Wilkinson DG, Peters G, Dickson C, McMahon AP (1988): Expression of the FGF-related proto-oncogene *int*-2 during gastrulation and neurulation in the mouse. EMBO J 7:691–695.

Wilkinson DG, Bhatt S, McMahon AP (1989): Expression pattern of the FGF-related proto-oncogene *int*-2 suggests multiple roles in fetal development. Development 105:131–136.

Xu J, Nakahara M, Crabb JW, Shi E, Matuo Y, Fraser M, Kan M, Hou J, McKeehan WL (1992): Expression and immunochemical analysis of rat and human fibroblast growth factor receptor (flg) isoforms. J Biol Chem 267:17792–17803.

Yamaguchi TP, Conlon RA, Rossant J (1992): Expression of the fibroblast growth factor receptor FGFR-1/flg during gastrulation and segmentation in the mouse embryo. Dev Biol 152:75–88.

Yayon A, Klagsbrun M, Esko JD, Leder P, Ornitz DM (1991): Cell surface, heparin-like molecules are require for binding of basic fibroblast growth factor to its high affinity receptor. Cell 64:841–848.

Yoshida T, Muramatsu H, Muramatsu T, Sakamoto H, Katoh O, Sugimura T, Terada M (1988): Differential expression of two homologous and clustered oncogenes, *Hst1* and *Int-2*, during differentiation of F9 cells. Biochem Biophys Res Comm 157:618–625.

Zhan X, Hu X, Friedman S, Maciag T (1992): Analysis of endogenous and exogenous nuclear translocation of fibroblast growth factor-1 in NIH 3T3 cells. Biochem Biophys Res Commun 188:982–991.

Zhu X, Komiya H, Chirino A, Faham S, Fox GM, Arakawa T, Hsu BT, Rees DC (1990): Three-dimensional structures of acidic and basic fibroblast growth factor. Science 251:90–93.

ABOUT THE AUTHORS

KEITH MILLER is a fifth-year M.D., Ph.D. student in the Department of Pathology and Microbiology at the University of Nebraska Medical Center in Omaha, Nebraska. After completing the first two years of medical school, Mr. Miller began his graduate training in Dr. Angie Rizzino's laboratory in 1991. The continuing focus of his research is the function of fibroblast growth factors during embryonic development and tumorigenesis. He received his B.S. from Cornell University in 1989.

ANGIE RIZZINO is a Professor in the Eppley Institute for Cancer Research and in the Department of Pathology and Microbiology at the University of Nebraska Medical Center in Omaha, Nebraska. Dr. Rizzino is chairman of the Cancer Research Training Program of the Eppley Institute and he teaches graduate courses dealing with topics in the cell cycle, signal transduction mechanisms, development biology, and molecular biology. After receiving a B.S. in biology from the State University of New York at Stony Brook, he pursued doctoral research in the Department of Biochemistry at SUNY Stony Brook in the laboratory of Martin Freundlich where he earned a Ph.D. in 1974. This was followed by postdoctoral training in the laboratories of Brian McCarthy at UC San Francisco and Gordon Sato at UC San Diego where he worked on histone gene expression and the growth factor requirements of embryonal carcinoma cells, respectively. Since 1980, Dr. Rizzino's research has focused on the effects of differentiation on the expression of growth factors and their receptors. More recently, his research has focused on the transcriptional regulation of the growth factor genes, FGF-4 and TGF-β2. Dr. Rizzino has been a visiting scientist at the Roche Institute of Molecular Biology (Nutley, NJ) and an exchange scientist at the College De France (Paris, France). He serves on the Human Embryology and Development study section of the National Institutes of Health. In addition, he is an associate editor for *In Vitro Cellular and Developmental Biology* and he is a consulting editor for the journal *Cytotechnology.*

Growth Factors and Signal Transduction in Development: 51–73

Platelet-Derived Growth Factor in Development

Daniel F. Bowen-Pope and Ronald A. Seifert

I. INTRODUCTION

Platelet-derived growth factor (PDGF) was first recognized as an active substance released from blood platelets that could stimulate the proliferation of fibroblasts and vascular smooth muscle cells [reviewed in Ross et al., 1986]. This suggested that a normal function of PDGF might be to stimulate connective tissue cells to proliferate at sites of injury, and that chronic release of PDGF could play a role in the etiology of atherosclerotic lesions [Ross et al., 1974]. Development of specific methods for detection of PDGF from other sources led, during the mid 1980s, to the realization that many cultured cells can secrete PDGF. This growth factor has now been proposed to play a role in almost every situation in which connective tissue cells migrate and/or proliferate, including oncogenic transformation, tissue response to injury, and normal embryogenesis.

In this review, we will first summarize basic information about PDGF and PDGF receptors. We will then consider advantages and problems with various types of evidence that have been used to evaluate the role that PDGF may play in embryogenesis and/or development. Finally, we will consider evidence for PDGF involvement in the development of specific organ systems.

II. PDGF AND ITS RECEPTORS

A. The PDGF/PDGF Receptor System

The PDGF system (Fig. 1) consists of two ligand genes (PDGF A-chain and PDGF B-chain) and two receptor genes (PDGFRα and PDGFRβ). Within this family, the ligand proteins dimerize covalently with each other to form three possible biologically active homo- and heterodimers (AA, AB, and BB) [Heldin and Westermark, 1990]. PDGFRα can bind both PDGF A-chain and PDGF B-chain, but PDGFRβ can bind only PDGF B-chain [Hammacher et al., 1989; Seifert et al., 1989]. The receptor proteins dimerize noncovalently with each other to form the high-affinity biologically active dimeric forms of the PDGF receptor ($\alpha\alpha$, $\alpha\beta$, $\beta\beta$) [Seifert et al., 1989; Kanakaraj et al., 1991; Heidaran et al., 1991; Raj et al., 1992]. These specificity rules mean that PDGF-AA can bind only to cells which express PDGFRα. By contrast, PDGF-BB can bind to cells which express either PDGFRβ *or* PDGFRα.

The interaction between PDGF and its receptors is very highly conserved in evolution. Human PDGF binds to receptors on cultured chicken and fish cells almost as well as to receptors on human fibroblasts [Bowen-Pope et al., 1985] and elicits a response by hydra, a coelenterate [Hanai et al., 1987]. This conser-

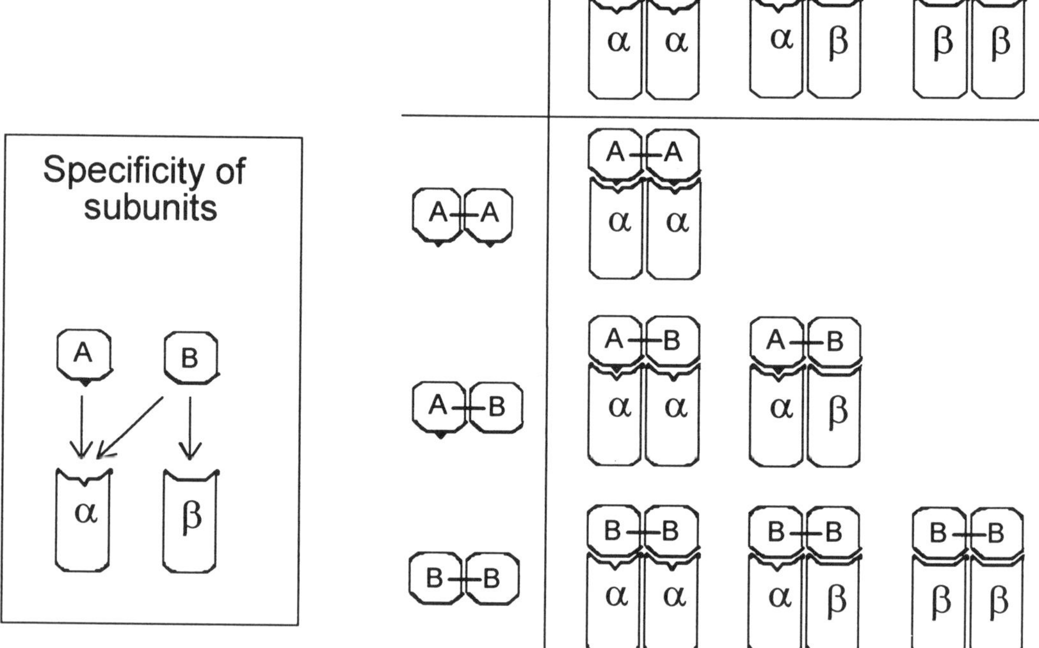

Fig. 1. *PDGF and its receptors. The panel on the left shows the two PDGF subunit chain polypeptide chains (A-chain and B-chain) and indicates the receptor subunit proteins (PDGFRα and PDGFRβ) to which they can bind. The panel on the right shows the biologically active forms of PDGF and its receptors. Secreted PDGF is always found as one of the three possible disulfide-bonded dimers shown on the column at the left of this panel. High-affinity binding sites are created when receptor subunit proteins dimerize noncovalently to form one (or more) of the three possible receptor dimers shown on the top row. Using the rules for subunit specificity shown in the left panel, the high-affinity dimeric receptors are predicted to have the specificity shown in the right panel.*

vation is convenient experimentally, because it means that readily available recombinant human PDGF can be used to test for responses in any convenient animal species model system.

B. Differences in Splicing, Processing, Cell Retention, and Extracellular Sequestration Could Be Involved in Regulation of PDGF Function

Both A-chain and B-chain have N-terminal propeptide sequences which are cleaved before secretion to generate the mature protein [Robbins et al., 1985]. The function of the propeptides is not clear. Recombinant PDGF folds and dimerizes normally without them. Their presence does not seem to inhibit PDGF activity, i.e., unprocessed PDGF is not an inactive zymogen. Unlike the FGF family, in which some members have secretory signal peptides and others do not, both PDGF A-chain and PDGF B-chain are secreted. PDGF A-chain is much more efficiently secreted than is PDGF B-chain. From studies with transfected cells in culture, this difference has been attributed to a sequence present in PDGF A-chain, and missing in PDGF B-chain, which somehow facilitates secretion or release into the culture medium [LaRochelle et al., 1990]. If this difference in secretion has biological significance it could be to optimize PDGF-AA for functioning as a *paracrine* factor, where diffusion to a nearby target is necessary, and to optimize PDGF-BB for functioning as an *autocrine* factor by minimizing the amount of PDGF which diffuses away. Beckmann et al. [1988] found that 3T3 cells, which expressed both PDGFRα and PDGFRβ, could be oncogenically transformed by transfection with either PDGF A-chain or PDGF B-chain, but that PDGF B-chain was far more potent. At the time, it seemed that this might be explained through the difference in efficiency of secretion. However, subsequent studies with chimeric PDGF molecules demonstrated that efficacy of transformation could be dissociated from cell retention, and that it correlated with receptor specificity instead [LaRochelle et al., 1990].

Alternative splicing of the A-chain transcript generates two possible sequences at the C-terminus: a long form which includes exon 6 and a short form which does not [Collins et al., 1987]. Exon 6 in PDGF B-chain is not alternatively spliced but the presence or absence of exon 6-encoded sequences in PDGF B-chain may be regulated via posttranslational cleavage. LaRochelle et al. [1991] reported that the exon 6 sequence in both A-chain and B chain are able to mediate retention of PDGF within the cell. It has also been proposed that exon 6 sequences may interact with extracellular matrix components and contribute to sequestration or localization of PDGF [Raines and Ross, 1992]. However, the relative abundance of long and short forms of PDGF A-chain transcript do not change during early mouse development [Mercola et al., 1990]. This suggests that PDGF A-chain exon 6–mediated interactions with matrix or membrane components are not regulated at the level of alternative splicing. Possibly they are regulated via proteolysis as suggested for PDGF B-chain. It should be noted that exon 6 sequences are not the only possible mediators of cell or matrix sequestration. For example, PDGF can bind to SPARC, a secreted glycoprotein which is abundant in many tissues undergoing morphogenetic changes. PDGF bound to soluble SPARC is unable to bind to its receptor [Raines et al., 1992].

C. The Biological Activities and Functions of PDGF: General Insights and Prejudices Derived From Studies With Cultured Cells

1. Connective tissue cell types express PDGF receptors. Studies with a vast range of cultured cell types have indicated that expression of PDGF receptors is characteristic of "connective tissue" cell types [reviewed in Bowen-Pope et al., 1985]. The pattern of PDGF receptor expression is thus narrower than for epidermal growth factor (EGF) and fibroblast growth factor (FGF) receptors, which are expressed by epithelial cells as well as by connective tissue cells.

2. PDGF can be synthesized by a very wide range of cultured cells. PDGFs (A-chain, B-chain, or both) are expressed by a broader range of cultured cells than are PDGF receptors. For example, all of the cell types cultured from large arteries can synthesize PDGF [reviewed in Ross et al., 1986]. These include the endothelial cells which line the vessels, the smooth muscle cells which constitute the vessel wall, the platelets which adhere and degranulate in damaged regions, and the macrophages which enter into developing atherosclerotic lesions. Of these, the smooth muscle cells also express PDGF receptors while the endothelial cells and platelets do not. These studies with cultured cells suggested that PDGF can function in the vessel wall as either an autocrine factor or as a paracrine factor.

In some cases, PDGF expression seems to be constitutive. An example of this would be cultured vascular smooth muscle cells from newborn rat aortas [Seifert et al., 1984]. In other cases, expression is transiently induced in response to some acute external stimulus. For example, fibroblasts express PDGF A-chain transcripts in response to exposure to exogenous PDGF or to serum [Paulsson et al., 1987]. This endogenously synthesized PDGF might serve to accentuate the response to a brief exposure to exogenous PDGF. PDGF transcripts are also induced in response to other growth factors, for example transforming growth factor β (TGF-β) [Leof et al., 1985] and interleukin-1 (IL-1) [Raines et al., 1989]. It has been proposed that this induced PDGF, acting as an autocrine stimulant, is responsible for the ability of TGF-β (and IL-1) to stimulate proliferation of connective tissue cells. To the extent that TGF-β is mitogenic through PDGF induction, TGF-β would function as a mitogen much like a member of the PDGF family: it would only be mitogenic for cells which express PDGF receptors.

3. PDGF is mitogenic, chemotactic, and affects matrix production. PDGF was purified on the basis of its ability to stimulate cell proliferation. In certain culture systems (e.g., myoblasts and glial precursors) the mitogenic stimulation of progenitor cells is accompanied by an inhibition, or delay, of terminal differentiation [Jin et al., 1990; Yablonka-Reuveni et al., 1990; Jin et al., 1991; Richardson et al., 1988; Noble et al., 1988; Raff et al., 1988]. Whether the inhibition of differentiation is the cause of the continued proliferation or vice versa is not clear. The result is the same: the number of differentiated cells ultimately obtained is increased. Cell types which respond to PDGF as a mitogen also respond to it as a chemotactic agent [reviewed in Ross et al., 1986].

All of the above processes have obvious potential roles in embryogenesis. If PDGF produced by a source tissue "S" were stimulating proliferation in adjacent receptor-positive target tissue "T", it seems likely that it would also stimulate chemotaxis of cells from T to S. If S were an epithelial sheet, this could help establish and maintain apposition of S and T, e.g., apposition of dermis to epidermis. If S were a point source, it would become surrounded by T. What if it were desirable to stimulate proliferation *without* attracting the target cells to the source? This might be possible through the PDGFRα if PDGF-AA can be nonchemotactic (or antichemotactic) as reported for some assays with cultured cells [see next section]. Alternatively, it may be that the ability of cells to migrate in response to a chemotactic gradient can be regulated independently, e.g., by factors affecting cell–cell and cell–matrix adhesion. Another factor which contributes to differential regulation of chemotaxis and proliferation is that these two responses often show very different dose–response relationships. As the concentration of PDGF is increased, the mitogenic response plateaus, while the chemotactic response peaks, then declines to baseline. As discussed below, the local concentration of PDGF *in vivo* is not known.

4. Different isoforms of PDGF may stimulate different processes. In attempting to evaluate what roles PDGF might play in different developmental processes it would be helpful to know whether the biological activities of the different isoforms of PDGF are intrinsically different. This question has not yet been clearly answered. The PDGFRα and PDGFRβ receptor proteins differ significantly in the sequence of their intracellular domains, especially the interkinase domain, which has been implicated in interacting with second messenger systems [reviewed in Williams, 1989]. This suggests that they might have evolved to generate different intracellular signals. Several studies have demonstrated differences in the responses of specific cell types to different isoforms of PDGF and these differences seem to reflect the ability of the different isoforms to bind to different forms of PDGF receptor. For example, Nister et al. [1988] reported that PDGF-AA is not chemotactic for fibroblasts. This was interpreted as demonstrating that PDGFRα (which is the only receptor subunit to which PDGF-AA can bind) is not competent to mediate chemotaxis. Koyama et al. [1992] reported that PDGF-AA inhibits chemotaxis stimulated by PDGF-BB or by fibronectin. These results would indicate that PDGFRα and PDGFRβ are *qualitatively* different in their signal transduction pathways. Although this is possible, there is a simpler explanation for the data. The lower potency of PDGF-AA may reflect *quantitative* differences in levels of expression of the two receptor subunits. This is supported by the observations that although PDGF-AA is indeed less mitogenic than PDGF-BB for human fibroblasts, these cells express many more PDGFRβ than PDGFRα and thus bind much more PDGF-BB than PDGF-AA. Cell types which express large numbers of both subunits, and bind comparable amounts of the two PDGF isoforms, are *equally* mitogenically responsive to the two isoforms [Seifert et al., 1989]. When the two subunits are expressed by transfection into receptor-negative host cells they both confer mitogenic and chemotactic responsiveness [Matsui et al., 1989; Ferns et al., 1990].

The hypothesis that PDGFRα and PDGFRβ mediate different biological effects on target cells is attractive because it would "explain" why more than one form of PDGF receptor has evolved. How can one justify the evolution of two receptors if they do not mediate different effects? Differing specificities of the two receptor subunits and the secretion of different isoforms of PDGF by different cells at different times would allow the PDGF system to function as a set of partially overlapping and partially distinct regulatory networks. For example, a cell expressing PDGFRα could respond to any isoform of PDGF. A second cell type expressing only PDGFRβ could respond to PDGF-BB, but not to PDGF-AA or PDGF-AB. PDGF-BB thus functions as a "universal ligand" and stimulates all PDGF-responsive cells, while PDGF-AA and PDGF-AB target subsets of this population.

III. APPROACHES TO EVALUATING THE ROLES OF THE PDGF/PDGF RECEPTOR SYSTEM *IN VIVO*

A. Deducing the Biological Roles of PDGF From *In Vivo* Patterns of Expression: Practical and Theoretical Considerations

One approach to evaluating when and where PDGF may be playing a biological role is to determine when and where PDGF and its receptors are expressed. Unlike classical hormones, it appears that PDGF may usually act locally in short-range autocrine and paracrine pathways which are spatially restricted within a tissue or organ. To be really useful, information about PDGF/PDGF receptor expression needs to include anatomical information about spatial relationships between putative producing and responding cells. Localization of PDGF in the vicinity of cells expressing receptors would be taken to indicate that the receptor-expressing cells are probably responding to the PDGF. This inference is supported by two arguments: First, inferring a response is the simplest prediction; lack of response requires postulating an additional inhibitory regulatory factor or condition. Second, specific patterns of ligand and receptor expression presumably evolve to serve a function; expression of nonfunctioning signalling systems seems unlikely.

We feel that these arguments are valid as far as they go, and we continue to pursue localization approaches in our own laboratory. Nevertheless, we have come to believe that there are both technical and theoretical limitations to the type of information that can be derived in practice. First of all, it is very difficult to convincingly quantify receptor and ligand levels. This is a problem because reliable quantitative information should be an integral part of evaluating whether there are adequate levels of receptor to drive a significant stimulation. Secondly, it seems likely from the information presented later in this review, that the PDGF/PDGF receptor system usually functions through a close juxtaposition (or identity) of cells which express PDGF receptors and cells which secrete PDGF. The receptor-expressing target cells are *chronically* exposed to PDGF which they bind, respond to, and degrade along with the receptor. We do not know the rate at which PDGF is entering the system and we do not know how many receptors are initially available to be occupied by this PDGF. Because receptor–ligand complexes are degraded, the functioning of the PDGF/PDGF receptor system erases evidence of its existence. What we need to evaluate the *in vivo* situation is thus *kinetic* data (number of receptors occupied per cell per time) but what we can detect is *steady-state* levels (how many receptors remain *un*occupied and how much PDGF remains *un*bound).

Ultimately, the only safe conclusion from localization of expression data is merely that cells which express PDGF receptor protein or transcript are potential targets for PDGF and that cells which express PDGF transcripts are potential sources of PDGF.

B. Evaluating the Role of PDGFRα Expression Using a Receptor-Deficient Mutant Mouse

In order to determine the *magnitude* or *nature* of the roles suggested from localization data, one needs to be able to experimentally test the effects of enhanced or diminished receptor or ligand expression. To date, little information of this sort is available. One approach which has been used is to evaluate the consequences of receptor absence in a mouse mutant lacking the PDGFRα gene. This system is introduced next. Additional specific findings are described separately under sections describing the development of organ systems.

1. The phenotype of *Patch* homozygotes demonstrates that PDGFRα expression is necessary for normal development. One potentially useful approach to evaluating gene function *in vivo* is to determine the consequences of gene disruption. Mapping the PDGFRα gene by *in situ* hybridization to metaphase chromosomes showed that it is located very close to *c-kit* [Gronwald et al., 1990] and Southern blotting demonstrated that the PDGFRα is deleted in a genetically defined mouse mutation called *Patch* (Ph) [Stephenson et al., 1991]. No other known genes are affected by the *Patch* mutation but this is still a formal possibility. When the Ph mutation was first described, it was noted that Ph/Ph homozygotes die before birth after exhibiting certain gross abnormalities, including subepidermal blebs and general edema [Gruneberg and Truslove, 1960]. Truslove and Grüneberg calculated that about two-thirds of Ph/Ph embryos die before embryonic day 12 (E12). The remaining one-third die before birth. Ph/Ph embryos are considerably smaller than normal littermates at all stages after about E9. Nevertheless, on E14 both wild type and Ph/Ph are at comparable developmental stages, as evaluated by a complete separation of the digits and rotation of the hand. This indicates that the small size of Ph/Ph embryos results from a reduction in the growth rate rather than a retardation in the timing of development.

Once the *Patch* mutation was known to involve the PDGFRα gene, the phenotype of the mutants was examined more thoroughly and compared with the pattern of expression of PDGFRα transcripts observed by *in situ* analysis of normal development [Schatteman et al., 1992; Morrison-Graham et al., 1992; Orr-Urtreger et al., 1992]. Many additional specific defects in development were noted in tissues which would normally express PDGFRα, usually mesoderm derivatives or tissues dependent upon the nonneurogenic neural crest. Many of these are described in more detail in the sections below. For convenience and reference, the pattern of expression of PDGFRα transcripts throughout development is summarized in Figure 2.

2. The phenotype of *Patch* heterozygotes demonstrates that some developmental processes are perturbed by even a 50% reduction in PDGFRα levels. *Patch* is considered to be a "dominant" mutation because heterozygotes have a discernible phenotype. The reported phenotype is very mild: an area of white fur in the midsection and a larger prefrontal bone. Localized pigmentation defects are characteristic of heterozygotes of mutations in other genes (e.g., *c-kit* and *steel*) and have been shown to result from defects in the migration or differentiation of melanocyte precursors. It seems that the melanocyte pathway may be particularly demanding and requires full functioning of its components. It is also likely that pigmentation defects will be noted because even small defects are obvious visually. Little attention has been paid to *Patch* heterozygotes and many less obvious mild defects may have gone unnoticed.

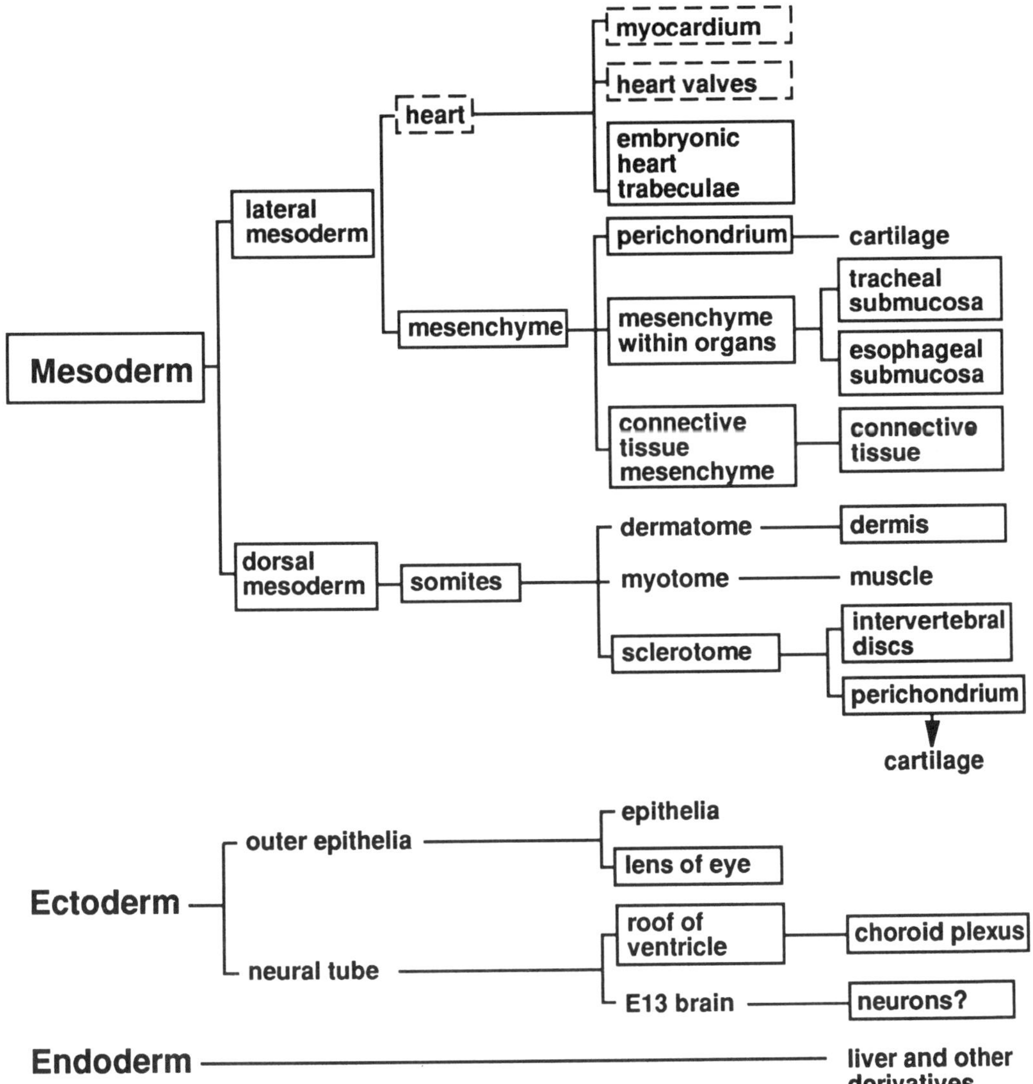

Fig. 2. *Expression of PDGFRα transcripts during mouse development. In this simplified schema of development of different tissue types from the three original germ layers, cell types which express PDGFRα transcripts are shown within boxes. The box is broken if expression is transient or nonuniform.*

Since the *Patch* mutation is a deletion, the phenotype of heterozygotes cannot be due to synthesis of a mutant form of PDGFRα. Instead, it presumably reveals those processes which are so dependent on full levels of expression of PDGFRα that they cannot function adequately in heterozygotes (expressing 50% of normal levels). Further study of the heterozygous phenotype may shed light on the role that regulation of PDGFRα expression plays in regulating PDGF function *in vivo*. Changes in the level of PDGF receptors (usually PDGFRβ) have been noted in many proliferative pathologies. We and others have suggested that levels of PDGF receptor in normal adult tissues are not adequate to support a full stimulation by PDGF and that upregulation of receptor levels is therefore necessary. Ph/+ heterozygotes provide a test system for determining whether wild type PDGF receptor levels are at least twofold higher than required for various developmental processes and whether a twofold reduction in the upregulated levels is enough to significantly affect pathological processes in which the PDGF receptor has been implicated.

3. The severity of the defects in Ph/Ph homozygotes is influenced by additional factors. Considerable variation in the severity of the *Patch* phenotype has been noted. The Ph/Ph embryos which we obtained from colonies in Eugene or Seattle showed the same abnormalities, and died at the same stages, as described by Grüneberg and Truslove [1960]. The Ph/Ph embryos studied by Orr-Urtreger et al. [1992] showed abnormalities earlier and usually died before showing the characteristic later abnormalities. Even within individual colonies or within single litters, some Ph/Ph individuals show more severe abnormalities than others. Multiple regulatory pathways undoubtedly affect the development of every tissue. The extent to which other signal transduction pathways can compensate for defects in the PDGFRα/PDGF A-chain pathway would depend on the "activity" of those other pathways. The activity of these pathways could be influenced by differences in the environment of different embryos *in utero* and by small genetic differences between individuals. These observed differences are useful rather than disturbing, because the very early defects observed by Orr-Urtreger et al. [1992] reveal an important role of PDGFRα in early development. We were not able to observe this early effect most likely because the absence of PDGFRα may have been largely compensated by some other process(es). By the same token, because Ph/Ph embryos in our colony frequently survive until late gestation, they are useful for evaluating roles of the PDGFRα in later developmental processes.

4. Limitations to the utility of studying the Ph/Ph mutant. Ph/Ph embryos do not survive until birth. This prevents them from being used for studies of the role of PDGFRα in late development (e.g., in the maturation of the nervous system) or in adult homeostasis and pathogenesis. The utility of the *Patch* deletion mutant is also limited by our inability to control the *timing* of the loss of receptor function. This has two consequences: (1) As development proceeds, abnormalities could be compounded due to the increasingly abnormal *general* embryonic environment, and (2) we cannot distinguish when PDGFRα expression is required—whether at the time the defect is observed or at some earlier stage. These limitations need to be kept in mind when evaluating the results noted in the following sections.

IV. SURVEY OF EXPRESSION AND POSSIBLE ROLES OF THE PDGF/PDGF RECEPTOR SYSTEM IN DEVELOPMENT

A. Oocytes and Very Early Embryos

Transcripts for the A-chain of PDGF have been detected as a maternally encoded transcript in Xenopus oocytes [Mercola et al., 1988] and in mouse blastocysts [Rappolee et al., 1988], although *in situ* hybridization did not detect PDGFRα in ovarian oocytes [Orr-Urtreger et al., 1992]. Cultured embryonal carcinoma cell lines have been used as models for blastocyst inner cell mass cells to permit cell culture analysis of PDGF and PDGFR expression during preimplantation development. The "undifferentiated" embryonal carcinoma cells secrete PDGF and do not bind significant amounts of PDGF [Rizzino and Bowen-Pope, 1985; Gudas et al., 1983]. Upon induction of differentiation by retinoic acid, the cells assume characteristics of primitive endoderm, cease to accumulate detectable PDGF in the conditioned medium, and express PDGF receptors. The interpretation that the undifferentiated cells expressed both PDGF A-chain and PDGF receptors (and were driven to proliferate through a PDGF autocrine loop), that the differentiation-induced increase in PDGF binding and the apparent loss of PDGF secretion reflected the cessation of autocrine PDGF synthesis and response, is probably not correct. PDGF A-chain transcripts are expressed both before and after differentiation, but PDGFRα transcripts are not expressed until differentiation is induced [Mercola et al., 1990; Wang et al., 1990]. The cessation of PDGF accumulation in the medium presumably reflects the expression of PDGFRα which are able to bind and degrade (and presumably respond to) the secreted PDGF. It therefore seems that the cessation of accumulation of PDGF protein in the growth medium reflects the *initiation* of an autocrine PDGF loop rather than its *cessation*. The PDGF B-chain/PDGFRβ system does not seem to function at a significant level in this system. PDGF B-chain is expressed at low levels and PDGFRβ is not induced until 6 days of induction by retinoic acid (vs. 3 days for PDGFRα).

Embryonal stem (ES) cell lines (e.g., the ES-D3 line) retain even more of the properties of an uncommitted inner cell mass cells than do embryonal carcinoma cells. Under the proper conditions, they can give rise to every cell type in the body. Undifferentiated ES-D3 cells do not express PDGFRα that can be detected by Northern blot analysis. When ES-D3 cells are cultured as aggregates in suspension, they form simple embryoid bodies which resemble the inner cell mass of an E4.5 murine blastocyst. *In situ* hybridization demonstrates that the outer cells express PDGFRα transcripts while the inner cells do not [Schatteman et al., 1992]. If these simple embryoid bodies are maintained in suspension culture for 8 days or longer, cystic embryoid bodies start to form. These consist of a fluid-filled cavity surrounded by a PDGFRα expressing cell layer with the properties of extraembryonic endoderm, and in some cases, an additional ectoderm-like layer. *In situ* hybridization to early embryos *in vivo* showed a pattern of PDGFRα expression that is consistent with the *in vitro* ES cell results.

We have derived ES cell lines from PDGFRα-negative Ph/Ph blastocysts and find that they proceed through *in vitro* embryogenesis in a way indistinguishable from wild-type ES cell lines [unpublished observations]. In suspension culture they give rise to simple embryoid bodies, which develop into cystic embryoid bodies. After attachment, they give rise to differentiated cell types. Ph/Ph embryos also develop normally *in vivo* during this early (presomite) period [Schatteman et al., 1992]. There is no decrease in the number of conceptuses which could be attributed to early lethality in matings between heterozygotes. Thus, despite the evidence that PDGF A-chain and PDGFRα are expressed in early embryos, they do not seem to play an "essential" role during this period. What does this indicate about the role of PDGF in early murine development? Is PDGFRα only one of many genes which make small additive contributions? Or does PDGFRα normally play a significant role but, in its absence, may a different gene(s) assume these functions? The product of this different gene might be expressed in normal embryos but play an unrelated role. Alternatively, expression of the different gene could be induced or augmented in Ph/Ph mutants through some sort of general regulatory network which tends to maintain developmental pathways. An example of this seems to be elevated expression of myf 5 in embryos whose MyoD gene has been disrupted [Rudnicki et al., 1992].

B. Placenta and Extraembryonic Tissues

Mouse blastocysts implant at about E4.5 and human blastocysts at about E6.5. Some of the cells in the developing placenta and extraembryonic tissues are of maternal origin but the whole ensemble develops and functions as an integrated physiological unit in concert with the developing embryo proper. Implanted mouse embryos can be located and dissected by E6.5, although it can be difficult to completely remove placental tissue. At these very early stages, nuclease protection assays indicated that PDGF B-chain is expressed strongly by the placenta and weakly by the embryo. In fact, the low levels apparently detected in the embryo could be due to residual placental tissue [Mercola et al., 1990]. Northern blot analysis confirms that PDGF B-chain is expressed by the human placenta, but there is disagreement as to whether expression peaks at mid gestation [Taylor and Williams, 1988] or is highest during the first trimester and declines thereafter [Goustin et al., 1985]. In either case, knowledge of the average level of expression in this complex tissue is not as informative as the pattern of expression by the component cell types.

During the initial stages of human pregnancy, trophoblast cells from the blastocyst proliferate and invade the uterine endometrium. They establish villi which constitute the interface with the maternal blood. The villi become filled with embryonic mesenchyme and blood vessels. Goustin et al. [1985] found that cytotrophoblasts expressed PDGF B-chain transcripts *in vivo* and bound ^{125}I-PDGF in culture and concluded that proliferation of these cells was driven through a PDGF B-chain/PDGF receptor autocrine loop. The highest level of expression of PDGF A-chain in the placenta is by cytotrophoblasts at the base of the growing tips of the villi [Holmgren et al., 1991]. Cytotrophoblasts did not express detectable PDGFRα. This makes it unlikely that this PDGF A-chain is acting as an autocrine stimulant for cytotrophoblasts. Presumably, the PDGF A-chain, and any B-chain not bound by PDGFRβ on the cytotrophoblasts, stimulate the surrounding PDGFRα-positive stromal cells. A paracrine system of this sort could help ensure that the developing villi contain enough mesodermal and/or mesenchymal cells to develop an adequate supporting stroma and vasculature. Holmgren et al. (1991) also provided evidence for a expression of the PDGF/PDGF receptor system during placental angiogenesis. This is described in Section IV.G on blood vessel development.

At the egg-cylinder stage of the mouse embryo, PDGFRα transcripts are detectable in visceral and parietal endoderm [Orr-Urtreger et al., 1992]. In normal E6.5 mouse embryos, the extraembryonic region of the visceral endoderm expresses higher levels of PDGFRα transcripts than does the embryonic region. Orr-Urtreger et al. [1992] observed that Ph/Ph mutants at this stage began to display a hypertrophic and convoluted visceral endoderm which develops into a hypertrophic "cauliflower-like" yolk sac. This remained one of the most dramatic aspects of the Ph/Ph phenotype in their colony and most of the mutants died before E11.5. In our colonies, Ph/Ph mutants survived considerably longer and yolk sac development was much closer to normal. As noted above, we suspect that this reflects the influence of small uncontrolled variables (possibly including differences in genetic background or maternal health) on compensatory or regulatory processes.

C. Early Postimplantation Embryogenesis

1. PDGF A-chain/PDGFRα, but not PDGF B-chain or PDGFRβ, are expressed in early postimplantation embryos. Between E6.5 and E8.5 the mouse embryo begins as a simple egg cylinder of about 800 cells, gastrulates, and develops to the early somite stage with a beating heart. Using nuclease protection assays to quantitate PDGF and PDGF receptor transcripts from E6.5–E8.5 mouse embryos we found that the relative abundance of the two PDGF receptor transcripts changed dramatically during this period [Mercola et al., 1990]. PDGFRα and PDGF A-chain transcripts were abundant throughout the period whereas PDGFRβ transcripts were not detected until the end of this period, and PDGF B-chain was not detected. This suggests that a PDGF A-chain/PDGFRα pathway could be operating, but that a PDGF B-chain/PDGFRβ pathway is unlikely, or is spatially very restricted. There has not been a systematic survey of PDGF B-chain/PDGFRβ expression during later stages of development that can be compared with the surveys done with PDGF A-chain/PDGFRα for developmental stages up to E8.5. We do know, however, that PDGFRβ is much more highly and widely expressed in normal and pathological adult tissues than is PDGFRα. Therefore, some time after E8.5, the relative abundances of the receptors must change.

The transcripts that were quantitated by nuclease protection have also been localized *in vivo* by *in situ* hybridization [Schatteman et al., 1992; Morrison-Graham et al., 1992; Orr-Urtreger et al., 1992; Orr-Urtreger and Lonai, 1992]. In the embryo proper, expression is first detected in the mesoderm as this germ layer first forms (E6.5) but is also expressed at low levels in the primitive streak, the precursor to mesoderm. PDGFRα expression remains characteristic of many mesodermal derivatives during later development.

2. The absence of PDGFRα expression does not detectably affect development until E9–9.5. Despite the observation that normal embryos do express PDGF A-chain and PDGFRα early in development, our colony of Ph/Ph mice did not show defects in the embryo proper until E9–9.5. The mesoderm seemed to form normally. This would be consistent with the observation that PDGF is not competent to induce mesoderm in the *Xenopus* animal cap system. Alternatively, it may be that development *is* affected during this early period, but without obvious *visible* consequences. External defects are first detectable in Ph/Ph homozygotes by about E9–9.5. These probably reflect defects in the development of sclerotome derivatives and are discussed in Section IV.E on the musculoskeletal system.

D. Between E10 and E11 Some of the Defects in *Patch* Embryos Can Be Transiently Compensated for by Other Systems

Although observed earlier in development, embryos with subepidermal blisters are not observed between E10 and E11. After E12 they are observed again, this time accompanied by additional internal and external defects. Almost all of our information about the development of Ph/Ph embryos derives from serial sacrifice after different gestational periods, rather than by observation of individual embryos over time. Using this means we could not distinguish between the following two explanations for the temporary disappearance of a recognizable mutant phenotype: First, that severely affected individuals show the phenotype at E9–9.5 and die before E10, while less severely affected individuals show no phenotype until after E12; and second, that individuals with a defective phenotype at E9–9.5 recover temporarily between E10–E11 then relapse. The second possibility was shown to be correct by culturing E9.5 mutant embryos *in vitro* and observing that they lose, then later regain, the mutant phenotype [Morrison-Graham et al., 1992]. It appears that there is a period from about E10 to E12 when the defects produced by prior lack of PDGFRα expression can be compensated for by some process which is independent of PDGFRα function. The period of compensation is relatively brief but its existence does illustrate the phenomenon of compensatory systems. These compensatory systems are presumably very advantageous to the embryo, but obscure the analysis of mutant phenotypes.

E. Development of the Musculoskeletal System

1. The PDGF A-chain/PDGFRα system is expressed during the formation of muscle-cartilage functional units. The early somites express PDGFRα transcripts throughout. As regions of the somites differentiate, PDGFRα expression is lost in the myotome region and enhanced in the sclerotome region. Cells from the sclerotome region migrate away and express particularly high levels of PDGFRα transcripts at sites in which they are condensing to form the intervertebral discs and the cartilage precursors of the axial skeleton. PDGFRα expression by chondrocytes is lost as they differentiate into mature hypertrophic chrondrocytes and as ossification begins. Expression continues in the perichondrium. It appears that the PDGFRα system is much more important for the initial proliferation of chondrocyte precursors than for their subsequent differentiation. We have observed [unpublished observations] that high levels of PDGFRα are expressed by the osteoblasts that later invade the cartilage, but we have not evaluated what role this expression plays in osteoblast function.

As cells from the myotome differentiate into muscle, they remain PDGFRα-negative, but they do express PDGF A-chain transcripts [Orr-Urtreger and Lonai, 1992]. This suggests that the PDGFRα/PDGF A-chain system could be involved in the initial formation of coherent muscle-cartilage-bone functional units. PDGF-AA secreted by developing muscles would promote the initial recruitment and/or proliferation of chondrocyte precursors to form the adjacent cartilage and perichondral tissues, and muscle attachment sites.

Limb buds form from mesenchymal cells which migrate away from the lateral plate mesoderm. The pattern of expression of PDGFRα and PDGF A-chain in the developing limbs is similar to that in the axial structures; PDGFRα is expressed by mesenchyme and lost during chondrogenesis and ossification, while PDGF A-chain is expressed by the surface ectoderm and by muscle. Both PDGFRα and PDGF A-chain transcripts are expressed by the apical ectodermal ridge (AER), an area of thickened ectoderm at the tip of the developing limb bud [Orr-Urtreger and Lonai, 1992]. The AER forms in response to some signal from the underlying mesenchyme and thereafter directs the further differentiation of the limb, for example by inhibiting the induction of chondrogenesis in cells directly below it. The AER is a self-sustaining population of cells which does not exchange cells with its surroundings. Although a role for the PDGFRα/PDGF A-chain system as an autocrine stimulator within the AER would be reasonable, it is difficult to imagine that this system could account for the inhibition by the AER of chondrogenesis, a process normally stimulated by PDGF.

2. The phenotype of Ph/Ph embryos indicates that the PDGFRα/PDGF A-chain system plays an important role in early axial cartilage development. The earliest visible defect in Ph/Ph embryos is that the neural tube in the region between the otic vesicle and the first few somites is distorted and may be flanked by one or more subepidermal blisters [Grüneberg and Truslove, 1960; Morrison-Graham et al., 1992]. This reflects the virtual absence of presumptive sclerotomal cells (which express PDGFRα in normal embryos) in the vicinity of the notochord in E9.5 embryos [Schatteman et al., 1992], and suggests that the initial migration or proliferation of sclerotomal precursors requires PDGFRα expression. Eventually, some cartilage and bone does form in Ph/Ph embryos, but the dorsal portions of many vertebrae result in spina bifida, and absence of many intervertebral discs results in fusion of adjacent vertebrae. Back muscles in Ph/Ph embryos are poorly developed or missing [Schatteman et al., 1992]. This may be a secondary consequence of the malformation of the vertebral attachment sites for these muscles. Other bones formed by endochondral ossification of a cartilage precursor are also occasionally abnormal but these abnormalities are not as severe as the defects in the vertebral column.

The exact cause of the subepidermal blisters in Ph/Ph embryos is not known. Electron microscopy after fixation with agents that aggregate proteoglycans demonstrates clear differences in proteoglycan structure between Ph/Ph and normal embryos. Proteoglycans are a major determinant of tissue hydration. This difference in extracellular matrix structure could result from differences in the type of cell which populates the affected regions in normal and Ph/Ph embryos. For example, sclerotomal derivatives may be replaced by dermatomal derivatives in Ph/Ph embryos. Alternatively, the cell type could be normal but the type of extracellular matrix components which they secrete could be affected by their inability to respond to stimulation through PDGFRα.

3. Cell culture data also suggests a possible role for PDGFRβ and PDGF B-chain in muscle development. Consistent with the *in vivo* expression data, cultured chondroblast-like cells express relatively high levels of

PDGFRα and cultured myoblasts express very low levels of this receptor. A comparable body of information about PDGFRβ and PDGF B-chain expression *in vivo* is not available; however, some cultured myoblasts do express PDGFRβ. Myoblasts isolated from chicken embryos of different ages expressed higher levels of PDGFRβ as development progressed; PDGF-BB is active as a mitogen for these cells [Yablonka-Reuveni and Seifert, 1993]. Mouse C2 and mm-14 myoblast cell lines were derived from adult mouse muscle and are thus thought to represent the residual undifferentiated myoblasts ("satellite cells") which give adult muscle some proliferative potential. The mm-14 line does not express either PDGFRα or PDGFRβ and does not respond to either PDGF-AA or PDGF-BB. In contrast, C2 cells do express PDGFRβ and exposure to PDGF-BB stimulates proliferation and delays differentiation, ultimately resulting in the formation of larger myotubes [Yablonka-Reuveni et al., 1990]. Is the phenotype of one of these cell lines an artifact of culture, or do the two cell lines represent different myoblast subtypes or stages? In order to evaluate muscle responsiveness to growth factors in a more *in vivo*-like environment, without losing the advantages of control over the environment gained with cell culture, Bischoff et al. [1986] have isolated intact muscle fibers from the adult rat toe. The satellite cells within the fibers proliferated when exposed to FGF but not when exposed to PDGF (isoform not known). Z. Yablonka-Reuveni [unpublished observations] repeated this study using all three isoforms of PDGF and confirmed that none of them elicited a proliferative response. This would indicate that the PDGFRβ is either not expressed by adult satellite cells *in vivo* or is not sufficient (by itself) to mediate a mitogenic response. The latter possibility would not rule out a role of the PDGF/PDGFR system in injury or in embryogenesis. PDGF responsiveness may require a concomitant alteration in other aspects of the cell's local environment which could serve to prevent mature muscle fibers from responding to minor disturbances. This secondary constraint may be relieved during wounding and during the period in embryogenesis during which myoblast proliferation occurs.

F. PDGFRα Expression Is Essential for a Non-Neurogenic Subset of Neural Crest Cells

Neural crest cells are ectodermal derivatives which form in the dorsal region of the neural tube. Before closure of the neural tube, they begin to migrate from the neural tube and ultimately participate in a remarkably wide range of developmental processes. The developmental fates of neural crest cells can be divided into neurogenic and nonneurogenic subsets. The neurogenic subset develops into components of the sensory, autonomic, and enteric ganglia. No defects in these tissues have been observed in Ph/Ph embryos. The nonneurogenic subset contributes to many tissues, all of which are abnormal in Ph/Ph embryos. A very close facsimile of the defects in these tissues can be produced by surgical ablation of specific regions of the neural crest in avian embryos [Kirby et al., 1983]. This argues that the normal functioning of nonneurogenic neural crest cells depends on PDGFRα expression. Tissues which are abnormal after these neural crest ablations have been described by Morrison-Graham et al. [1992].

1. Melanocytes. Ph/Ph homozygotes die before melanocytes can be identified. However, melanocyte behavior is likely to be severely affected because Ph/+ heterozygotes, which express half the normal level of PDGFRα transcripts, lack functional melanocytes in the skin of the belly and back; this results in the pathognomonic white "patch" [Grüneberg and Truslove, 1960]. It seems as if the neural crest cell/melanocyte pathway is so dependent on expression of PDGFRα that even the 50% reduction found in Ph/+ heterozygotes is detrimental to their development.

2. Thyroid and thymus. These organs form through an interaction of cranial neural crest cells with epithelia or with mesodermal derivatives. *In situ* hybridization demonstrates high levels of PDGFRα expression in the surrounding stromal component and no expression in the epithelial components of the thyroid and thymus glands. In Ph/Ph homozygotes these organs are rudimentary or absent.

3. Head and face. Neural crest cells in the head and face differentiate into some of the same cell types as are derived from the mesoderm in the lower body. These include connective tissue and bones of the face and jaw. A characteristic abnormality in Ph/Ph homozygotes is a "cleft face" in which the two sides of the face do not meet in the midline. Ph/+ heterozygotes have a widened prefrontal bone. This could be related to the split-face phenotype of Ph/Ph homozygotes, in which the prefrontal bone widens to fill in a region that should have been filled by the adjacent cranial bones.

G. The Formation and Maturation of Blood Vessels

1. Formation of placental blood vessels. Based on their studies of the developing human placenta, Holmgren et al. [1991] proposed that the endothelial cells of developing capillaries express both PDGFRβ and PDGF-BB and that their proliferation is driven through this autocrine loop. Later, the endothelial cells themselves cease to express PDGFRβ and the PDGF-BB acts as a paracrine stimulator of the adjacent PDGFRβ-positive mesenchymal cells. As the capillaries develop into arterioles and arteries, the mesenchymal cells immediately around the endothelial tube differentiate into vascular smooth muscle cells which continue to express PDGFRβ and now also express PDGF A-chain. PDGF-AA that is secreted by the smooth muscle cells presumably acts on the surrounding mesenchymal cells, because these are the only cells in which the authors detected PDGFRα transcripts.

2. Formation of embryonic blood vessels. The development of large vessels in the mouse embryo may proceed through a pathway similar to that described in the previous section for the human placenta, although we have information only about PDGFRα. We observed PDGFRα transcripts in the loose mesenchyme surrounding the endothelial tubes which constitute the primitive vessels [Schattemann et al. 1992]. Orr-Urtreger et al. [1992] noted that embryonic mouse blood vessels are positive for the PDGFRα at E9.9–10.5. As the cells around the endothelium organize into what begins to look like a muscular wall, their expression of PDGFRα declines compared with expression by the cells which remain in the surrounding mesenchyme.

The *Patch* mutant sheds some light on the role that PDGFRα may play in vascular development. Dilation of the heart and vasculature, along with edema, were the major defects reported in the initial analysis of Ph/Ph embryos [Grüneberg and Truslove, 1960]. Despite these suggestive observations, we observed a structural defect in a large artery only in the *truncus arteriosis* (next section). Although further analysis may reveal other differences between blood vessels in Ph/Ph mutants and +/+ animals, it is clear that the basic structure of most blood vessels can be established in the absence of signalling through PDGFRα.

Ph/Ph embryos also show abnormalities in the pattern of distribution and branching of vessels, especially in the neck and head. The distribution network of these blood vessels is confused, and the density of vessels seems to be higher than normal [G.C. Schatteman et al., manuscript in preparation]. But these changes could be secondary to other abnormalities in the development of head mesenchyme.

3. Remodeling of the *truncus arteriosis.* The most obvious defect in the organization of the vascular system in Ph/Ph embryos is "persistent *truncus arteriosis*". In a normal E11 mouse heart, blood from both ventricles exits via a single large vessel, the *truncus arteriosis*. By E13, this vessel has divided (septated) to form the aorta and pulmonary artery. The septation of the outflow tracts requires the participation of neural crest cells which migrate out from the cranial neural crest into the region of the truncal cushion. Some migration begins by E8 but most occurs between E9.5 and E10.5 [J.A. Weston, personal communication]. When the cranial crest region of an avian embryo is surgically removed, septation does not occur and the embryos display defects in other organs (thymus, thyroid, etc., see above) which are also defective in Ph/Ph embryos. This suggests that failure of the neural crest component accounts for the aberrant development of the *truncus arteriosis*. This is consistent with the observation that other major blood vessels, which do not involve neural crest cells, develop relatively normally. *In situ* hybridization reveals that the truncal cushion tissue in a normal embryo expresses relatively high levels of PDGFRα transcripts. It seems likely that at least some of these receptor-positive cells are neural crest cell–derived and that the absence of PDGFRα expression in neural crest cells of Ph/Ph embryos directly compromises their migration or proliferation.

Persistent *truncus arteriosis* is a relatively common human developmental abnormality, but it is unlikely that these cases result from defective PDGFRα expression because the defects in other organs which characterize Ph/Ph mice are not observed in these patients. Proper septation seems to require the normal function of many genes and is easily perturbed by many different forms of genetic and environmental insult.

4. Changes in PDGF/PDGF receptor expression may accompany postnatal maturation of arteries. Changes in PDGF and PDGF-receptor expression may also occur during the postembryonic period of vascular development. Smooth muscle cells cultured from the aorta of newborn rats, but not from 2-month-old rats, secrete large amounts of PDGF, display very few PDGF binding sites, and proliferate rapidly in the absence of exogenous PDGF [Seifert et al., 1984]. When this was initially observed, we proposed that autocrine stimulation by PDGF drives the proliferation of smooth muscle cells during the neonatal period, and that growth stops when PDGF production is shut off. More recent studies have suggested that this hypothesis may not explain smooth muscle cell behavior *in vivo*. Majesky et al. [1988] found that rat pup and adult smooth muscle cells *in vivo* express comparable levels of PDGF A-chain and B-chain transcripts. The differences between the two cell cultures show up during subculturing *ex vivo* as a dramatic increase in PDGF B-chain transcripts by pup-derived cells, but not by the adult cells. This suggests that conclusions from studies with cultured smooth muscle cells may not be directly applicable to smooth muscle cells *in vivo*. Nevertheless, the cell culture environment, although not identical to the *in vivo* environment, probably reveals some underlying differences between the regulation of the PDGF/PDGF receptor system in pup and adult smooth muscle cells.

H. Development of the Heart

In E8 mouse embryos, PDGFRα transcripts are expressed by mesenchymal cells surrounding the heart tube endothelium. The primitive heart begins to beat at E9. By E10.5, PDGFRα expression becomes prominent in the pericardium. By E11.5, the endocardium no longer expresses detectable PDGFRα transcripts; neither does the myocardium, although there seem to be small pockets of cells that express PDGFRα transcripts that are observed in some sections. PDGFRα are also found in the primitive trabeculae and pericardium. By E15, rudimentary valves which express high levels of PDGFRα can be distinguished.

Muslin and Williams [1991] found that PDGF-BB and TGF-β (but not basic FGF) increased the frequency with which heart tissue developed in explants of axolotl prospective cardiac mesoderm. This indicates that PDGF-BB (PDGF-AA was not tried) could be involved in the early induction of a portion of the mesoderm to become cardiac tissue. If PDGF plays a critical role in mouse cardiac development it must be through the PDGFRβ, because early heart development has not been observed to be disrupted in Ph/Ph embryos. The most obvious cardiac defects in Ph/Ph embryos, e.g., an absence of aortic valves, occur in the outflow tract region. It seems likely that the same deficit in neural crest cell participation accounts for both the lack of septation of the *truncus arteriosis* (Section IV.G.3) and the lack of formation of valves at the base of this blood vessel. In general, however, the heart itself is relatively well formed in Ph/Ph embryos.

I. Development and Repair of Skin and Connective Tissue

Skin consists of an outer epithelial layer (keratinocytes) and an inner mesenchymal layer (dermis). Transplantation studies have indicated that the dermis regulates the proliferation and differentiation of the overlying epidermis; this regulation may partly be mediated by keratinocyte growth factor produced by dermal fibroblasts [Finch et al., 1989]. It seems that the PDGF system may mediate a reciprocal interaction: a stimulation of the dermis by the epidermis. Evidence for this comes from the pattern of expression of PDGF A-chain and PDGFRα and from the phenotype of Ph/Ph embryos. *In situ* hybridization detects PDGF A-chain transcripts in the embryonic ectoderm (developing epidermis) throughout mouse development [Orr-Urtreger and Lonai, 1992]. PDGFRα transcripts are detected in the dermatome portion of the somites. The cells in the dermatome migrate toward the epidermis and proliferate under it to form the dermis. Some of this migration and/or proliferation seems to be driven through the PDGF A-chain/PDGFRα system because the thickness of the dermis is greatly reduced in Ph/Ph embryos which do not express PDGFRα [Schatteman et al., 1992]. The numbers of fibroblasts between organs is also reduced and the overall cohesiveness of Ph/Ph embryos compared to normal embryos is reduced to such a degree that the organs frequently break through the skin and spill into the extraembryonic space.

In adult animals, PDGF has been proposed to play a role in stimulating proliferation of dermal cells during would healing [reviewed in Ross et al., 1986]. This hypothesis initially assumed that the PDGF would be delivered by blood platelets (known to be a rich source of PDGF) and that the major target would be dermal fibroblasts (known to express PDGF receptors). Because the healing of a full-thickness wound involves the concomitant development of a new dermis and epidermis, some of the same processes involved in the embryogenesis of the skin might be reactivated during would healing where the PDGF might come from epithelial cells at the wound margin.

The role of PDGF in wound healing has not been directly experimentally tested. Ph/Ph animals are not useful for these studies because they die before birth. The addition of exogenous PDGF does not accelerate wound healing under conditions in which the normally rapid rate of healing is compromised, e.g., in diabetic animals [Knighton et al., 1985; Grotendorst et al., 1986; Greenhalgh et al., 1990]. Under these conditions, the most obvious effect of the PDGF is to increase the influx of leukocytes followed by augmentation of neovascularization. The initial target for PDGF in wound healing is probably the leukocyte (which expresses very low levels of PDGF receptors), rather than the fibroblast. Although this observation alone does not shed much light on the role of PDGF in development, it does remind us that the dermis is a complex tissue consisting of more than just fibroblasts. Thus, the virtual absence of a dermis in Ph/Ph embryos might not result exclusively, or even predominantly, from lack of PDGFRα expression by fibroblasts.

J. Development and Repair of the Kidney

1. The PDGF B-chain/PDGFRβ system plays autocrine and paracrine roles in kidney development. The kidney develops from an inductive interaction between two mesodermal derivatives, the ureteric duct and the blastema. In the human, all of the stages in the development of the glomerulus can be seen in a single section of E100 kidney. The outermost

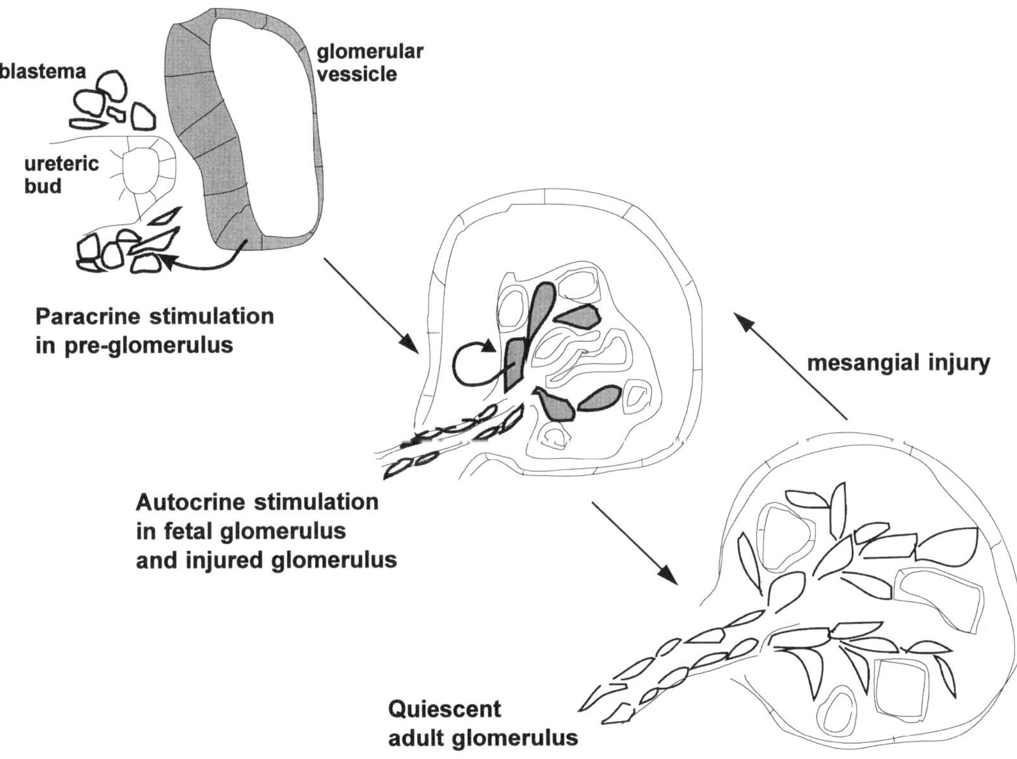

Fig. 3. *Expression of PDGFRβ and PDGF B-chain during development of the human kidney and in response to injury. Cells producing PDGF B-chain are shaded. Cells expressing PDGFRβ are shown with a darker outline. The outlines of mesangial cells in the adult glomerulus are intermediate in thickness to indicate detectable levels of PDGFRβ. Arrows within a diagram lead from the source to the target of PDGF action.*

zone (blastema) is the least mature, while essentially mature glomeruli are present in the innermost areas. At the earliest stages of glomerulogenesis, PDGF B-chain is expressed by the glomerular vesicle, while PDGFRβ is expressed only outside the vesicle by blastemal and connective tissue elements [Alpers et al., 1992] (Fig. 3). This suggests a possible paracrine interaction within the kidney, in which PDGF from the vesicle may modulate the processes of glomerular induction, or stimulate proliferation or migration of receptor-positive cells from the vascular and/or connective tissue into the glomerular cleft, where they develop into the mesangial tuft. Once the mesangial-like cells have entered the glomerulus they continue to express PDGFRβ but now also express PDGF B-chain. This provides the possibility of an autocrine stimulus for further mesangial cell proliferation. In adult glomeruli, expression of both PDGF B-chain and PDGFRβ is greatly reduced, further proliferation virtually ceases, and the glomerulus has achieved its mature, adult phenotype.

The final, autocrine phase of glomerular development appears to be reactivated in adults under certain pathological conditions (Fig. 3). Cultured mesangial cells produce PDGF when stimulated by a variety of mediator substances known to participate in the response to renal injury [Silver et al., 1989]. Expression of PDGF B-chain and PDGFRβ is greatly increased *in vivo* following mesangiolytic injury in the rat [Iida et al., 1991; Yoshimura et al., 1991]. This could mediate the compensatory proliferation of mesangial cells which follows the injury by reactivating the PDGF/PDGFR autocrine loop that originally established the mature glomerulus. In this case there is some direct experimental evidence of function: infusion of a neutralizing anti-PDGF B-chain antibody blocks about 40% of the proliferative response to mesangiolytic injury [Johnson et al., 1991].

2. Suggestive evidence that the PDGF A-chain/PDGFRα system is also involved in kidney development. PDGFRα transcripts in the kidney are detected by *in situ* hybridization, and PDGFRα protein can be detected by immunocytochemistry [R.A. Seifert, unpublished data]. Unlike PDGFRβ, which is expressed by both undifferentiated blastema and by stroma, PDGFRα expression is detected only by stromal elements. The pattern of expression of PDGF A-chain during kidney development *in vivo* has not been reported.

Abnormalities in Ph/Ph kidneys have not been noted. This argues that the PDGF A-chain/PDGFRα system cannot be necessary for normal kidney development. In adult glomeruli, PDGFRα and PDGF A-chain transcript levels are low, even after glomerular injury, when the levels of PDGF B-chain and PDGFRβ transcripts increase dramatically [Iida et al., 1991]. This argues against a role for the PDGF A-chain in the kidney's response to injury. Cell culture data does suggest one potential role for PDGF A-chain/PDGFRα in *abnormal* kidney development. Wilms' tumor (nephroblastoma) is one of the most common childhood tumors. The tumor resembles a fetal kidney in that it contains an undifferentiated component (blastema) as well as various differentiated components (tubules, glomeruli, stroma). Stroma-like cells cultured from Wilms' tumors express substantial levels of PDGF A-chain transcripts [Fraizer et al., 1987]. Gashler et al. [1992] found that in transient expression assays, the PDGF A-chain promoter is repressed by the WT-1 protein encoded by the gene which is mutated in Wilms' tumors. These observations suggest that PDGF A-chain might contribute to cell proliferation during early kidney development and that the WT-1 protein represses PDGF A-chain expression at some point during later development. Continued expression of PDGF A-chain in individuals with mutations in the WT-1 gene might contribute to abnormal proliferation. It must be noted, however, that the WT-1 protein influences transcription of other genes as well, and misregulation of those other genes might be equally, or more, important for regulation of proliferation.

K. Development of the Lens

Brewitt and Clark [1988] reported that pulsatile exposure to PDGF maintained transparency and normal rates of protein synthesis in the organ culture of newborn rat lens. Although the lens is entirely constituted of ectoderm-derived epithelial cells (lens fiber cells) and PDGF receptor expression had not previously been demonstrated by an epithelial cell type, subsequent *in situ* hybridization studies confirmed that the lens is one of the few ectodermal derivatives which does express PDGFRα [Schatteman et al., 1992]. In an E11.5 embryo, PDGFRα transcripts can be clearly seen in the developing lens. The receptor-positive lens is surrounded by a receptor-negative optic cup extending from the neural tube. In Ph/Ph embryos the lens is substantially smaller than normal. This demonstrates that PDGF, through the PDGFRα, plays an important role in lens development *in vivo* as well as in organ culture. Why PDGFRα should be expressed, and function, in the lens but not in most other epithelial cells is a complete mystery.

L. Development of the Nervous System

There is a considerable body of information implicating PDGF in the development of the central nervous system (CNS). The most detailed studies have been of the optic nerve, because it is one the simplest components of the CNS. Studies with cultured rat optic nerve glial cells have led to the hypothesis that PDGF secreted by type-1 astrocytes controls the proliferation, survival, and differentiation of other glial cell types [reviewed in Raff, 1989]. Around the time of birth, type-1 astrocytes differentiate in the optic nerve. The hypothesis proposes that these cells secrete PDGF, which promotes the proliferation of a population of undifferentiated O2-A progenitor cells and prevents their premature differentiation. Eventually, several weeks after birth, the O2-A precursor cells differentiate into oligodendrocytes and type-2 astrocytes. The evidence for this hypothesis is that culture medium conditioned by type-1 astrocytes (1) is able to stimulate proliferation, and delay differentiation, of cultured O-2A cells, (2) contains PDGF-AA, (3) is inactivated by antisera against PDGF, and (4) can be mimicked by pure PDGF-AA. Although PDGF can prevent the death of newly formed oligodendrocytes, it does not prevent the death of mature oligodendrocytes [Barres et al., 1992]. This can be accomplished by insulin-like growth factors and ciliary-neurotrophic factor [Barres et al., 1993]. The role of PDGF in this system thus seems to be restricted to a relatively brief period in development.

Surprisingly, *in situ* hybridization and immunocytochemistry detect very low levels of expression of PDGF A-chain [Yeh et al., 1991] and B-chain [Sasahara et al., 1991] transcripts in glial cells and high-level expression of these transcripts in neurons throughout the nervous system. However, expression in the optic nerve was not evaluated in those reports. In mouse embryos, PDGF A-chain expression can be detected at E15, before the differentiation of most glial cells types, and persists at high levels in the adult brain. This suggests that the neurons may be an important source of PDGF in the early developing brain. This hypothesis is attractive because it could explain how the number and location of glial cells is coordinated with the neurons which they support.

In the peripheral nervous system (PNS), PDGF and PDGF receptors are expressed in the neonatal rat dorsal root ganglia and sciatic nerve [Eccleston et al., 1993; and references therein]. The neurons, as well as the Schwann cells which myelinate these neurons, express PDGF A-chain, PDGF B-chain, PDGFRβ, and much lower levels of PDGFRα. This would be consistent with autocrine and/or paracrine stimulatory pathways in the PNS. As the rats mature, PDGF receptor expression declines in myelin-forming, but not in non-myelin-forming Schwann cells. The PDGF/PDGF receptor system may thus play some role in maintenance of non-myelin-forming Schwann cells in the PNS. When the chick sciatic nerve is injured, there is an induction of PDGF binding sites (presumably receptors) on the fibroblast-like cells of the adjacent endoneurium, but not on the neurons [Raivich and Kreutzberg, 1987].

There is little evidence for PDGFRα expression or function during the early stages of development of the nervous system. PDGFRα transcript expression is not detected in the neural tube. At E9.5, the only portion of the CNS that expresses PDGFRα transcripts is the roof of the fourth ventricle. The choroid plexuses develop from these areas and continue to express PDGFRα throughout later development. In Ph/Ph mice, the choroid plexus is both small and disorganized. After E13, the ventricles of most Ph/Ph embryos appear collapsed, perhaps due to the inability of the choroid plexus to produce sufficient cerebrospinal fluid. By *in situ* hybridization, PDGFRα transcripts begin to be detected in the brain by E14–E16, which is somewhat earlier than the appearance of glial cells, and this expression is by neurons rather than by glial cells. Most glial cell proliferation, and much other differentiation and proliferation in the brain, occurs just before and after birth. Ph/Ph mutants rarely survive to these stages so that the relative importance of PDGFRα expression for central nervous system development cannot be evaluated using Ph/Ph mutants.

V. SUMMARY

Specific proposals for biological roles for PDGF in development have expanded immensely since PDGF was first discovered in platelets and was proposed to mediate the proliferative response of local vascular smooth muscle cells and fibroblasts [reviewed in Ross et al., 1986]. The following general themes can be recognized which characterize the PDGF/PDGF receptor system during the initial development of tissues, as well as during their response to pathological perturbation in adult animals.

1. PDGF stimulates connective tissue proliferation and function via short-range (paracrine or autocrine) interactions. During development, PDGF and its receptors are often expressed by adjacent cell layers, with epithelial layers secreting the PDGF and adjacent mesenchymal layers expressing receptors for PDGF and responding to it. Occasionally, PDGF and PDGF receptors are both expressed by the same connective tissue cell types. These cells are always proliferating and a PDGF autocrine is proposed to contribute toward stimulating this proliferation.

2. Regulation of the PDGF/PDGF receptor system occurs by regulation of receptor expression as well as regulation of ligand availability. In adult animals, connective tissue cell types frequently respond to tissue injury by upregulating the expression of both PDGF and PDGF receptors.

3. The pattern of expression of PDGF and PDGF receptors during development and in response to injury is clearly concordant with a role for the PDGF/PDGF receptor system in driving connective tissue proliferation. Nevertheless, there is little direct experimental evidence concerning the magnitude of the role played by PDGF versus by other growth factor systems. The phenotype of the PDGFRα-deficient mutant Ph/Ph mouse indicates that signalling through PDGFRα plays an essential role in the development of dermis and other connective tissue cell types. Much less is known about the function of PDGFRβ during development.

REFERENCES

Alpers CE, Seifert RA, Hudkins KL, Johnson RJ, Bowen-Pope DF (1992): PDGF B-chain, PDGF receptor and α-actin in human glomerulogenesis: Developmental patterns recapitulated in glomerulonephritis. Kidney Int 42:390–399.

Barres BA, Hart IK, Coles HSR, Burne JF, Voyvodic JT, Richardson WD, Raff MC (1992): Cell death and control of cell survival in the oligodendrocyte lineage. Cell 70:31–46.

Barres BA, Schmidt R, Sendnter M, Raff MC (1993): Multiple extracellular signals are required for long-term oligodendrocyte survival. Development 118: 283–295.

Beckmann MP, Betsholtz C, Heldin C-H, Westermark B, Di Marco E, Di Fiore PP, Robbins KC, Aaronson SA (1988): Comparison of biological properties and transforming potential of human PDGF-A and PDGF-B chains. Science 241:1346–1349.

Bischoff R (1986): Proliferation of muscle satellite cells on intact myofibers in culture. Dev Biol 115:129–139.

Bowen-Pope DF, Seifert RA, Ross R (1985): The platelet-derived growth factor receptor. In Boynton AL, Leffert HL (eds): Control of Animal Cell Proliferation. New York: Academic Press, pp 281–312.

Braun T, Rudnicki MA, Arnold HH, Jaenisch R (1992): Targeted inactivation of the muscle regulatory gene Myf-5 results in abnormal rib development and perinatal death. Cell 71:369–382.

Brewitt B, Clark JI (1988): Growth and transparency in the lens, an epithelial tissue, stimulated by pulses of PDGF. Science 242:777–779.

Collins T, Bonthron DT, Orkin SH (1987): Alternative RNA splicing affects function of encoded platelet-derived growth factor A chain. Nature 328:621–624.

Eccleston PA, Funa K, Heldin C-H (1993): Expression of platelet-derived growth factor (PDGF) and PDGF α- and β-receptors in the peripheral nervous system: an analysis of sciatic nerve and dorsal root ganglia. Dev Biol 155:459–470.

Ferns GAA, Sprugel KH, Seifert RA, Bowen-Pope DF, Kelly JD, Murray M, Raines EW, Ross R (1990): Relative platelet-derived growth factor receptor subunit expression determines cell migration to different dimeric forms of PDGF. Growth Factors 3:315–324.

Finch PW, Rubin JS, Miki T, Ron D, Aaronson SA (1989): Human KGF is FGF-related with properties of a paracrine effector of epithelial cell growth. Science 245:752–755.

Fraizer GE, Bowen-Pope DF, Vogel AM (1987): Production of platelet derived growth factor by cultured Wilms' tumor cells and fetal kidney cells. J Cell Physiol 133:169–174.

Gashler AL, Bonthron DT, Madden SL, Rauscher FJ III, Collins T, Sukhatme VP (1992): Human platelet-derived growth factor A chain is transcriptionally repressed by the Wilms' tumor suppressor WT1. Proc Natl Acad Sci USA 89:10984–10988.

Goustin AS, Betsholtz C, Pfeifer-Ohlsson S, Persson H, Rydnert J, Bywater M, Holmgren G, Heldin C, Westermark B, Ohlsson R (1985): Coexpression of the sis and myc proto-oncogenes in developing human placenta suggests autocrine control of trophoblast growth. Cell 41:301–312.

Greenhalgh DG, Sprugel KH, Murray MJ, Ross R (1990): PDGF and FGF stimulate wound healing in the genetically diabetic mouse. Am J Pathol 136:1235–1246.

Gronwald RGK, Adler DA, Kelly JD, Disteche CM, Bowen-Pope DF (1990): The human PDGF receptor α-subunit gene maps to chromosome 4 in close proximity to c-kit. Hum Genet 85:383–385.

Grotendorst GR, Martin GR, Pencev D, Sodek J, Harvey AK (1986): Stimulation of granulation tissue formation by platelet-derived growth factor in normal and diabetic rats. J Clin Invest 76:2323–2329.

Grüneberg H, Truslove GM (1960): Two closely linked genes in the mouse. Genet Res 1:69–90.

Gudas LJ, Singh JP, Stiles CD (1983): Secretion of growth regulatory molecules by teratocarcinoma stem cells. In Silver LM, Martin GR, Strickland S (eds): Teratocarcinoma Stem Cells, Cold Spring Harbor Conferences on Cell Proliferation, Vol 10. Cold Spring Harbor, NY Cold Spring Harbor Laboratory, pp 229–236.

Hammacher A, Mellström K, Heldin CH, Westermark B (1989): Isoform-specific induction of actin reorganization by platelet-derived growth factor suggests that the functionally active receptor is a dimer. EMBO J 8:2489–2495.

Hanai K, Kato H, Matsuhashi S, Morita H, Raines EW, Ross R (1987): Platelet proteins, including platelet-derived growth factor, specifically depress a subset of the multiple components of the response elicited by glutathione in Hydra. J Cell Biol 104:1675–1681.

Heidaran MA, Pierce JH, Yu J-C, Lombardi D, Artrip JE, Fleming TP, Thomason A, Aaronson SA (1991): Role of αβ receptor heterodimer formation in β platelet-derived growth factor (PDGF) receptor activation by PDGF-AB. J Biol Chem 266:20232–20237.

Heldin C-H, Westermark B (1990): Platelet-derived growth factor: mechanism of action and possible *in vivo* function. Cell Regul 1:555–566.

Holmgren L, Glaser A. Pfeifer-Ohlsson S, Ohlsson R (1991): Angiogenesis during human extraembryonic development involves the spatiotemporal control of PDGF ligand and receptor gene expression. Development 113:749–754.

Iida H, Seifert R, Alpers CE, Gronwald RGK, Philips PE, Pritzl P, Gordon K, Gown AM, Ross R, Bowen-Pope DF, Johnson RJ (1991): Platelet-derived growth factor (PDGF) and PDGF receptor are induced in mesangial proliferative nephritis in the rat. Proc Natl Acad Sci USA 88:6560–6564.

Jin P, Rahm M, Claesson-Welsh L, Heldin C-H, Sejersen T (1990): Expression of PDGF A-chain and β-receptor genes during rat myoblast differentiation. J Cell Biol 110:1665–1672.

Jin P, Sejersen T, Ringertz NR (1991): Recombinant platelet-derived growth factor-BB stimulates growth and inhibits differentiation of rat L6 myoblasts. J Biol Chem 266:1245–1249.

Johnson RJ, Raines EW, Floege J, Yoshimura A, Pritzl P, Alpers C, Ross R (1991): Inhibition of mesangial cell proliferation and matrix expansion in glomerulonephritis in the rat by antibody to platelet derived growth factor. J Exp Med 175:1413–1416.

Kanakaraj P, Raj S, Khan A, Bishayee S (1991): Ligand-induced interaction between A- and B-type platelet-derived growth factor (PDGF) receptors: role of receptor heterodimers in kinase activation. Biochemistry 30:1761–1767.

Kirby ML, Gale TF, Stewart DE (1983): Neural crest cells contribute to normal aorticopulmonary septation. Science 220:1059–1061.

Knighton DR, Fiegel VD, Austin LL, Ciresi KF, Butler EL (1985): Classification and treatment of chronic nonhealing wounds: successful treatment with autologous platelet-derived wound healing factors. Ann Surg 204:322–330.

Koyama N, Morisaki N, Saito Y, Yoshida S (1992): Regulatory effects of platelet-derived growth factor-AA homodimer on migration of vascular smooth muscle cells. J Biol Chem 267:22806–22812.

LaRochelle WJ, Giese N, May-Siroff M, Robbins KC, Aaronson SA (1990): Molecular localization of the transforming and secretory properties of PDGF A and PDGF B. Science 248:1541–1544.

LaRochelle WJ, May-Siroff M, Robbins KC, Aaronson SA (1991): A novel mechanism regulating growth factor association with the cell surface: identification of a PDGF retention domain. Genes Dev 5:1191–9.

Leof EB, Proper JA, Goustin AS, Shipley GD, DiCorleto PE, Moses HL (1985): Induction of c-sis mRNA and platelet-derived growth factor–like material by transforming growth factor; type-β: a proposed model for indirect mitogenesis involving autocrine activity. Proc Natl Acad Sci USA 83:2453–2457.

Majesky MW, Benditt EP, Schwartz SM (1988): Expression and developmental control of platelet-derived growth factor A-chain and B-chain/Sis genes in rat aortic smooth muscle cells. Proc Natl Acad Sci USA 85:1524–1528.

Matsui T, Pierce JH, Fleming TP, Greenberger JS, LaRochelle WJ, Ruggiero M, Aaronson SA (1989): Independent expression of human α or β platelet-derived growth factor receptor cDNAs in a naive hematopoietic cell leads to functional coupling with mitogenic and chemotactic signaling pathways. Proc Natl Acad Sci USA 86:8314–8318.

Mercola M, Melton DA, Stiles CD (1988): Platelet-derived growth factor A chain is maternally encoded in Xenopus embryos. Science 241:1223–1225.

Mercola M, Wang C, Kelly J, Brownless C, Jackson-Grusby L, Stiles C, Bowen-Pope D (1990): Selective expression of PDGF A and its receptor during early mouse embryogenesis. Dev Biol 138:114–122.

Morrison-Graham K, Schatteman GC, Bork T, Bowen-Pope DF, Weston JA (1992): A PDGF receptor mutation in the mouse (*Patch*) perturbs the development of a non-neuronal subset of neural crest-derived cells. Development 115:133–142.

Muslin AJ, Williams LT (1991): Well-defined growth factors promote cardiac development in axolotl mesodermal explants. Development 112:1095–1101.

Nister M, Hammacher A, Mellström K, Siegbahn A, Rönnstrand L, Westermark B, Heldin C-H (1988): A glioma-derived PDGF A chain homodimer has different functional activities than a PDGF AB heterodimer purified from human platelets. Cell 52:791–799.

Noble M, Murray K, Stroobant P, Waterfield MD, Riddle P (1988): Platelet-derived growth factor promotes division and motility and inhibits premature differentiation of the oligodendrocyte/type-2 astrocyte progenitor cell. Nature 333:560–562.

Orr-Urtreger A, Bedfor MT, Do M, Eisenbach L, Lonai P (1992): Developmental expression of the α receptor for platelet-derived growth factor, which is deleted in the embryonic lethal *Patch* mutation. Development 115:289–303.

Orr-Urtreger A, Lonai P (1992): Platelet-derived growth factor-A and its receptor are expressed in separate, but adjacent cell layers of the mouse embryo. Development 115:1045–1058.

Paulsson Y, Hammacher A, Heldin C-H, Westermark B (1987): Possible positive autocrine feedback mechanism in the prereplicative phase of human fibroblasts. Nature 328:715–717.

Raj S, Kanakaraj P, Khan SA, Bishayee S (1992): Type-specific antibodies to the platelet-derived growth factor receptors: role in elucidating the structural and functional characteristics of receptor types. Biochemistry 31:1774–9.

Raff MC (1989): Glial cell diversification in the rat optic nerve. Science 243:1450–1455.

Raff MC, Lillien LE, Richardson WD, Burne JF, Noble MD (1988): Platelet-dervied growth factor from astrocytes drives the clock that times oligodendrocyte development in culture. Nature 333:562–565.

Raines EW, Dower SK, Ross R (1989): Interleukin-1 mitogenic activity for fibroblasts and smooth muscle cells is due to PDGF-AA. Science 243:393–396.

Raines EW, Lane TF, Iruela-Arispe ML, Ross R, Sage EH (1992): The extracellular glycoprotein SPARC interacts with platelet-derived growth factor (PDGF)-AB and -BB and inhibits the binding of PDGF to its receptors. Proc Natl Acad Sci USA 89:1281–1285.

Raines EW, Ross R (1992): Compartmentalization of PDGF on extracellular binding sites dependent on exon-6-encoded sequences. J Cell Biol 116:533–543.

Raivich G, Kreutzberg GW (1987): Expression of growth factor receptors in injured nervous tissue, II. Induction of specific platelet-derived growth factor binding in the injured PNS is associated with a breakdown in the blood-nerve barrier. J Neurocytol 16:701–711.

Rappolee DA, Mark D, Banda MJ, Werb Z (1988): Wound macrophages express TFT-α and other growth factors *in vivo:* analysis by mRNA phenotyping. Science 241:708–712.

Richardson WD, Pringle N, Mosley MJ, Westermark B, Dubois-Dalcq M (1988): A role for platelet-derived growth factor in normal gliogenesis in the central nervous system. Cell 53:309–319.

Rizzino A, Bowen-Pope DF (1985): Production of PDGF-like growth factors by embryonal carcinoma cells and binding of PDGF to their endoderm-like differentiated cells. Dev Biol 110:15–22.

Robbins KC, Leal F, Pierce JH, Aaronson S (1985): The v-sis/PDGF-2 transforming gene product localizes to cell membranes but is not secretory protein. EMBO J 4:1783–1792.

Ross R, Glomset J, Kariya B, Harker L (1974): A platelet-dependent serum factor that stimulates the proliferation of arterial smooth muscle cells *in vitro*. Proc Natl Acad Sci USA 71:1207–1210.

Ross R, Raines EW, Bowen-Pope DF (1986): The biology of platelet-derived growth factor. Cell 46:155–169.

Rudnicki MA, Braun T, Hinuma S, Jaenisch R (1992): Inactivation of MyoD in mice leads to upregulation of the myogenic HLH gene Myf-5 and results in apparently normal muscle development. Cell 71:383–390.

Sasahara M, Fries JWU, Raines EW, Gown AM, Westrum LE, Frosch MP, Bonthron DT, Ross R, Collins T (1991): PDGF B-chain in neurons of the central nervous system, posterior pituitary, and in a transgenic model. Cell 64:217–227.

Schatteman GC, Morrison-Graham K, Koppen AV, Weston JA, Bowen-Pope DF (1992): Regulation and role of PDGF receptor α-subunit expression during embryogenesis. Development 115:123–131.

Seifert RA, Schwartz SM, Bowen-Pope DF (1984): Developmentally regulated production of platelet-derived growth factor-like molecules. Nature 311: 669–671.

Seifert RA, Hart CE, Phillips PE, Forstrom JW, Ross R, Murray MJ, Bowen-Pope DF (1989): Two different subunits associate to create isoform-specific platelet-derived growth factor receptors. J Biol Chem 264:8771–8778.

Silver BJ, Jaffer FE, Abboud HE (1989): Platelet-derived growth factor synthesis in mesangial cells: induction by multiple peptide mitogens. Proc Natl Acad Sci USA 86:1056–1060.

Stephenson DA, Mercola M, Anderson E, Wang C, Stiles CD, Bowen-Pope DF, Chapman VM (1991): Platelet-derived growth factor receptor α-subunit gene (PDGFRα) is deleted in the mouse patch (Ph) mutation. Proc Natl Acad Sci USA 88:6–10.

Taylor RN, Williams LT (1988): Developmental expression of platelet-derived growth factor and its receptor in the human placenta. Mol Endocrinol 2:627–631.
Wang C, Kelly J, Bowen-Pope DF, Stiles CD (1990): Retinoic acid promotes transcription of the platelet-derived growth factor α-receptor gene. Mol Cell Biol 10:6781–6784.
Williams LT (1989): Signal transduction by the platelet-derived growth factor receptor. Science 243:1564–1570.
Yablonka-Reuveni Z, Balestreri TM, Bowen-Pope DF (1990): Regulation of proliferation and differentiation of myoblasts derived from adult mouse skeletal muscle by specific isoforms of PDGF. J Cell Biol 111:1623–1629.
Yablonka-Reuveni Z, Seifert RA (1993): Proliferation of chicken myoblasts is regulated by specific isoforms of PDGF: evidence for differences between myoblasts from mid and late stages of development. Dev Biol 156:307–318.
Yeh H-J, Ruit KG, Wang YX, Parks WC, Snider WD, Deuel TF (1991): PDGF A-chain gene is expressed by mammalian neurons during development and in maturity. Cell 64:209–216.
Yoshimura A, Gordon K, Alpers CE, Floege J, Pritzl P, Ross R, Couser WG, Bowen-Pope DF, Johnson RJ (1991): Demonstration of PDGF B-chain mRNA in glomeruli in mesangial proliferative nephritis by *in situ* hybridization. Kidney Int 40:470–476.

ABOUT THE AUTHORS

DANIEL F. BOWEN-POPE is Professor of Pathology at the University of Washington in Seattle, Washington. After receiving a B.A. *summa cum laude* from the University of California at San Diego, he pursued doctoral research at the University of California at Berkeley in the laboratory of Harry Rubin, where he earned a Ph.D. in 1979 for work on the regulation of sugar transport and cell proliferation by cations. This was followed by postdoctoral research at the University of Washington in the laboratory of Russell Ross. This research began with the identification of a cell surface receptor for platelet-derived growth factor (PDGF) and broadened into a general investigation of the structure and physiological functions of the PDGF/PDGF receptor system in normal development and in pathological conditions.

RONALD A. SEIFERT is Lecturer in the Department of Pathology at the University of Washington in Seattle, Washington. Despite his title, he has very little teaching responsibilities but rather conducts research on the structure, function, and *in vivo* expression and role of the receptor for platelet-derived growth factor. After receiving a B.S. in Biological Oceanography, he entered into a graduate program in the School of Fisheries at the University of Washington with an interest in invertebrate pathology. At the suggestion of his advisor, he transferred to the Department of Pathology where he received his Ph.D. in 1983 for work on the mechanisms of oyster amoebocyte aggregation. Since that time he has carried out research on the platelet-derived growth factor receptor in the laboratory of Daniel Bowen-Pope.

Growth Factors and Signal Transduction in Development: 75–95

T-Cell Signalling and Activation Through the Interleukin-2 Receptor Complex

Naohiro Terada, Richard A. Franklin, Joseph J. Lucas, and Erwin W. Gelfand

I. INTRODUCTION

Interleukin-2 (IL-2) is produced by T cells in response to activation by mitogens, alloantigens, or antigens presented in the context of appropriate major histocompatibility complex (MHC) molecules. The most important effect of IL-2 is the autocrine or paracrine stimulation of T cells expressing specific cell surface receptors (IL-2 receptors). T-cell activation fits a competence and progression model [Smith, 1988; Pardee, 1989]. Competence is first conferred on T cells by stimulation of the T-cell antigen receptor, which promotes entry of resting T cells into the G_1 phase of the cell cycle. This stage comprises T-cell activation, including the transient induction of IL-2 messenger RNA (mRNA) and a more sustained increase in IL-2 receptor mRNA expression. However, by itself, competence induction is not sufficient to result in T-cell proliferation. *In vitro,* this phase can be accomplished by a short exposure of T cells to mitogens [Kumagai et al., 1988]. Progression will follow, but is dependent on IL-2/IL-2 receptor interaction, which transduces additional signals required for entry into S phase and sustained IL-2 gene transcription [Kumagai et al., 1988; Terada et al., 1992]. Thus, IL-2 plays a central role in the immune system and is primarily responsible for inducing clonal expansion of activated T cells.

In addition to its potent activity for T-cell proliferation, IL-2 has multiple, additional biologic activities. These include induction of proliferation and differentiation of B cells [Waldmann et al., 1984; Jung et al., 1984], induction of proliferation and maturation of oligodendroglial cells [Benvenisto and Merrill, 1986], induction of lymphokine-activated killer (LAK) cells [Grimm et al., 1982] and activation of natural killer (NK) cells [Henney et al., 1981]. Further, IL-2 can induce apoptosis of T cells under some conditions, suggesting that IL-2 may be involved in thymic or extrathymic selection [Lenardo, 1991].

In this review, current knowledge of the structure and regulation of the production of IL-2 and expression of its receptors will be outlined; this will be followed by an overview of signalling events resulting from IL-2/IL-2 receptor interaction, with emphasis on the role this interaction plays in promoting cell cycle progression.

II. INTERLEUKIN-2

IL-2 was originally discovered in tissue culture supernatants from human peripheral blood lymphocytes stimulated with T-cell-specific mitogens [Morgan et al., 1976]. This cytokine was initially described as T-cell growth factor (TCGF), T-cell mitogenic factor (TMF), thymocyte stimulation factor (TSF), by others names; subsequent studies have revealed that B cells, natural killer (NK) cells, and lymphokine-activated killer (LAK) cells are also responsive to IL-2. Human IL-2 and mouse IL-2 have a molecular weight of 14–16 kD and 15–30 kD, respectively, as determined by sodium dodecyl sulfate polyacrylamide gel electrophoresis (SDS-PAGE) [Robb and Smith, 1981]. Complete nucleotide sequence analysis of the cDNA encoding human and mouse IL-2 suggests that the precursors are translated as 153 and 169 amino acids, respectively, with the signal peptide sequence [Taniguchi et al., 1983; Yokota et al., 1985; Kashima et al., 1985]. Cleavage of the 20-amino-acid leader peptide yields 133- and 149-amino-acid proteins with a predicted molecular weight of 15.4 kD and 17–19 kD. Mouse and human IL-2 share a 63% homology at the amino acid level. Discrepancies in the apparent molecular weights of mature human and murine IL-2 are due to differential O-glycosylation at a conserved threonine residue at position 3. The carboxy-terminus (amino acids 131–133) and the two cysteine residues at positions 58 and 105, which form an intramolecular disulfide bond, are required for biological activity and binding. These cysteine residues are also conserved in human and mouse sequences, and reduction of single disulfide bonds results in the loss of biological activity of IL-2 [Robb et al., 1984; Wang et al., 1984]. The human IL-2 gene is localized on chromosome 4q [Fujita et al., 1983; Holbrook et al., 1984; Siegel et al., 1984]. Its 5′-flanking region is highly conserved in human and mouse IL-2 genes. In the human IL-2 gene, the upstream region spanning nucleotides −129 to −319 from the transcription initiation site contains enhancer-binding sequences required for transcription of IL-2 [Fujita et al., 1986; Siebenlist et al., 1986]. A T-cell-specific transcription factor, nuclear factor of activated T cells (NF-AT), plays a central role in the initiation of transcription of IL-2 [Durand et al., 1988; Shaw et al., 1988]. The immunosuppressive drugs, cyclosporin A(CsA) and FK506, selectively inhibit the activation of this transcription factor, and completely inhibit the initiation of IL-2 transcription [Emmel et al., 1989].

III. INTERLEUKIN-2 RECEPTOR

A. General

IL-2 exerts its effects through a specific saturable receptor system found primarily on T cells [Morgan et al., 1976; Smith, 1984; Robb et al., 1984; Taniguchi et al., 1986; Smith, 1988], B lymphocytes [Waldmann et al., 1984; Jung et al., 1984; Bich-Thuy and Fauci, 1985; Miyawaki et al., 1987], NK cells [Brooks and Henney, 1985; Henney et al., 1981], and LAK cells [Rosenberg et al., 1985; Grimm et al., 1982]. The IL-2 receptor is a heterotrimeric complex consisting of single α, β, and γ subunits [Taniguchi and Minami, 1993; Takeshita et al., 1992; Takeshita et al., 1990]. The α-chain (Tac, p55) binds to IL-2 with low affinity ($Kd \cong 10^{-8}$ M) [Hatakeyama et al., 1985; Greene et al., 1985], but binding of IL-2 solely to the α-chain does not trigger any known second messenger pathways [Mills et al., 1990; Kondo et al., 1987]. Association of the γ-chain and β-chains seems to form the heterodimeric intermediate-affinity ($Kd \cong 10^{-9}$ M) form of the receptor [Takeshita et al., 1992; Minami et al., 1992]. Association of the α-, β-, and γ-chains forms the high-affinity IL-2 receptor ($Kd \cong 10^{-11}$ M) [Takeshita et al., 1992; Minami et al., 1992]. The complexity of the roles that each of these chains plays in signal transduction is beginning to be unraveled.

B. The α-Chain

The α-chain of the IL-2 receptor was first identified and originally described as the Tac antigen. The gene encoding the α-chain is located on chromosome 10p14-15 and is only expressed in activated lymphoid cells. This gene spans 35 kilobases (kb) and, like the γ-chain gene, contains 8 exons separated by 7 introns. The mature form of IL-2 receptor α-chain contains 251 amino acids, following cleavage of the 19-amino-acid signal sequence, and encodes a protein of approximately 55 kD. Amino acids 1–219, 220–238, and 239–251 make up the extracellular, transmembrane, and cytoplasmic portions, respectively [Minami et al., 1992]. The α-chain has potential sites for carbohydrate addition and glycosylation, as well as sites for sulfation and phosphorylation [Leonard et al., 1983; Leonard et al., 1985; Shakelford and Trowbridge, 1984]. Two sites of phosphorylation have been identified in the cytoplasmic tail (Ser 247 and Thr 250); however, phosphorylation of these sites does not seem to modulate receptor function [Hatakeyama et al., 1986]. Transcription and expression of the α-chain can be increased by treating cells with an agent that activates protein kinase C (PKC), such as phorbol ester.

C. The β-Chain

The β-chain is present at low levels on resting $CD8^+$ T lymphocytes and is absent from resting $CD4^+$ lymphocytes [Taniguchi and Minami, 1993]. Following stimulation, the β-chain gene is expressed in $CD4^+$ cells and is markedly upregulated in $CD8^+$ cells [Taniguchi and Minami, 1993; Siegel et al., 1987]. The gene encoding the β-chain is present on chromosome 22q11.2-12. The gene is approximately 24 kb long and contains 10 exons separated by 9 introns. The mature form of the IL-2 receptor β-chain is 525 amino acids long, following the cleavage of the 26-amino-acid signal sequence, and encodes a protein of approximately 75 kD. Amino acids 1–214, 215–239, and 240–525 make up the extracellular, transmembrane, and cytoplasmic portions, respectively [Gnarra et al., 1990; Leonard et al., 1984; Nikaido et al., 1984; Cosman et al., 1984]. Like the γ-chain, described below, but unlike the α-chain, the β-chain also contains the WS motif (Trp-Ser-nonconserved amino acid-Trp-Ser) and four conserved cysteine residues, characteristic of the hematopoietic family of cytokine receptors that includes the interleukin-3, -4, -5, and -6 receptors, among others [Taga and Kishimoto, 1992]. The β-chain clearly has the largest intracellular region (286 amino acids) of any of the three chains that make up the IL-2 receptor. However, the cytoplasmic tail of the β-chain, like those of the γ- and α-chains, does not seem to contain any similarity to known catalytic motifs [Hanks et al., 1988; Hatakeyama et al., 1989]. Several functionally important regions of the cytoplasmic portion of the β-chain have been defined [see Taniguchi and Minami, 1993, for review]. It seems that two distinct cytoplasmic regions of the β-chain are involved in IL-2 induced cellular signalling. The "serine-rich" region was identified as a region critical for IL-2-induced mitogenic signalling. Another cytoplasmic region of the β-chain, the "acidic" region, is responsible for its physical association with the src-family protein tyrosine kinase, $p56^{lck}$ and is critical for activation of $p56^{lck}$, following IL-2 stimulation.

D. The γ-Chain

The γ-chain is the most recently discovered chain in the IL-2 receptor complex [Noguchi et al., 1993] and is encoded on chromosome Xq13 [Noguchi et al., 1993]. The genetic mutation that leads to X-linked severe combined immunodeficiency, a disease characterized by diminished cell-mediated and humoral immunity [Gelfand and Dosch, 1983], maps to this same position [Noguchi et al., 1993]. The gene is approximately 4.2 kb in size and contains 8 exons separated by 7 introns [Noguchi et al., 1993]. The gene is constitutively expressed in T cells and encodes a 64-kD protein that spans the plasma membrane [Takeshita et al., 1992; Takeshita et al., 1990]. Following cleavage of the 22-amino-acid signal sequence, the mature protein contains 347 amino acids [Takeshita et al., 1990]. Amino acids 1–232 make up the extracellular portion of the protein, while amino acids 233–261 and 262–347 make up the transmembrane and cytoplasmic portions of the protein, respectively [Takeshita et al., 1992]. The extracellular region of this chain contains the WS motif and cysteines at positions 62, 72, 102, 115, a feature shared with other members of the hematopoietic family of cytokine receptors [Taga and Kishimoto, 1992; Takeshita et al., 1992]. Although the γ-chain does not seem to contain any kinase domain it does contain an SH2 (Src homology domain) [Conley, 1992] which could potentially bind to phosphoproteins with catalytic activity [Koch et al., 1991].

IV. SIGNALLING INDUCED BY IL-2/IL-2 RECEPTOR INTERACTION

A. IL-2 as a Progression Factor for T-Cell Proliferation

T-cell proliferation occurs through a two-stage mechanism, similar to those seen with other systems studied, most notably with growth-controlled fibroblasts [Pardee, 1989]. Although the particular extracellular signals or ligands and the receptors vary greatly for cells of different lineages, there is a remarkable similarity in the overall pattern of events which occur during cellular activation following receptor ligation. Interaction of the competence-inducing factor with the cell initiates entry into the cell cycle from a resting state, the G_0 to G_1 phase transition, and is followed by binding of a second growth factor with an appropriate receptor to initiate the progression phase, that is, commitment to proceed through G_1 phase, replication of the cellular genome, and cell division. The point of commitment to enter into the proliferative phase has been most clearly defined through analysis of yeast and is called the "start point" [see Forsburg and Nurse, 1991 for review]. The less rigorously defined "restriction point" in animal cells appears to be analogous to this point, although the molecular mechanism involved remains to be identified. In any case, it seems that after passage through this point, completion of a cell cycle is independent of further mitogen–receptor interaction. Virtually all animal cell systems which have been studied in sufficient detail follow this two-stage mechanism of cell cycle entry and proliferation. It is suspected that resting cell systems which proliferate after application of a single stimulus are likely responding to unresolved, but different, components of the mitogen.

Possible reasons for this dual-level control of cell proliferation can be readily imagined. Keeping cells in a true resting or G_0 state, such as that seen with normal lymphocytes, minimizes the organism's expenditure of energy, as the overall biosynthetic rate of such cells is greatly depressed. However, the complete "awakening" of such cells requires many biochemical steps over many hours' time. Under certain conditions, it may thus be advantageous to maintain cells in a semiresting or competent state once they have entered or reentered the cell cycle and accomplished the first series of events needed for proliferation. This includes expression of the receptors needed for interaction with the progression-inducing factor. This state of competence is usually characterized by a high level of expression of such receptors but yet the cells have not begun the doubling in overall protein and RNA content which occurs primarily during the G_1 phase. Furthermore, the dual level of control may also be involved in the mechanism(s) which permit cells to quickly leave the proliferative state as needed, for example, under conditions of nutrient deprivation. As first elaborated by Pardee [Pardee, 1974, 1989], the concept of the "restriction point" was envisaged as a window of opportunity through which cells could assess the external environment and "decide" whether or not division was appropriate.

The two-phase process of activation has been clearly delineated using normal human lymphocytes [Smith, 1988]. Stimulation of the cells through the T-cell antigen receptor/CD3 complex initiates entry into the cell cycle, and the production of IL-2 and expression of its receptor. Interaction of the cytokine with its receptor then initiates the series of events, many of which are only now being elucidated, which permit progression through G_1 phase and commitment to genome replication and cell division. Detailed analyses of the sets of specific biochemical events induced by the two sets of signals (i.e., induction of competence, triggering of progression) have revealed that they are highly redundant, in that many appear to occur quite similarly at both regulatory points. As expected, key differences in the two sets of events have also emerged, although the significance of many of these differences remains to be explained at a functional level.

At both points, entry into the cell cycle after antigen–receptor interaction and the initiation of the progression phase triggered by IL-2/IL-2 receptor interaction, there occurs the rapid induction of a pattern of gene expression. As for other basic processes studied, these patterns are overlapping. At both points, transcription of several genes, most notably a set of protooncogenes, including the cellular *fos, jun, myc* and *raf-1* genes, is induced [see Smith, 1988 for review]. However, comparative analysis of the "immediate early" genes (functionally defined as those which are induced after stimulation in the presence of cycloheximide and thus not dependent on prior protein synthesis) that are expressed at the two phases has revealed numerous differences. For example, in a recently described random isolation and characterization of eight IL-2-induced, immediate early genes, it was shown that seven were unique to the cytokine-induced phase and not expressed during the G_0 to G_1-phase transition [Beadling et al., 1993]. One of the seven unique gene products encoded was the previously identified 33-kD Pim-1 kinase, whose role in IL-2-induced proliferation remains to be determined, as do the identities of the other IL-2 induced transcripts.

As for transcription, other metabolic processes show overlapping patterns of activity at the two transitions. As described in detail below, activation through the T-cell receptor induces a series of events, including tyrosine phosphorylation of a number of substrates, hydrolysis of phosphatidylinositol, elevation of intracellular calcium concentrations, and activation of PKC. Subsequent stimulation by IL-2 also involves both tyrosine and serine/threonine phosphorylation of numerous substrates and, for example, activation of the important serine-threonine protein kinase, MAP kinase, but seems to occur independently of the latter three events. Other events specific to the series of events initiated by IL-2 are only now being identified. One event under intensive study is the IL-2-induced activation of the Raf-1 kinase, which repeatedly did not occur at the earlier (G_0 to G_1) transition [Zmuidzinas et al., 1991; Turner et al., 1991]. However, recently Franklin et al. [1994] demonstrated Ras/Raf-1 activation after ligation of the T-cell receptor.

B. Ionic and Biochemical Responses Induced by IL-2

As a progression signal, it is likely that the IL-2 receptor is associated with different signalling pathways than those coupled with the T-cell receptor (TCR). In contrast to ligation of the TCR, activation of the cells by IL-2 does not lead to turnover of membrane phosphatidylinositols, the liberation of inositol phosphates, or increases in intracellular concentrations of Ca^{2+} ($[Ca^{2+}]_i$) [Mills et al., 1985a; Mills et al., 1986a; Gelfand et al., 1987a, b; Gelfand, 1990]. Further, IL-2-induced proliferation is unaffected if extracellular Ca^{2+} is removed or $[Ca^{2+}]_i$ is clamped, i.e., maintained at a particular value [Mills et al., 1985a, b; Gelfand et al., 1987b].

One of the earliest responses to IL-2 is an increase in cytosolic pH, the result of activation of the Na^+/H^+ antiporter [Gelfand et al., 1987b; Gelfand, 1990; Mills et al., 1986b]. Although increases in intracellular pH have been causally linked to cell proliferation in many other systems, blocking the antiporter with amiloride analogs or clamping intracellular pH, does not affect IL-2 induced proliferation *in vitro* [Gelfand et al., 1987b; Mills et al., 1986b]. These findings indicate that neither activation of the Na^+/H^+ antiporter nor cytosolic alkalinization are critical processes in IL-2-dependent signalling. Interestingly, stimulation of the β chain of the IL-2 receptor with high concentrations of IL-2, in the absence of the α chain, is sufficient to activate Na^+/H^+ exchange [Mills and May, 1987].

Activation of the Na^+/H^+ exchanger may result from activation of PKC, and IL-2 has been reported to induce the translocation of PKC from the cytosol to the plasma membrane [Farrar and Anderson, 1985]. Activation of PKC following IL-2/IL-2 receptor interaction, however, has not been consistently observed [Mills et al., 1986]. Further, IL-2-dependent cell proliferation is not prevented by blocking PKC activity [Mills et al., 1988] or by depleting cells of PKC [Valge et al., 1988]. In addition, IL-2 is capable of inducing cell proliferation even in cells which constitutively lack functional PKC [Mills et al., 1988].

C. Tyrosine Kinases

Triggering of the IL-2 receptor results in the tyrosine phosphorylation of a number of proteins ranging in size from 38 to 120 kD [Saltzman et al., 1990]. Some of these proteins have been identified; for example, the β-chain of the IL-2 receptor appears to be the 75-kD protein phosphorylated following IL-2 stimulation [Mills et al., 1990]. In addition, the Raf-1 kinase is known to be tyrosine phosphorylated following IL-2 stimulation [Turner et al., 1991]. Although increased tyrosine phosphorylation clearly occurs following IL-2 stimulation of T lymphocytes, there does not appear to be homology with any of the known tyrosine kinases in the cytoplasmic regions of the IL-2 receptor components [Hatakeyama et al., 1989; Hanks et al., 1988; Leonard et al., 1984]. Perhaps a portion of this activity is due to $p56^{lck}$, or another src kinase, that is associated with the acid-rich region in the cytoplasmic tail of the β-chain. Activation of $p56^{lck}$, which can occur following IL-2 stimulation, requires the dephosphorylation of tyrosine residue 505 [Seft and Campbell, 1991]. This suggests that not only does tyrosine phosphorylation occur following IL-2 stimulation, but that tyrosine phosphatases may also be activated. Moreover, it is unlikely that $p56^{lck}$ activation is an absolute requirement for proliferation because IL-2-induced proliferation occurs in a T-cell line that lacks $p56^{lck}$, and in cells that lack the portion of the β-chain that binds $p56^{lck}$ [Taniguchi and Minami, 1993; Sawami et al., 1992]. Recently a tyrosine kinase gene that is not a member of the *src* family of kinases was found to be induced following IL-2 stimulation. This novel tyrosine kinase is a 72-kD protein that has been named Itk in mouse [Siliciano et al., 1992] and Emt in human [Gibson, Leung, Squire, Hill, Arima, Goss, Hogg and Mills, submitted]. Since $p56^{lck}$ and potentially other *src* kinases are also activated following T-cell receptor triggering [Saltzman et al., 1990], this enzyme may be responsible for some of the differences in tyrosine phosphorylation that occur following both IL-2 and T-cell receptor signalling.

D. Phosphatidylinositol 3-Kinase and Glycosylphosphatidylinositol Metabolism

Phosphatidylinositol 3-kinase (P13K) phosphorylates phosphatidylinositol, phosphatidylinositol 4-phosphate, and phosphatidylinositol 4,5-bisphosphate at the D-3 position of the inositol ring [Whitman et al., 1988; Remillard et al., 1991]. This enzyme is rapidly activated after IL-2-stimulation [Remillard et al., 1991]. P13K is composed of two distinct subunits with approximate molecular weights of 110 kD and 85 kD. The p85 subunit contains both *src* homology region 2 (SH2) and SH3 domains and seems to regulate the activity of the 110-kD subunit. P13K can be immunoprecipitated with antibodies directed to the β-chain of the IL-2 receptor, as well as with antibodies to phosphotyrosine and the *src* protein tyrosine kinase, Fyn [Remillard et al., 1991; Augustine et al., 1991]. These data suggest that P13K is able to form complexes and potentially be regulated by, or itself regulate, the activities of these proteins. The phosphoinositol products formed by P13K activation may represent new second messenger systems. In addition to P13K-generated phosphatidylinositols, other phosphatidylinositols may play a role in IL-2-induced signalling. Following IL-2 stimulation, glycophosphatidylinositol lipids are hydrolized and inositolphosphoglycan and myristylated diacylglycerol are formed [Merida et al., 1990; Eardley and Koshland, 1991]. The rate of formation of these compounds parallels the dose requirements for IL-2-induced growth [Eardley and Koshland, 1991]. Inositolphosphoglycans have strong effects on glucose utilization [Saltiel and Sorbara-Kazan, 1987; Kelly et al.,

1987], will stimulate the proliferation of several insulin-dependent cell lines [Witters and Watts, 1988], and augment the proliferation of IL-2-stimulated T cells [Merida et al., 1990].

E. p21^ras^

p21^ras^ was originally identified as a protein which is capable of inducing cell transformation. p21^ras^ activity is modulated by the binding of guanosine triphosphate (GTP; stimulatory) or guanosine diphosphate (GDP; inhibitory). Activation of p2l^ras^ (p21^ras^-GTP) can be mediated via the inhibition of GTPase activating protein (GAP) or the activation of a protein which exchanges GTP for p21^ras^-bound GDP [Downward, 1991; Trahey and McCormick, 1987; Shou et al., 1992]. IL-2 is known to activate p21^ras^ [Izquierdo et al., 1992]. Although the route by which IL-2 binding to its receptor mediates p21^ras^ activation is not known, it seems that both the "acidic" and "serine-rich" regions of the β-chain of the IL-2 receptor are required for this activation [Satoh et al., 1992]. In addition, IL-2 promotes some of the events that are presumed to be downstream of p21^ras^ activation such as the activation of the 70–75-kD serine-threonine kinase, Raf-1 [Turner et al., 1991].

F. Raf-1 Kinase/MAP Kinase/p90 S6 Kinase

Raf-1, the product of the ubiquitously expressed c-*raf*-1 gene, is a 70–75-kD cytoplasmic phosphoprotein with intrinsic serine-threonine kinase activity [Bonner et al., 1985; Moelling et al., 1984]. Ligand-induced activation of the platelet-derived growth factor (PDGF) and epidermal growth factor (EGF) receptor tyrosine kinases leads to the association of Raf-1 with the receptor and, in the case of the PDGF receptor, results in the phosphorylation of Raf-1 on tyrosine residues [Morrison et al., 1989; Morrison et al., 1988]. Interaction of IL-2 with the IL-2 receptor, as well as that of IL-3 or granulocyte-macrophage colony stimulatory factor and their receptors, results in the activation of Raf-1 kinase, possibly as a result of its phosphorylation [Turner et al., 1991; Carroll et al., 1990]. Raf-1 is a candidate for the activation of MEK1, or MAP kinase kinase, the enzyme which activates MAP kinase [Lange-Carter et al., 1994].

MAP kinase is a 42-kD serine-threonine kinase which is activated by many different stimuli in most cells; it is a member of the family of extracellular regulated kinases (erk) [Boulton and Cobb, 1991]. The phosphorylation and activation of MAP kinase are rapidly induced by IL-2 within 1 min in the IL-2-dependent human T-cell line Kit225 [Franklin, Sawami, Terada, Lucas, and Gelfand, submitted; Terada et al., 1993a]. One of the known downstream events of MAP kinase is activation of p90 S6 kinase (p90^rsk^). MAP kinase can phosphorylate p90^rsk^ *in vitro*, and the activity of MAP kinase and phosphorylation of p90^rsk^ correlate well *in vivo* [Sturgill et al., 1988]. IL-2 can also rapidly induce phosphorylation and activation of p90^rsk^ in T cells, within 5 min of the addition of the cytokine [Franklin, Sawami, Terada, Lucas, and Gelfand, submitted; Sawami et al., 1992; Terada et al., 1993a]. Although this serine-threonine kinase p90^rsk^ can phosphorylate ribosomal S6 protein *in vitro*, it seems not to phosphorylate S6 *in vivo* [Ballou et al., 1991; Blenis et al., 1991]. The substrate for p90^rsk^ in the intact cell remains unknown; however, p90^rsk^ can phosphorylate other substrates such as glycogen synthetase and c-Fos in vitro [Lavoinne et al., 1991; Dent et al., 1990; Chen et al., 1992]. In a number of cell types, including T cells, p21^ras^ and Raf-1 activation are associated with MAP kinase activation [deVries-Smits et al., 1992; Leevers et al., 1992; Campbell et al., 1992; Franklin et al., 1994] and p56^lck^ has been shown to directly phosphorylate and activate MAP kinase as well [Ettenhadieh et al., 1992]. It is likely that either of these pathways (p21^ras^/Raf-1 or p56^lck^), or a combination of the two, leads to MAP kinase activation following IL-2 stimulation as well.

G. p70 S6 Kinase/Ribosomal S6 Phosphorylation

A serine-threonine kinase, p70 S6 kinase (p70^s6k^), is also induced by IL-2 within 5 min of addition of the growth factor, and its activity reaches a maximum within 30–60 min [Kuo et al., 1992; Calvo et al., 1992; Terada et al., 1993a]. Although the upstream events leading to activation of this kinase are not well defined, they are distinct from the Raf-1/MEK/MAP/p90^{rsk} kinase pathway [Ballou et al., 1991; Blenis et al. 1991]. p70^{s6k} is responsible for *in vivo* phosphorylation of ribosomal S6 protein in mammalian cells, and seems to augment the translational rate of mRNAs [see Proud, 1992 for review]. Inhibition of p70^{s6k} activity and S6 phosphorylation by rapamycin results in a marked reduction of protein synthesis and inhibition of cell growth induced by IL-2. The inhibition of cell proliferation with rapamycin was most potent when resting cells were cultured with mitogens [Terada et al., 1993a, b]. In contrast, once cells have entered the cell cycle, rapamycin had only marginal inhibitory activity on cell cycle progression. Indeed when rapamycin was added to cultures of continuously cycling cells, cell proliferation for an additional two cycles proceeded without any impedance [Terada et al., 1993a; Terada et al., submitted]. These results indicate that activity of p70^{s6k} and phosphorylation of S6 are more likely required for resting cells entering into the cell cycle than for already cycling cells. In T cells, mitogenic stimulation by phytohemagglutinin (PHA) of resting cells or IL-2/IL-2 receptor interaction can induce p70^{s6k} activity [Terada et al., 1992b]. Thus, stimulation of p70^{s6k} activity by IL-2 may facilitate competent T-cell entry into the progression phase, perhaps by upregulating translational activity.

H. Cyclin-Dependent Kinases (Cdks) and Cyclins

The outcome of the series of events initiated by the interaction of IL-2 with its receptor is "progression" through the G_1 phase of the cell cycle, genome replication and cell division. Analysis of this series of cell cycle events in a variety of systems, including yeast, amphibians, marine invertebrates, and mammals, indicates that they are, at least in part, regulated by a family of kinases, the cyclin-dependent kinases (cdks), and their associated cyclins [reviewed in Maller, 1991; Forsburg and Nurse, 1991]. In mammals, five cdks and five *families* of cyclins have been so far identified, but the specific roles of very few of these molecules have yet been defined. The mechanism(s) by which the chain of signal transduction events triggered by the interaction of IL-2 with its receptor leads to progression through the cell cycle remains especially enigmatic. Links between the biochemical events listed above and the family of cdks and cyclins which seem to drive the cell cycle remain to be identified.

Identification of the first member of the cdk family, p34$^{cdc2/cdk1}$, was derived from isolation and characterization of temperature-sensitive cell division cycle (cdc) mutants in yeast [Forsburg and Nurse, 1991; Hartwell, 1973; Nurse and Bissett, 1981]. Genetic and molecular studies have established that the enzyme, a serine-threonine protein kinase, plays a central role at two points in the yeast cell cycle: at the G_2/M phase transition and at passage through the point called "Start" at which time the cell becomes committed to progression through the cycle [Maller, 1991; Forsburg and Nurse, 1991]. At mitosis, p34^{cdc2}, which is present in large amounts, is directly responsible for the breakdown of cellular structure that is essential for cell division, by phosphorylating structural components of the cytoskeleton, the nuclear envelope, and chromatin [Dessev et al., 1991; Moreno and Nurse, 1990]. The mechanisms of action at other points in the cycle remain to be established, because *in vivo* substrates for the kinase have not been definitively identified. The p34^{cdc2} kinase was also identified by a second line of research involving analysis of the cell cycle of activated amphibian oocytes. The amphibian homolog of p34^{cdc2} was identified as the catalytic subunit of MPF (mitosis or maturation promoting factor). When complexed with cyclin B, the kinase plays a similar role at the G_2/M phase transition as its yeast counterpart [Dunphy et al., 1988; Gautier et al., 1988]. Analysis of the mammalian homolog of yeast

$p34^{cdc2}$ established its mitotic function as well [Lee and Nurse, 1987; Riabowol et al., 1989; Hamaguchi et al., 1992]. Whether it plays a role at other points in the cell cycle remains unclear. Furthermore, in higher eukaryotic systems, $p34^{cdc2}$ is but one member of a family of cdk kinases each of which may play a distinct role in the cell cycle [Meyerson et al., 1991; Reed, 1992]. So far, five members of the cdk family have been identified.

Also of importance here is that proper functioning of the kinases requires association with a cyclin. Five families of cyclins (A–E) have been so far identified in mammalian cells [reviewed in Reed, 1992]. As noted above, $p34^{cdc2(cdk1)}$, complexed with cyclin B, functions at mitosis, but suggested role(s) for $p34^{cdc2}$, complexed with cyclin A, at the G_1/S transition or during S phase itself have not been firmly established [Pagano et al., 1992; Marraccino et al., 1992]. Indirect evidence suggests that $p33^{cdk2}$, complexed with cyclin E, has a role at the G_1/S phase transition and/or in S phase [Dulic et al., 1992; Koff et al., 1992]. A possible mitotic function for $p33^{cdk2}$ has also been suggested [Rosenblatt et al., 1992]. The $p36^{cdk4}$ kinase, associated with cyclins of the D family, may play a role, earlier in the cell cycle, mediating the effects of growth factors in inducing cells to enter the cell cycle [Matsushime et al., 1992]. Again, however, evidence for these functions is indirect; the substrates for the kinases have not been identified. Finally, the possible roles for $p36^{cdk3}$ [Meyerson et al., 1992], cyclin C [Lew et al., 1991], and for the recently discovered $p31^{cdk5}$ remain unknown. The $p31^{cdk5}$ [Xiong et al., 1992; Hellmich et al., 1992] kinase was identified by virtue of its association with D-type cyclins and with proliferating cell nuclear antigen (PCNA) [Xiong et al., 1992], a component of DNA polymerase δ which appears to be involved in DNA synthesis and repair.

$p34^{cdc2}$ and $p33^{cdk2}$ are so far the best-studied kinases and the regulation of their activities is similar. Posttranslational regulation of these protein kinases over a range of at least three-fold is achieved by several means [Dalton, 1992; Welch and Wang, 1992].

1. Within the ATP binding sites of the enzymes, a threonine (residue 14) and tyrosine (residue 15) must be dephosphorylated for activity. This is accomplished at least in part by a homologue of the yeast Cdc25 phosphatase [Millar and Russell, 1992; Gabrielli et al., 1992; Atherton-Fessler et al., 1993]. As for the cdks, there is a family of these phosphatases, with three members (Cdc25A,B,C) identified to date [Sadhu et al., 1990; Galaktionov and Beach, 1991].

2. For maximum activation, a threonine residue near the carboxyl terminus (161 for Cdc2; 160 for Cdk2) must be phosphorylated. This may be accomplished by a homolog of the newly described enzyme CAK, or cyclin-dependent kinase activating kinase [Gu et al., 1992; Solomon et al., 1992]. Whether or not this enzyme, identified in *Xenopus,* will also be one member of a family of kinases is unknown.

3. Finally, the kinase must be complexed with a cyclin [Reed, 1992]. So far, a complex pattern of associations has been described, with each kinase forming complexes with several different cyclins. However, because most conclusions have been drawn from studies of complexes *in vitro,* it is still unclear whether the level of combinatorial complexity which has emerged actually exists *in vivo.*

Remarkably, the precise role(s) of the cyclins in promoting activity of the Cdk/cyclin complexes remains unclear. Recent results showing that the cytoplasmic to nuclear distribution ratio of some cyclins changes during the cell cycle has led to the proposal that cyclins target or "carry" the kinases to their appropriate sites in the cell [Pines and Hunter, 1991; Hubbard and Cohen, 1993]. The discoveries of Cdk/cyclin duplexes in multiprotein complexes also containing the transcription factor E2F (which regulates transcription of genes encoding proteins needed for S-phase entry and progression) [Nevins, 1992], or with components of the DNA-replication machinery like PCNA [Xiong et al., 1992], immediately suggests possible ways in which the cdks may exert their control over cell cycle transitions: for example, by altering the phosphorylation state of chromatin proteins involved in regulating cell cycle–dependent transcription and/or the replication of the genome.

In summary, analysis of a variety of systems from fungi to mammals has revealed both the outlines and many details of the mechanisms which regulate the major cell cycle transitions. However, acquisition of specific knowledge concerning these events in the system under study here, the IL-2-induced proliferation of human lymphocytes, has only recently begun. Furukawa et al. [1990] first demonstrated the presence of $p34^{cdc2}$ kinase in normal human T cells and showed that it was first synthesized in activated T cells very late in G_1 phase, just prior to S phase entry. Lucas et al. [1992] subsequently described the differentially phosphorylated forms of the kinase which were present throughout the T-cell cycle and identified a second, related kinase which was made earlier in the cell cycle than $p34^{cdc2}$. Synthesis of the kinase and also of cyclin A, which is the major p34-associated cyclin apart from cyclin B, which appears later in the cell cycle, was inhibited both by CsA and by rapamycin [Terada et al., 1993b], indicating that these events are ultimately dependent on prior IL-2/IL-2 receptor interaction. The second, novel kinase noted above was subsequently shown to be $p33^{cdk2}$ [Lucas et al., submitted]. This kinase was present, in very low amounts and in an enzymatically inactive form, in resting T cells isolated from peripheral blood. Transcription of the *cdk2* gene, increased accumulation of the protein, and, most importantly, activation of the $p33^{cdk2}$ protein as a serine-threonine protein kinase was first detected in mid G_1 phase. Comparative studies of the activities of the two kinases indicated that the major role for the $p34^{cdc2}$ kinase was restricted to the G_2/M phase transition in T-cell proliferation, whereas the $p33^{cdk2}$ kinase likely functioned primarily in late G_1 phase and S phase, although the precise roles for the kinases are yet to be determined.

Analysis of the patterns of transcription of the families of cyclin genes in normal T lymphocytes has also been described [Ajchenbaum et al., 1993]. Resting T cells express, at a very low level, the cyclin D2 and cyclin C genes. Increased transcription of these genes is observed after T-cell activation, with increased D2 transcripts being detectable as early as 2 hr after stimulation. Transcripts encoding cyclins A, D3, and E were not detected in resting cells and began to appear about 18–24 hr, 12–16 hr, and 18–24 hr, respectively. The cyclin D1 gene seems not to be expressed in T cells. The induction of transcription of all of the genes clearly occurred in G_1 phase; except for a partial reduction in the accumulation of cyclin A transcripts, all were expressed normally in cells arrested prior to S-phase entry. Increased expression of the cyclin D2 and C genes and the induction of expression of the cyclin A, D3, and E genes were all inhibited by CsA under conditions which completely prevented production of IL-2 mRNA.

For those members of the *cdk* and *cyclin* gene families which have been identified and studied to date, i.e., the *cdc2* and *cdk2* genes and all of the known T-cell-expressed *cyclin* genes (*A, B, C, D2, D3,* and *E*), these results indicate that transcription of the genes is, because of the kinetics of gene induction and/or the gene sensitivity to inhibition by CsA and/or rapamycin, dependent on IL-2/ IL-2 receptor interaction, at least in normal human T lymphocytes. Similar findings have also been reported for the IL-2-mediated induction of *cyclin* and *cdk* genes using the hematopoietic cell line BAF-B03, after transfection with IL-2 receptor cDNA clones [Shibuya et al., 1992]. An elucidation of the mechanisms by which the chain of events initiated by IL-2/IL-2 receptor interaction culminates in activation of the Cdk/cyclin complexes needed for further cell-cycle transitions thus becomes a major focus of research. However, the findings that low levels of the $p33^{cdk2}$ protein and of cyclin C and D2 transcripts are present in purified populations of resting T cells leaves open the possibility that these molecules may also play some role prior to IL-2/IL-2 receptor interaction.

One member of the cdk family, $p36^{cdk4}$, has been implicated in events associated with the mitogen-activated entry of resting cells into the cell cycle [Matsushime et al., 1991, 1992]. Since $p36^{cdk4}$ is the earliest inducibly transcribed member of the cdk family in activated T lymphocytes [Lucas et al., unpublished observation], its activity may be independent of IL-2/IL-2 receptor interaction, so that $p36^{cdk4}$ is in a position to mediate the effects of the cytokine in inducing cell cycle progression. It is also noteworthy that this kinase was identified through its association with cyclins of the D family and that *cyclin D2* seems to be the first cyclin gene activated after T-cell stimulation. Further analysis of $p36^{cdk4}$ and its associated cyclin(s), especially cyclin D2, in normal human T cells is therefore currently an area of intense investigation.

Whether $p36^{cdk4}$ is activated as part of the acquisition of the competent state, and is thus in a position to *mediate* the action of the cytokine, or is, in fact, activated as part of the early progression phase remains unknown. However, forging the link between initial signal transduction events and the network of cell cycle regulatory kinases and cyclins will likely come through analysis of this protein complex.

V. IMMUNOSUPPRESSANTS

A. General

Delineation of the actions of the immunosuppressants, CsA, FK506, and rapamycin have contributed significantly to our understanding of the regulation of IL-2 production and the role of IL-2 receptor signalling in T-cell activation. As described in detail elsewhere [Morris, 1991; Schreiber, 1991], CsA, FK506, and rapamycin share many common characteristics and also distinct differences in their actions. All three drugs suppress T-cell proliferation [Morris, 1991; Schreiber, 1991], prolong graft survival in organ transplantation, and prevent the onset of autoimmune disease in animals [Morris, 1991]. CsA and FK506 are particularly important therapeutic agents that have found widespread use in preventing graft rejection. It is of interest that all three compounds bind to distinct families of cellular receptors, the immunophilins, which exhibit peptidyl-prolyl isomerase (PPIase) activity [Takahashi et al., 1989; Fisher et al., 1989; Siekierka et al., 1989; Harding et al., 1989]. CsA binds to the family of cyclophilins; FK506 and rapamycin bind to the group of FK506-binding proteins (FKBPs). In fact, FK506 is structurally very similar to rapamycin [Schreiber, 1991]. Although the drugs can inhibit the enzymatic activity of their receptors *in vitro*, it has been shown that inhibition of PPIase activity is unrelated to their immunosuppressive activity [Sigal et al., 1991]. However, for their activity, the drugs must bind to the immunophilins [Wiederrecht et al., 1991; Tropshung et al., 1989]. Elucidation of the structures of FKBP and cyclophilins through the use of X-ray crystallography and nuclear magnetic resonance (NMR) techniques demonstrates that the drugs undergo a dramatic alteration in conformation when bound to immunophilins [Van Duyne et al., 1991; Kallen et al., 1991; Fesik et al., 1991]. Thus, the drug—receptor complex forms a unique structural entity that is represented by neither the drug nor the receptor alone.

B. Cyclosporin A and FK506

CsA and FK506 show virtually identical inhibitory properties during T-cell activation. Both of the drugs inhibit T-cell proliferation by blocking the transcription of early T-cell activation genes, especially those of IL-2 and IL-4 [Morris, 1991]. This effect seems to be mediated, in part, through inhibition of NF-AT activity, which is essential for the initiation of IL-2 transcription [Durand et al., 1988; Shaw et al., 1988; Emmel et al., 1989; Terada et al., 1992a]. A detailed proposed mechanism of inhibition of NF-AT activity by CsA or FK506 has been recently presented

[Flanagan et al., 1991]. For the full activation of NF-AT, there appears to be a requirement for at least two components of NF-AT. One is NF-ATc, a constitutively expressed cytoplasmic component, which is translocated into the nucleus following cell activation via a calcium-dependent pathway. The other component is NF-ATn, a nuclear component. NF-ATn is induced by PKC-dependent signals such as phorbol esters [Flanagan et al., 1991] and has been demonstrated to contain AP-1 (Jun/Fos) [Jain et al., 1992; Northrop et al., 1993]. CsA or FK506 seems to inhibit the translocation of NF-ATc to the nucleus. Furthermore, the CsA/cyclophilin or FK506/FKBP complexes directly bind to calcineurin/calmodulin complexes, and inhibit the activity of calcineurin (a Ca^{2+}-dependent serine-threonine phosphatase, classified as PP2B) [Liu et al., 1991]. Although there is no direct evidence that calcineurin regulates the translocation of NF-ATc, the results of several studies have suggested that calcineurin affects the IL-2 promoter in an FK506- and/or CsA-sensitive manner [Fruman et al., 1992; O'Keefe et al. 1992]; in addition the translocation of NF-ATc to the nucleus is a calcium-dependent event [Flanagan et al., 1991].

The immunosuppressive action of CsA and FK506 is therefore likely explained by the selective inhibition of IL-2 or IL-4 (especially IL-2) transcription, which is regulated at least in part by calcineurin. However, it remains to be determined whether calcineurin will play a similar role in other CsA- and/or FK506-sensitive events, such as mast cell degranulation [Morris, 1991]. In addition, CsA and FK506 can inhibit cell proliferation of other cell types, including B lymphocytes [Morris, 1991]. Since B cell proliferation does not involve activation of NF-AT or the production of IL-2 or IL-4, the further mechanisms of action of the drugs remain to be defined.

C. Rapamycin

Rapamycin inhibits T-cell proliferation through a different mechanism from that of CsA and FK506. Although rapamycin and FK506 are structurally related and are active following binding to identical cellular proteins (FKBPs), rapamycin does not affect transcription of IL-2 or IL-3 [Morris, 1991], activation of NF-AT [Schreiber, 1991], or activity of calcineurin [Liu et al., 1991]. Earlier studies have shown that rapamycin, but not CsA or FK506, can inhibit T-cell proliferation, even that triggered by mitogens that initiate activation in a Ca^{2+}-independent manner [Morris, 1991]. Rapamycin, but not CsA or FK506, has been demonstrated to inhibit IL-2- or IL-4-induced proliferation of cytokine-dependent cell lines [Morris, 1991]. Further, rapamycin has been shown to inhibit IL-3- or granulocyte-macrophage colony stimulation factor (GM-CSF)-dependent proliferation [Morris, 1991]. Therefore, an attractive hypothesis derived from these observations is that rapamycin may inhibit the unique events induced by a common signalling pathway through cytokine receptors. However, one pathway targeted by rapamycin has been $p70^{s6k}$ activation [Chung et al., 1992; Kuo et al., 1992; Price et al., 1992; Calvo et al., 1992; Terada et al., 1992b, 1993a], a pathway activated by general mitogenic stimuli as well [Sturgill and Wu, 1991]. For example, in T cells, $p70^{s6k}$ is activated not only by IL-2 receptor signalling but also by PHA or phorbol ester/calcium ionophore [Terada et al., 1992b]. Rapamycin does not inhibit $p70^{s6k}$ directly [Price et al., 1992], but it seems to target an unknown upstream event. Inhibition of the $p70^{s6k}$ pathway by rapamycin may be selective, because the drug does not affect a number of other early signalling events or kinase activities (including MAP kinase, p90 S6 kinase ($p90^{rsk}$), tyrosine phosphorylation of proteins detected using antiphosphotyrosine antibodies, and transcription of c-fos or c-myc). Similarly rapamycin does not inhibit the activity of $p34^{cdc2}$ once the kinase has been activated [Morice et al., 1993]. Although the upstream event(s) for $p70^{s6k}$ activation is (are) not known, a downstream event has been partially characterized. $p70^{s6k}$ activation is the major (if not exclusive) kinase that regulates phosphorylation of ribosomal

S6 protein *in vivo* in mammalian cells; a related molecule p90^{rsk} does not affect S6 phosphorylation *in vivo* [Ballou et al., 1991; Blenis et al., 1991]. In fact, inhibition of p70^{s6k} by rapamycin is always accompanied by the inhibition of S6 phosphorylation, despite the persistence of activation of p90^{rsk}. Inhibition of S6 phosphorylation by rapamycin results in inhibition of protein synthesis. Interestingly, rapamycin inhibits translation of selective mRNAs, including mRNA encoding ribosomal proteins, suggesting that phosphorylation of S6 might be involved in selective translational control in higher eukaryotic cells [Terada et al., submitted]. Rapamycin inhibits p70^{s6k} activity and the S6 phosphorylation pathway in diverse mammalian cell lines, derived from different tissues, including lymphocytes, neuronal tissues, hepatocytes, and fibroblasts [Chung et al., 1992; Kuo et al., 1992; Price et al., 1992; Calvo et al., 1992; Terada et al., 1992b, 1993a]. Rapamycin inhibits proliferation of these cells to various degrees, and it is not clear that the differential sensitivity to rapamycin, in terms of proliferation, is the result of a differential sensitivity to p70^{s6k} inactivation in a given tissue, or of the particular growth status of the cell line.

Resting cells entering the cell cycle exhibit a greater sensitivity to rapamycin compared to continuously cycling cells [Terada et al., 1992a]. Resting T cells stimulated with PHA, or IL-2-deprived Kit225 cells stimulated with IL-2, cannot enter the cell cycle in the presence of rapamycin [Terada et al., 1993a, b]. However, rapamycin does not inhibit Kit225 cell proliferation when it is added to cycling cells [Terada et al., 1993a]. Rapamycin-treated cycling Kit225 cells continue to divide for two to three cycles in the absence of apparent p70^{s6k} activity; cell size nonetheless becomes progressively smaller and the G_1 phase of the cell cycle is gradually prolonged in subsequent cycles (unpublished observations). Because the increase in protein amount is necessary for normal resting cells to enter the cell cycle [Brooks, 1977], inhibition of protein synthesis by rapamycin may explain these differential effects of rapamycin on resting and cycling cells. Although rapamycin has little or no effect on most initial events triggered by IL-2 binding to the high-affinity IL-2 receptor complex, the drug provides an effective means for studying the role of p70^{s6k} and ribosomal S6 phosphorylation in IL-2-dependent cell proliferation and activation of the translational machinery.

VI. CONCLUSION

The interaction between IL-2 and its receptor critically regulates the magnitude and duration of the T-cell immune response following mitogen- or antigen-induced activation. The regulation of IL-2 production, the molecular characterization of the receptor and the signalling events which follow ligand–receptor binding are rapidly being elucidated. Figure 1 illustrates a model of IL-2 receptor–mediated signalling which incorporates much of the information discussed in this review. Many aspects of this descriptive model remain to be clarified, most notably:

1. The different roles of common signalling events which are induced both by a competence-inducing stimulus (via the T-cell receptor) and by a progression-triggering factor (IL-2).
2. Identification of the nature and role of unique signalling molecule(s) associated with the IL-2 receptor (and cytokine receptor family in general), especially their role in transducing signals for G_1 progression to S phase.
3. Identification of the molecules associated with the γ-chain of the IL-2 receptor.
4. Elucidation of the role of the regulation of protein synthesis by p70^{s6k} activity in G_1 progression to S phase.
5. Identification of the molecules and/or events which can transduce a signal leading to the activation of Cdk/cyclin complexes following IL-2/IL-2 receptor interaction.

Despite the obvious complexity of these events and the involvement of an increasing

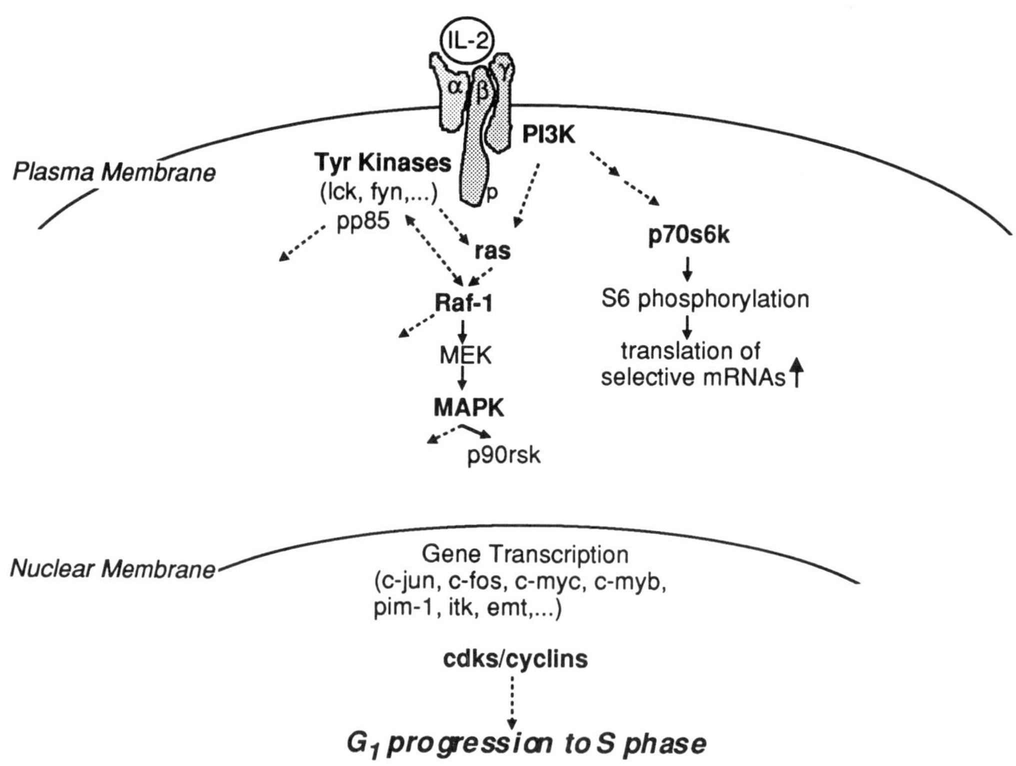

Fig. 1. Signaling events induced following IL-2/IL-2 receptor interaction. *Binding of IL-2 to the high-affinity IL-2 receptor complex (α, β, and γ chains) leads to activation of IL-2 receptor-associated src family tyrosine kinases such as Lck and Fyn. Activation of these kinases results in their autophosphorylation and the tyrosine phosphorylation of multiple substrates, including the IL-2 receptor β, Raf-1, and an 85-kD substrate (pp85), which seems to be distinguished from the p85 subunit of P13K [Satoh et al., 1992; Taniguchi and Minami, 1993]. P13K may also be associated with the activated IL-2 receptor complex, and P13K activity is rapidly induced following IL-2 stimulation. IL-2 also triggers Ras activation by an as-yet-undefined pathway; activation of Ras may be involved in Raf-1 activation. Activation of the serine-threonine kinase Raf-1 leads to the activation of a series of serine-threonine kinases, including MEK (MAP kinase kinase), MAP kinase, and* $p90^{rsk}$. *MAP kinase and* $p90^{rsk}$ *potentially phosphorylate substrates such as the Jun, Myc, and Fos proteins, thus transducing signals from the cytoplasm to the nucleus. IL-2 also induces mRNA transcription of the* jun, fos, myc, myb, pim-1, itk, *and* emt *genes. Another serine-threonine kinase,* $p70^{s6k}$, *is also activated by IL-2, and phosphorylates ribosomal S6 protein. This pathway seems to result in increased translation of ribosomal protein mRNAs which support protein synthesis required for* G_1 *progression. This complex cascade of signal transduction events induced by IL-2 ultimately leads T cells to DNA synthesis (S phase). Whether or not cells enter S phase is likely controlled by Cdk/cyclin complexes. The relationship between the IL-2-induced kinases described above and the activation of Cdk/cyclin complexes is now being defined.*

array of proteins and enzymatic activities, knowledge already gained about IL-2/IL-2 receptor interaction and the pathways involved has provided important insights into the understanding of T-cell immunodeficiency and the mechanism of action of certain immunosuppressive drugs.

ACKNOWLEDGMENT

We are grateful to Agota Szepesi, Kozo Takase, Joanne Domenico, Hiren R. Patel, and Attila Tordai for helpful discussions and for their assistance in the work performed in the authors' laboratory. This work was supported by grants AI-26490 and AI-29704 from the NIH.

REFERENCES

Ajchenbaum F, Ando K, DeCaprio JA, Griffin JD (1993): Independent regulation of human D-type cyclin gene expression during G1 phase in primary human T lymphocytes. J Biol Chem 268:4113–4119.

Atherton-Fessler S, Parker LL, Geahlen RL, Piwnica WH (1993): Mechanisms of p34cdc2 regulation. Mol Cell Biol 13:1675–1685.

Augustine JA, Sutor SL, Abraham RT (1991): Interleukin-2 and polyomavirus middle T antigen-induced modification of phosphatidylinositol 3-kinase activity in activated T lymphocytes. Mol Cell Biol 11:4431–4440.

Ballou LM, Luther H, Thomas G (1991): MAP2 kinase and 70K S6 kinase lie on distinct signalling pathways. Nature 349:348–350.

Beadling C, Johnson KW, Smith KA (1993): Isolation of interleukin 2–induced immediate-early genes. Proc Natl Acad Sci USA 90:2719–2723.

Benvenisto EN, Merrill JE (1986): Stimulation of oligodendroglial proliferation and maturation by interleukin-2. Nature 321:610–613.

Bich-Thuy LT, Fauci AS (1985): Direct effect of interleukin 2 on the differentiation of human B cells which have not been preactivated in vitro. Eur J Immunol 15:1075–1079.

Blenis J, Chung J, Erikson E, Alcorta DA, Erickson RL (1991): Distinct mechanisms for the activation of the RSK kinases/MAP2 kinase/pp90^{rsk} and pp70-S6 kinase signaling systems are indicated by inhibition of protein synthesis. Cell Growth Differ 2:279–285.

Bonner TI, Kerby SB, Sutrave P, Gunnell MA, Mark G, Rapp UR (1985): Structure and biological activity of human homologs of the *raf/mil* oncogene. Mol Cell Biol 51:1400–1407.

Boulton TG, Cobb MH (1991): Identification on multiple extracellular signal-regulated kinases (ERKs) with antipeptide antibodies. Cell Regul 2:357–371.

Brooks RF (1977): Continuous protein synthesis is required to maintain the probability of entry into S phase. Cell 12:311–317.

Brooks CG, Henney GS (1985): Interleukin-2 and regulation of natural killer cell activity in cultured cell populations. Contemp Top Mol Immunol 10:63–92.

Calvo V, Crews CM, Vik TA, Bierer BE (1992): Interleukin 2 stimulation of p70 S6 kinase activity is inhibited by the immunosuppressant rapamycin. Proc Natl Acad Sci USA 89:7571–7575.

Campbell GS, Pang L, Miyasaka T, Saltiel AR, Carter-Su C (1992): Stimulation by growth hormone of MAP kinase activity in 3T3-F442A fibroblasts. J Biol Chem 267:6074–6080.

Carroll M, Clark-Lewis I, Rapp UR, May WS (1990): Interleukin-3 and granulocyte-macrophage colony-stimulating factor mediate rapid phosphorylation and activation of cystosolic c-raf. J Biol Chem 265:19812–19820.

Chen R-H, Sarnecki C, Blenis J (1992): Nuclear localization and regulation of erk- and rsk-encoded protein kinases. Mol Cell Biol 12:915–927.

Chung J, Kuo CJ, Crabtree GR, Blenis J (1992): Rapamycin-FKBP specifically blocks growth-dependent activation of and signaling by the 70 kd S6 protein kinases. Cell 69:1227–1236.

Conley ME (1992): Molecular approaches to analysis of x-linked immunodeficiencies. Annu Rev Immunol 10:215–238.

Cosman D, Ceretti DP, Larsen A, Park L, March C, Dower S, Gillis S, Urdall D (1984): Cloning, sequence and expression of human interleukin-2 receptor. Nature 312:768–771.

Dalton S (1992): Cell cycle regulation of the human cdc 2 gene. EMBO J 11:1797–1804.

Dent P, Lavoinne A, Nakielny S, Caudwell FB, Watt P, Cohen P (1990): The molecular mechanisms by which insulin stimulates glycogen synthesis in skeletal muscle. Nature 348:302–308.

Dessev G, Iovcheva-Dessev C, Bischoff JR, Beach D, Goldman R (1991): A complex containing p34^{cdc2} and cyclin B phosphorylates the nuclear lamin and disassembles nuclei of clam oocytes in vitro. J Cell Biol 112:523–533.

deVries-Smits AMM, Burgering BMTh, Leevers SJ, Marshall CJ, Bos JL (1992): Involvement of p21ras in activation of extracellular signal-regulated kinase 2. Nature 357:602–604.

Downward J (1991): Exchange rate mechanisms. Nature 352:282–283.

Dulic V, Lees E, Reed SI (1992): Association of human cyclin E with a periodic G1-S phase protein kinase. Science 257:1958–1961.

Dunphy WG, Brizuela L, Beach D, Newport J (1988): The Xenopus cdc2 protein is a component of MPF, a cytoplasmic regulator of mitosis. Cell 54:423–431.

Durand DB, Shaw JP, Bush MR, Replogle RE, Belagaje R, Crabtree GR (1988): Characterization of antigen receptor response elements within the interleukin-2 enhancer. Mol Cell Biol 8:1715–1724.

Eardley DD, Koshland ME (1991): Glycosylphosphatidylinositol: a candidate system for interleukin-2 signal transduction. Science 251:78–81.

Emmel EA, Verweij CL, Durand DB, Higgins KM, Lacy E, Crabtree GR (1989): Cyclosporin A specifically inhibits function of nuclear proteins involved in T cell activation. Science 246:1617–1620.

Ettehadieh E, Saghera JS, Pelech SL, Hess-Bienz D, Watts J, Shastri N, Aebersold R (1992): Tyrosyl phosphorylation and activation of MAP kinases by p56^{lck}. Science 255:853–855.

Farrar WL, Anderson WB (1985): Interleukin 2 stimulates association of protein kinase C with plasma membrane. Nature 315:233–234.

Fesik SW, Gampe RT, Eaton HL, Gemmecker G, Olejniczak ET, Neri P, Holzman TF, Egan DA, Edalji R, Simmer R, Helfrich R, Hochlowski J, Jackson M (1991): NMR studies of [U-^{13}C] cyclosporin-A bound to cyclophilin. Bound conformation and portions of cyclosporin involved in binding. Biochemistry 30:6574–6583.

Fisher G, Wittmann-Liebold B, Lang K, Kiefhaber T, Schmid FX (1989): Cyclophilin and peptidyl-prolyl cis-trans isomerase are probably identical proteins. Nature 337:476–478.

Flanagan WM, Corthesy B, Bram RJ, Crabtree GR (1991): Nuclear association of a T-cell transcription factor blocked by FK-506 and cyclosporin A. Nature 352:803–807.

Flanagan WM, Crabtree GR (1992): In vitro transcription faithfully reflecting T-cell activation requirements. J Biol Chem 267:399–406.

Forsburg SL, Nurse P (1991): Cell cycle regulation in the yeasts *Saccharomyces cerevisiae* and *Saccharomyces pombe*. Annu Rev Cell Biol 7:227–256.

Franklin RA, Tordoi A, Patel H, Gardner AM, Johnson GL, Gelfand EW (1994): Ligation of the T-cell receptor complex results in activation of the Ras/Raf-1/MEK/MAPK cascade in human T lymphocytes. J Biol Chem (in press).

Fruman DA, Klee CL, Bierer BE, Burakoff SJ (1992): Calcineurin phosphatase activity in T lymphocytes is inhibited by FK506 and cyclosporin A. Proc Natl Acad Sci USA 89:3686–3690.

Fujita T, Takaoka C, Matsui H, Taniguchi T (1983): Structure of the human interleukin 2 gene. Proc Natl Acad Sci USA 80:7437–7441.

Fujita T, Shibuya H, Ohashi K, Yamanishi K, Taniguchi T (1986): Regulation of human IL-2 gene: functional DNA sequences in the 5′ flanking region for the gene expression in activated T lymphocytes. Cell 46:401–407.

Furukawa Y, Piwnica-Worms H, Ernst TJ, Kanakura Y, Griffin JD (1990): cdc2 gene expression at the G1 to S transition in human T lymphocytes. Science 250:805–808.

Gabrielli BG, Lee MS, Walker DH, Piwnica-Worms H, Maller JL (1992): Cdc25 regulates the phosphorylation and activity of the Xenopus cdk2 protein kinase complex. J Biol Chem 267:18040–18046.

Galaktionov K, Beach D (1991): Specific activation of cdc25 tyrosine phosphatases by B-type cyclins: evidence for multiple roles of mitotic cyclins. Cell 67:1181–1194.

Gautier J, Norbury C, Lohka M, Nurse P, Maller J (1988): Purified maturation-promoting factor contains the product of a Xenopus homolog of the fission yeast cell cycle control gene cdc2^{+}. Cell 54:433–439.

Gelfand EW, Dosch HM (1983): Diagnosis and classification of severe combined immunodeficiency disease. Birth Defects 19:65–72.

Gelfand EW, Mills GB, Cheung RH, Lee JW, Grinstein S (1987a): Contrasting requirements for IL-2 production, IL-2 receptor expression and lymphocyte proliferation. The role of calcium-dependent and calcium-independent signals. Lymphokines 14:155–157.

Gelfand EW, Mills GB, Cheung RK, Lee JW, Grinstein S (1987b): Transmembrane ion fluxes during activation of human T lymphocytes: role of Ca2+, Na+/H+ exchange and phospholipid turnover. Immunol Rev 95:59–87.

Gelfand EW (1990): Cytosolic calcium changes during T and B lymphocyte activation: biological consequences and significance. In Grinstein S, Rotstein (eds): Current Topics in Membranes and Transport, Vol 35. Orlando, FL: Academic Press, pp. 153–177.

Gibson S, Leung B, Squire JA, Hill M, Arima N, Goss P, Hogg D, Mills GB (1993): Identification, cloning, and characterization of a novel human T-cell-specific tyrosine kinase located at the hematopoietin complex on chromosome 5q. Blood 82:1561–1572.

Gnarra JR, Otani H, Wang MG, McBride OW, Sharon M, Leonard WJ (1990): Human interleukin 2 receptor β-chain gene: chromosomal localization and identification of 5′ regulatory sequences. Proc Natl Acad Sci USA 87:3440–3444.

Greene WC, Robb RJ, Svetlik PB, Rusk GM, Depper JM, Leonard WJ (1985): Stable expression of cDNA encoding the human interleukin-2 receptor in eukaryotic cells. J Exp Med 162:363–368.

Grimm EA, Mazumder A, Zhang HZ, Rosenberg SA (1982): Lymphokine-activated killer phenomenon. Lysis of Natural killer-resistant fresh solid tumor cells by interleukin 2-activated autologous human peripheral blood lymphocytes. J Exp Med 155:1823–1841.

Gu Y, Rosenblatt J, Morgan DO (1992): Cell cycle regulation of CDK2 activity by phosphorylation of Thr160 and Tyr15. EMBO J 11:3995–4005.

Hamaguchi JR, Tobey RA, Pines J, Crissman HA, Hunter T, Bradbury EM (1992): Requirement for p34^{cdc2} kinase is restricted to mitosis in the mammalian cdc2 mutant FT210. J Cell Biol 117:1041–1053.

Hanks SK, Quinn AM, Hunter T (1988): The protein kinase family: conserved features and deduced phylogeny of the catalytic domains. Science 241:758–760.

Hartwell LH (1973): Three additional genes required for DNA synthesis in *S. cerevisiae*. J Bacteriol 115:966–974.

Hatakeyama M, Minamoto S, Uchiyama T, Hardy RR, Yamada G, Taniguchi T (1985): Reconstitution of functional receptor for human interleukin-2 in mouse cells. Nature 318:467–470.

Hatakeyama M, Minamoto S, Taniguchi T (1986): Intracytoplasmic phosphorylation sites of Tac antigen (p55) are not essential for the conformation, function, and regulation of the human interleukin 2 receptor. Proc Natl Acad Sci USA 83:9650–9654.

Hatakeyama M, Tsudo M, Minamoto S, Kono T, Doi T, Miyata T, Miyasaka M, Taniguchi T (1989): Interleukin-2 receptor beta chain gene: generation of three receptor forms by cloned human alpha and beta chain cDNAs. Science 244:551–556.

Hellmich MR, Pant HC, Wada E, Battey JF (1992): Neuronal cdc2-like kinase: a cdc2-related protein kinase with predominantly neuronal expression. Proc Natl Acad Sci USA 89:10867–10871.

Henney CS, Kuribayashi K, Kern DE, Gillis S (1981): Interleukin-2 augments natural killer cell activity. Nature 291:335–338.

Holbrook NJ, Smith KA, Fornace AJ, Comeau CM, Wiskocil RL, Crabtree GR (1984): T-cell growth factor: completed nucleotide sequence and organization of the gene in normal and malignant cells. Proc Natl Acad Sci USA 81:1634–1638.

Hubbard MJ, Cohen P (1993): On target with a new mechanism for the regulation of protein phosphorylation. TIBS 18:172–177.

Izquierdo M, Downward J, Otani H, Leonard WJ, Cantrell DA (1992): Interleukin (IL)-2 activation of p21ras in murine myeloid cells transfected with human IL-2 receptor beta chain. Eur J Immunol 22:817–821.

Jain J, McCaffrey PG, Valge-Archer VE, Rao A (1992): Nuclear factor of activated T cells contains Fos and Jun. Nature 356:801–804.

Jung LK, Hara T, Fu SM (1984): Detection and functional studies of p60–65 (Tac antigen) on activated human B cells. J Exp Med 160:1597–1602.

Kallen J, Spitzfaden C, Zurini MGM, Wider G, Widmer H, Wuthrich K, Walkinshaw MD (1991): Structure of human cyclophilin and its binding site for cyclosporin-A determined by X-ray crystallography and NMR spectroscopy. Nature 353:276–279.

Kashima N, Nishi-Takaoka C, Fujita T, Taki S, Yamada G, Hamuro J, Taniguchi T (1985): Unique structure of murine interleukin-2 as deduced from cloned cDNAs. Nature 313:402–404.

Kelly KL, Mato JM, Merida I, Jarret L (1987): Glucose transport and antilipolysis are differentially regulated by the polar head group of an insulin-sensitive glycophospholipid. Proc Natl Acad Sci USA 84:6404–6407.

Koch CA, Anderson D, Moran MF, Ellis C, Pawson T (1991): SH2 and SH3 domains: elements that control interactions of cytoplasmic signaling proteins. Science 252:668–674.

Koff A, Giordano A, Desai D, Yamashita K, Harper JW, Elledge S, Nishimoto T, Morgan DO, Franza BR, Roberts JM (1992): Formation and activation of a cyclin E-cdk2 complex during the G1 phase of the human cell cycle. Science 257:1689–1694.

Kondo S, Kinoshita M, Shimizu A, Saito Y, Konishi M, Sabe H, Honjo T (1987): Expression and functional characterization of artificial mutants of interleukin-2 receptor. Nature 327:64–67.

Kumagai N, Benedict SH, Mills GB, Gelfand EW (1988): Induction of competence and progression signals in human T lymphocytes by phorbol esters and calcium ionophores. J Cell Physiol 137:329–336.

Kuo CJ, Chung J, Fiorentino DF, Flanagan WM, Blenis J, Crabtree GR (1992): Rapamycin selectively inhibits interleukin-2 activation of p70 S6 kinase. Nature 358:70–73.

Lange-Carter CA, Pleiman CM, Gardner AM, Blumer KJ, Johnson GL (1993): A divergence in the MAP kinase regulatory network defined by MEK kinase and Raf. Science 260:315–319.

Lavoinne A, Erikson E, Maller J, Price DJ, Avruch J, Cohen P (1991): Purification and characterization of the insulin-stimulated protein kinase from rabbit skeletal muscle; close similarity to S6 kinase II. Eur J Biochem 199:723–728.

Lee M, Nurse P (1987): Complementation used to clone a human homologue of the fission yeast cell cycle control gene *cdc2*. Nature 327:31–35.

Leevers SJ, Marshall CJ (1992): Activation of extracellular signal-regulated kinase, ERK2, by p21ras oncoprotein. EMBO J 11:569–574.

Lenardo MJ (1991): Interleukin-2 programs mouse αβ T lymphocytes for apoptosis. Nature 353:858–861.

Leonard WJ, Depper JM, Robb RJ, Waldmann TA, Greene WC (1983): Characterization of the human receptor for T-cell growth factor. Proc Natl Acad Sci USA 80:6957–6961.

Leonard WJ, Depper JM, Crabtree GR, Rudikoff S, Pumphrey J, Robb RJ, Kronke M, Svetlik PB, Peffer NJ, Waldmann TA, Greene WC (1984): Molecular cloning and expression of cDNAs for the human interleukin-2 receptor. Nature 311:626–631.

Leonard WJ, Depper JM, Kronke M, Robb RJ, Waldmann TA, Greene WC (1985): The human receptor for T-cell growth factor: evidence for variable post-translational processing, phosphorylation, sulfation and the ability of precursor forms of the receptor to bind TCGF. J Biol Chem 260:1872–1880.

Lew DJ, Dulic V, Reed SI (1991): Isolation of three novel human cyclins by rescue of G1 cyclin (Cln) function in yeast. Cell 66:1197–1206.

Liu J, Farmer JJ, Lane WS, Friedman J, Weissman I, Schreiber SL (1991): Calcineurin is a common target of cyclophilin-syclosporin A and FKBP-FK506 complexes. Cell 66:807–815.

Lucas JJ, Terada N, Szepesi A, Gelfand EW (1992): Regulation of synthesis of p34cdc2 and its homologues and their relationship to p110Rb phosphorylation during cell cycle progression of normal human T cells J Immunol 148:1804–1811.

Maller JL (1991): Mitotic control. Curr Opin Cell Biol 3:269–275.

Marraccino RL, Firpo EJ, Roberts JM (1992): Activation

of the p34 CDC2 protein kinase at the start of S phase in the human cell cycle. Mol Biol Cell 3:389–401.
Matsushime H, Roussel M, Ashman RA, Sherr CJ (1991): Colony-stimulating factor 1 regulates novel cyclins during the G1 phase of the cell cycle. Cell 65:701–713.
Matsushime H, Ewen ME, Strom DK, Kato JY, Hanks SK, Roussel MF, Sherr CJ (1992): Identification and properties of an atypical catalytic subunit ($p34^{PSK-J3/cdk4}$) for mammalian D type G1 cyclins. Cell 71:323–334.
Merida I, Pratt JC, Gaulton GN (1990): Regulation of the interleukin 2–dependent growth responses by glycosylphosphatidylinositol molecules. Proc Natl Acad Sci USA 87:9421–9425.
Meyerson M, Faha B, Su LK, Harlow E, Tsai LH (1991): The cyclin-dependent kinase family. Cold Spring Harb Symp Quant Biol 56:177–186.
Meyerson M, Enders GH, Wu CL, Su LK, Gorka C, Nelson C, Harlow E, Tsai LH (1992): A family of human cdc2-related protein kinases. EMBO J 11:2909–2917.
Millar JBA, Russell P (1992): The cdc25 M-phase inducer: an unconventional protein phosphatase. Cell 68:407–410.
Mills GB, May C (1987): Binding of interleukin 2 to its 75-kDa intermediate affinity receptor is sufficient to activate Na+/H+ exchange. J Immunol 139:4083–4087.
Mills GB, Cheung RK, Grinstein S, Gelfand EW (1985a): Interleukin 2–induced lymphocyte proliferation is independent of increases in cytosolic-free calcium concentrations. J Immunol 134:2431–2435.
Mills GB, Cragoe EJ, Gelfand EW, Grinstein S (1985b): Interleukin 2 induces a rapid increase in intracellular pH through activation of a Na+/H+ antiport. Cytoplasmic alkalinization is not required for lymphocyte proliferation. J Biol Chem 260:12500–12507.
Mills GB, Stewart DJ, Mellors A, Gelfand EW (1986a): Interleukin 2 does not induce phosphatidylinositol hydrolysis in activated T cells. J Immunol 136:3019–3024.
Mills GB, Cheung RK, Cragoe EJ, Grinstein S, Gelfand EW (1986b): Activation of the Na+/H+ antiport is not required for lectin-induced proliferation of human T lymphocytes. J Immunol 136:1150–1154.
Mills GB, Girard P, Grinstein S, Gelfand EW (1988): Interleukin-2 induces proliferation of T lymphocyte mutants lacking protein kinase C. Cell 55:91–100.
Mills GB, May C, McGill M, Fung M, Baker M, Sutherland R, Greene WC (1990): Interleukin 2-induced tyrosine phosphorylation. Interleukin 2 receptor β is tyrosine phosphorylated. J Biol Chem 265:3561–3567.
Minami Y, Takeshi K, Yamada K, Taniguchi T (1992): The interleukin-2 receptors: insights into a complex signalling mechanism. Biochim Biophys Acta 1114:163–177.
Miyawaki T, Suzuki T, Butler JL, Cooper MD (1987): Interleukin-2 effects on human B cells activated in vivo. J Clin Immunol 7:277–287.
Moelling K, Heimann B, Beimling P, Rapp UR, Sander T (1984): Serine-and threonine-specific protein kinase activities of purified gag-mil and gag-raf proteins. Nature 312:558–561.
Moreno S, Nurse P (1990): Substrates for $p34^{cdc2}$: in vivo veritas? Cell 61:549–551.
Morgan DA, Ruscetti FW, Gallo R (1976): Selective in vitro growth of T lymphocytes from normal human bone marrows. Science 193:1007–1008.
Morice WG, Brunn GJ, Wiederrecht G, Siekierka JJ, Abraham RT (1993): Rapamycin-induced inhibition of p34cdc2 kinase activation is associated with G1/S-phase growth arrest in T lymphocytes. J Biol Chem 268:3734–3738.
Morris RE (1991): Rapamycin: FK506's fraternal twin or distant cousin? Immunol Today 12:137–140.
Morrison DK, Kaplan DR, Rapp UR, Roberts TM (1988): Signal transduction from the membrane to the cytoplasm: growth factors and membrane bound oncogenes increase Raf-1 phosphorylation and associated protein kinase activity. Proc Natl Acad Sci USA 85:8855–8859.
Morrison DK, Kaplan DR, Escobedo JA, Rapp U, Roberts TL, Williams LT (1989): Direct activation of the ser/thr kinase activity of raf-1 through tyrosine phosphorylation by the PDGF β receptor. Cell 58:649–657.
Nevins JR (1992): E2F: A link between the Rb tumor suppressor protein and viral oncoproteins. Science 258:424–429.
Nikaido T, Shimizu A, Ishiida N, Sabeh, Teshigawara K, Maeda M, Uchiyama T, Yodoi J, Honjo T (1984): Molecular cloning of cDNA encoding human interleukin-2 receptor. Nature 311:631–635.
Noguchi M, Yi H, Rosenblatt HM, Filipovich AH, Adelstein S, Modi WS, McBride OW, Leonard WJ (1993): Interleukin-2 receptor γ chain mutation results in X-linked severe combined immunodeficiency in humans. Cell 73:147–157.
Northrop JP, Ullman KS, Crabtree GR (1993): Characterization of the nuclear and cytoplasmic components of the lymphoid-specific nuclear factor of activated T cells (NF-AT) complex. J Biol Chem 268:2917–2923.
Nurse P, Bissett Y (1981): Gene required in G1 for committment to cell cycle and in G2 for control of mitosis in fission yeast. Nature 292:558–560.
O'Keefe S, Tamura J, Kincaid RL, Tocci MJ, O'Neil EA (1992): FK506-and Cyclosporin A-sensitive activation of the interleukin-2 promoter by calcineurin. Nature 357:692–694.
Pagano M, Pepperkok R, Verde F, Ansorge W, Draetta G (1992): Cyclin A is required at two points in the human cell cycle. Embo J 11:961–971.
Pardee AB (1974): A restriction point for control of nor

mal animal cell proliferation. Proc Natl Acad Sci USA 71:1286–1290.

Pardee AB (1989): G_1 events and regulation of cell proliferation. Science 246:603–608.

Pines J, Hunter T (1991): Human cyclins A and B1 are differentially located in the cell and undergo cell cycle-dependent nuclear transport. J Cell Biol 115:1–17.

Price DJ, Grove JR, Calvo V, Avruch J, Bierer BE (1992): Rapamycin-induced inhibition of the 70-kilodalton S6 protein kinase. Science 257:973–977.

Proud CG (1992): Protein phosphorylation in translational control. Curr Top Cell Regul 32:243–369.

Reed SI (1992): The role of p34 kinases in the G1 to S-phase transition. Annu Rev Cell Biol 8:529–561.

Remillard B, Petrillo R, Maslinski W, Tsudo M, Strom TB, Cantley L, Varticovski L (1991): Interleukin-2 receptor regulates activation of phosphatidylinositol 3-kinase. J Biol Chem 266:14167–14170.

Riabowol K, Draetta G, Brizuela L, Vandre D, Beach D (1989): The cdc2 kinase is a nuclear protein that is essential for mitosis in mammalian cells. Cell 57:393–401.

Robb RJ, Smith KA (1981): Heterogeneity of human T-cell growth factor(s) due to variable glycosylation. Mol Immunol 18:1087–1094.

Robb RJ, Kutny RM, Panico M, Morris HR, Chowdhry V (1984): Amino acid sequence and post-translational modification of human interleukin 2. Proc Natl Acad Sci USA 81:6486–6490.

Rosenberg SA, Lotze MT, Muul LM, Leitman S, Chang AE, Ettinghausen SE, Matory YL, Skibber JM, Shiloni E, Vetto JT (1985): Observations on the systemic administration of autologous lymphokine-activated killer cells and recombinant interleukin-2 to patients with metastatic cancer. N Engl J Med 313:1485–1492.

Rosenblatt J, Gu Y, Morgan DO (1992): Human cyclin-dependent kinase 2 is activated during the S and G2 phases of the cell cycle and associates with cyclin A. Proc Natl Acad Sci USA 89:2824–2828.

Sadhu K, Reed SI, Richardson H, Russell P (1990): Human homolog of fission yeast cdc25 mitotic inducer is predominantly expressed in G2. Proc Natl Acad Sci USA 87:5139–5143.

Saltiel AR, Sorbara-Cazan L (1987): Inositol glycan mimics the action of insulin on glucose utilization in rat adipocytes. Biochem Biophys Res Commun 149:1084–1092.

Saltzman EM, White K, Casnellie JE (1990): Stimulation of the antigen and interleukin-2 receptors on T lymphocytes activates distinct tyrosine kinases. J Biol Chem 265:10138–10142.

Satoh T, Minami Y, Kono T, Yamada K, Kawahara A, Taniguchi T, Kaziro Y (1992): Interleukin 2-induced activation of Ras requires two domains of interleukin 2 receptor β subunit, the essential region for growth stimulation and lck-binding domain. J Biol Chem 267:25423–25427.

Sawami H, Terada N, Franklin RA, Okawa H, Uchiyama T, Lucas JJ, Gelfand EW (1992): Signal transduction by interleukin 2 in human T cells: activation of tyrosine and ribosomal S6 kinases and cell-cycle regulatory genes. J Cell Physiol 151:367–377.

Schreiber SL (1991): Chemistry and biology of the immunophilins and their immunosuppressive ligands. Science 251:283–287.

Sefton BM, Campbell MA (1991): The role of tyrosine phosphorylation in lymphocyte activation. Annu Rev Cell Biol 7:257–274.

Shackelford DA, Trowbridge IS (1984): Induction of expression and phosphorylation of the human interleukin-2 receptor by phorbol diester. J Biol Chem 259:11706–11712.

Shaw JP, Utz PJ, Durand DB, Toole JJ, Emmel EA, Crabtree GR (1988): Identification of a putative regulator of early T cell activation genes. Science 241:202–205.

Shibuya H, Yoneyama M, Ninomiya TJ, Matsumoto K, Taniguchi T (1992): IL-2 and EGF receptors stimulate the hematopoietic cell cycle via different signaling pathways: demonstration of a novel role for c-myc. Cell 70:57–67.

Shou C, Farnsworth CL, Neel BG, Feig LA (1992): Molecular cloning of cDNAs encoding a guanine-releasing factor for ras p21. Nature 358:351–354.

Siebenlist U, Durand DB, Bressler P, Holbrook NJ, Norris CA, Kamoun M, Kant JA, Crabtree GR (1986): Promoter region of interleukin-2 gene undergoes chromatin structure changes and confers inducibility on chloramphenicol acetyltransferase gene during activation of T cells. Mol Cell Biol 6:3042–3049.

Siegel LJ, Harper ME, Wong-Staal F, Gallo R, Nash WG, O'Brien SJ (1984): Gene for T-cell growth factor: location on human chromosome 4q and feline chromosome B1. Science 223:175–178.

Siegel JP, Sharon M, Smith P, Leonard WJ (1987): The IL-2 receptor β chain (p70): role in mediating signals for LAK, NK, and proliferative activities. Science 238:75–78.

Siekierka JJ, Hung SH, Poe M, Lin CS, Sigal NH (1989): A cytosolic binding protein for the immunosuppressant FK506 has peptidyl-prolyl isomerase activity but is distinct from cyclophilin. Nature 341:755–757.

Sigal NH, Dumont F, Durette P, Siekierka JJ, Peterson L, Rich DH, Dunlap BE, Staruch MJ, Melino MR, Koprak SL, Williams D, Witzel B, Pisano JM (1991): Is cyclophilin involved in the immunosuppressive and nephrotoxic mechanism of action of cyclosporin A? J Exp Med 173:619–628.

Siliciano JD, Morrow TA, Desiderio SV (1992): *itk*, a T-cell-specific tyrosine kinase gene inducible by interleukin 2. Proc Natl Acad Sci USA 89:11194–11198.

Smith KA (1984): Interleukin 2. Annu Rev Immunol 2:319–333.

Smith KA (1988): Interleukin-2: inception, impact, and implications. Science 240:1169–1176.

Solomon MJ, Lee T, Kirschner MW (1992): Role of phosphorylation in p34cdc2 activation: identification of an activating kinase. Mol Biol Cell 3:13–27.

Sturgill TW, Wu J (1991): Recent progress in characterization of protein kinase cascades for phosphorylation of ribosomal protein S6. Biochim Biophys Acta 1092:350–357.

Sturgill TW, Ray LB, Erikson E, Maller JL (1988): Insulin-stimulated MAP-2 kinase phosphorylates and activates ribosomal protein S6 kinase II. Nature 334:715–718.

Taga T, Kishimoto T (1992): Cytokine receptors and signal transduction. FASEB J 6:3387–3396.

Takahashi N, Hayano T, Suzuki M (1989): Peptidyl-prolyl *cis-trans* isomerase is the cyclosporin A-binding protein cyclophilin. Nature 337:473–475.

Takeshita T, Asao H, Suzuki J, Sugamura K (1990): An associated molecule, p64, with high-affinity interleukin 2 receptor. Int Immunol 2:477–480.

Takeshita T, Asao H, Ohtani K, Ishii N, Kumaki S, Tanaka N, Munakata H, Nakamura M, Sugamura K (1992): Cloning of the gamma chain of the human IL-2 receptor. Science 257:379–382.

Taniguchi T, Matsui H, Fujita T, Takaoka C, Kashima N, Yoshimoto R, Hamuro J (1983): Structure and expression of a cloned cDNA for human interleukin-2. Nature 302:305–310.

Taniguchi T, Matsui H, Fujita T, Hatakeyama M, Kashima N, Fuse A, Hamuro J, Nishi TC, Yamada G (1986): Molecular analysis of the interleukin-2 system. Immunol Rev 92:121–133.

Taniguchi T, Minami Y (1993): The IL-2/IL-2 receptor system: a current overview. Cell 73:5–8.

Terada N, Or R, Weinberg K, Domenico J, Lucas JJ, Gelfand EW (1992a): Transcription of IL-2 and IL-4 genes is not inhibited by cyclosporin A in competent T cells. J Biol Chem 267:21207–21210.

Terada N, Lucas JJ, Szepesi A, Franklin RA, Takase K, Gelfand EW (1992b): Rapamycin inhibits the phosphorylation of p70 S6 kinase in IL-2- and mitogen-activated human T cells. Biochem Biophys Res Commun 186:1315–1321.

Terada N, Franklin RA, Lucas JJ, Blenis J, Gelfand EW (1993a): Failure of rapamycin to block proliferation once resting cells have entered the cell cycle despite inactivation of p70 S6 kinase. J Biol Chem 268:12062–12068.

Terada N, Lucas JJ, Szepesi A, Franklin RA, Domenico J, Gelfand EW (1993b): Rapamycin blocks cell cycle progression of activated T cells prior to events characteristic of the middle to late G1 phase of the cycle. J Cell Physiol 154:7–15.

Trahey M, McCormick F (1987): A cytoplasmic protein stimulates normal N-ras p21 GTPase but does not affect oncogenic mutants. Science 238:542–545.

Tropshung M, Barthelmess IB, Neupert W (1989): Sensitivity to cyclosporin A is mediated by cyclophilin in *Neurospora crassa* and *Saccharomyces cerevisiae*. Nature 342:953–955.

Turner B, Rapp U, App H, Greene M, Dobashi K, Reed J (1991): Interleukin 2 induces tyrosine phosphorylation and activation of p72–74 Raf-1 kinase in a T-cell line. Proc Natl Acad Sci USA 88:1227–1231.

Valge VE, Wong JGP, Dtalof BM, Sinskey AJ, Rab A (1988): Protein kinase C is required for responses to T cell receptor ligand but not to interleukin-2 in T cells. Cell 55:101–110.

Van Duyne, GD, Standaert RF, Karplus PA, Schreiber SL, Clardy J (1991): Atomic structure of FKBP-FK506, an immunophilin–immunosuppressant complex. Science 252:839–842.

Waldmann TA, Goldman CK, Robb RJ, Depper JM, Leonard WJ, Sharrow SO, Bongiovanni KF, Korsmeyer SJ, Greene WC (1984): Expression of interleukin-2 receptors on activated human B-cells. J Exp Med 160:1450–1466.

Wang A, Lu SD, Mark DF (1984): Site-specific mutagenesis of the human interleukin-2 gene: structure-function analysis of the cysteine residues. Science 224:1431–1433.

Welch PJ, Wang JY (1992): Coordinated synthesis and degradation of cdc2 in the mammalian cell cycle. Proc Natl Acad Sci USA 89:3093–3097.

Whitman M, Downes CP, Keeler M, Keller T, Cantley L (1988): Type I phosphatidylinositol kinase makes a novel inositol phospholipid, phosphatidylinositol-3-phosphate. Nature 311:644–646.

Wiederrecht G, Brizuela L, Elliston K, Sigal NH, Siekierka JJ (1991): FKB1 encodes a nonessential FK 506-binding protein in *Saccharomyces cerevisiae* and contains regions suggesting homology to the cyclophilins. Proc Natl Acad Sci USA 88:1029–1033.

Witters LA, Watts TD (1988): An autocrine factor from Reuber hepatoma cells that stimulates DNA synthesis and acetyl-CoA carboxylase. J Biol Chem 263:8027–8036.

Xiong Y, Zhang H, Beach D (1992): D type cyclins associate with multiple protein kinases and the DNA replication and repair factor PCNA. Cell 71:505–514.

Yokota T, Arai N, Lee F, Rennick D, Mosmann T, Arai K-I (1985): Use of a cDNA expression vector for isolation of mouse interleukin 2 cDNA clones: expression of T-cell growth-factor activity after transfection of monkey cells. Proc Natl Acad Sci USA 82:68–72.

Zmuidzinas A, Mamon HJ, Roberts TM, Smith KA (1991): Interleukin-2-triggered Raf-1 expression, phosphorylation, and associated kinase activity increase through G1 and S in CD3-stimulated primary human T cells. Mol Cell Biol 11:2794–2803.

ABOUT THE AUTHORS

Naohiro Terada is a staff scientist in the Division of Basic Sciences, Department of Pediatrics, National Jewish Center for Immunology and Respiratory Medicine in Denver, Colorado. After receiving his M.D. from Osaka University School of Medicine, Japan in 1981, he worked with Drs. Keisei Kawa-Ha and Takeo Kakunaga on the oncogenesis and clonality of childhood leukemia at Osaka University. This was followed by research at the National Jewish Center since 1989, where he worked with Dr. Erwin W. Gelfand on signal transduction in lymphocytes. Since 1992 he has been a faculty member of the Institute, and has focused on a study of the regulation of lymphocyte proliferation using immunosuppressive agents.

Richard A. Franklin is a senior research scientist at Paradigm Biosciences in Salt Lake City, Utah. Dr. Franklin received his B.S. at the University of Illinois. He received both his M.S. and Ph.D. at the University of Illinois in the laboratory of Dr. Keith Kelley. His thesis work involved elucidating the immunological changes that occur with age. Following the completion of his Ph.D. he received a Chateaubrian fellowship to identify links between the neuroendocrine and immune systems in the laboratory of Dr. Robert Dantzer at INSERM (Bordeaux, France). He recently completed his postdoctoral studies in the laboratory of Dr. Erwin W. Gelfand at the National Jewish Center for Immunology and Respiratory Medicine in Denver, Colorado. His postdoctoral studies in Dr. Gelfand's laboratory involved the identification of signal transduction pathways activated following receptor ligation of T and B lymphocytes. His current research involves alterations in signal transduction pathways that occur with age and how these pathways can be manipulated.

Joseph J. Lucas is a staff scientist in the Basic Sciences Division of the Department of Pediatrics at the National Jewish Center for Immunology and Respiratory Medicine. After receiving a B.A. degree in chemistry at LaSalle College in Philadelphia, he pursued doctoral studies at the University of Pennsylvania in the laboratory of Harold S. Ginsberg. He was awarded a Ph.D. degree in biochemistry for his work on the elucidation of transcriptional control patterns in the productive and abortive replicative cycles of the human adenovirus. After postdoctoral work on transcriptional regulation of vaccinia virus and, subsequently, of the mammalian genome with Joseph R. Kates at the University of Colorado in Boulder and at SUNY at Stony Brook, he joined the faculty of the Department of Microbiology of SUNY at Stony Brook Medical School. While there, he was involved to a great extent in graduation education, and designed, organized, and taught courses in cellular, molecular, and developmental biology. He also served as an Established Investigator of the American Heart Association. His recent research is focused on the differential regulation of cell cycle entry and progression in normal and transformed human cells.

Erwin W. Gelfand is Chairman of the Department of Pediatrics and Staff Scientist in the Division of Basic Sciences at the National Jewish Center for Immunology and Respiratory Medicine in Denver, Colorado. After receiving his B.Sc. degree at McGill University in Montreal, Quebec, he served his internship and residency at Montreal General Hospital, Montreal; Montreal Children's Hospital, Montreal; Children's Hospital Medical Center, Boston; and Harvard Medical School, Boston. He then completed a fellowship in immunology at Harvard Medical School. In 1971–72, he was a visiting scientist at the Max Planck Institute for Immunobiology in Freiberg, Germany. His research interests include the control of human T and B lymphocyte activation and functional maturation, immunodeficiency diseases, and bone marrow transplantation.

Growth Factors and Signal Transduction in Development: 97–122

The Transforming Growth Factor β Family in Vertebrate Embryogenesis

Rosemary J. Akhurst

I. INTRODUCTION

The existence of transforming growth factor β (TGFβ) as a distinct molecular entity was established 10 years ago [Assoian et al., 1983; Frolik et al., 1983; Roberts et al., 1983]. In 1985, the first human cDNA encoding TGFβ1 was molecularly cloned by Derynck et al. [1985], and over the last decade, an ever-increasing body of knowledge concerning the diverse biological activities of TGFβ has grown from studies on cells in culture [Roberts and Sporn, 1990].

This family of molecules attracted the attention of developmental biologists for several reasons, the first being the multifaceted nature of its bioactivity. It is now widely established that TGFβ can stimulate or inhibit the growth and/or differentiation of virtually all cells, depending on cell type, stage of differentiation, and physiology of the culture conditions. TGFβ is also a potent modulator of extracellular matrix deposition which, in turn, might mediate several of its biological effects. Molecular biologists revealed that the TGFβs are the prototype of a superfamily of related molecules [Roberts and Sporn, 1990], many of which had been isolated on the basis that they might play important developmental functions. These include the *Drosophila* gene, *Decapentaplegic (dpp), Xenopus vg1,* and the mammalian Müllerian inhibitory substance, each of which had been implicated as important in determining cell fate. Finally, in the late 1980s, a number of laboratories definitively demonstrated the widespread expression of TGFβs during mammalian embryogenesis [Heine et al., 1987; Sandberg et al., 1988; Lehnert and Akhurst, 1988; Rappolee et al., 1988; Wilcox and Derynck, 1988].

The purpose of this chapter is to review some of the molecular features of the TGFβ family which are pertinent to our understanding of its role in embryonic processes, and secondly to review what is currently known about TGFβ involvement in specific developmental systems.

II. BIOCHEMISTRY OF THE TGFβ FAMILY

A. The TGFβ Family: Genes and Protein Structure

Human TGFβ was originally isolated as a 25-kD homodimeric protein, now known to be TGFβ1. Further biochemical purification and molecular cloning studies revealed that there are three mammalian *tgf*β genes, encoding TGFβ1, -β2, and -β3, respectively [reviewed by Roberts and Sporn, 1990]. The existence of a possible fourth *tgf*β gene in the chick has been questioned [Burt and Jakowlew, 1992; Burt and Paton, 1992]. Indeed, it is probable that in this vertebrate too, there are only three genes, and that chick *tgf*β*4* is the avian homologue of mammalian *tgf*β*1* [Burt and Jakowlew, 1992].

The TGFβs, like other molecules in this superfamily, are translated as large precursor proteins (pre-pro-TGFβs) varying in size from

390 (TGFβ1) to 412 (TGFβ2 and β3) amino acids. Each isoform has a short signal peptide. The mature bioactive form is generated by proteolytic cleavage, generating the amino-terminal latency-associated peptide (βLAP), and the carboxy-terminal 112-amino-acid bioactive polypeptide. Dimeric βLAP remains associated noncovalently with the mature dimer, keeping it in a latent state. The small bioactive peptide is highly conserved between isoforms (approximately 70% amino acid homology) [Roberts and Sporn, 1990]. Individual isoform conservation between species is extreme, being in the region of 100% amino acid homology within the mature bioactive domain, and 85% homology within the βLAP [Roberts and Sporn, 1990]. The strong isoform conservation between species has been taken as evidence that the βLAP serves an important isoform-specific function.

The mature bioactive region of the TGFβs possesses nine conserved cysteine residues. Eight of these are involved in tightly-associated intramolecular disulfide bonds within the core of the molecule, whereas only one disulfide bond cross-links the two monomeric units. Hydrophobic interactions are thus thought to be important in the stabilization of dimers [Daopin et al., 1992; Schlunegger and Grutter, 1992].

Important structural features of the βLAPs of TGFβ1, -β2, and -β3 include interisoform conservation of three cysteine residues, and the existence of several (isoform-specific) N-linked glycosylation sites. Mannose-6-phosphate residues are added posttranslationally at these sites. Both the cysteine residues and the addition of glycoprotein side chains are thought to be important in protein folding, secretion, and activation. β1LAP and β3LAP both possess potential integrin-binding (RGD) sites, whereas this is not the case for β2LAP [Roberts and Sporn, 1990]. Denhez et al. [1990] suggested that this could provide a functional distinction for TGFβ2 compared to the other two isoforms. Interestingly, however, all three βLAPs are capable of associating interchangably with each of the mature isoform homodimers [Miller et al., 1992].

B. Transcriptional and Posttranscriptional Control of TGFβ Gene Expression

All three human TGFβ gene promoters have been cloned and characterized [Kim et al., 1989; Lafyatis et al., 1990; Noma et al., 1991]. Despite high amino acid sequence homology between the three isoforms, there is little sequence similarity between the gene promoters, consistent with expectations from their diverse patterns of expression during development, and in response to exogenous stimuli. However, there is considerable interspecies homology for each gene promoter [Geiser et al., 1991], which would be consistent with the conservation of developmental gene expression patterns between mouse and human [Gatherer et al., 1990; Millan et al., 1991].

It is clear that posttranscriptional, translational, and posttranslational control are important in the biosynthesis of TGFβs. Assoian et al. [1987] demonstrated that both unstimulated and activated monocytes contain comparable quantities of TGFβ1 mRNA, but that the protein is only released following monocyte activation. There is an increasing number of reports of major discrepancies between the TGFβ mRNA content of a cell, and its ability to secrete TGFβ protein [Ikeda et al., 1987; Coletta et al., 1990; Glick et al., 1989]. Glick et al. [1989] demonstrated that enhanced secretion of TGFβ2 in response to retinoic acid was due to a combination of transcriptional activation, message stabilization, and increased translational efficiency.

More recently, cell type–specific translational control of human TGFβ1 production has been definitively demonstrated in several malignant cell lines [Kim et al., 1992]. The TGFβ1 mRNA transcript possesses an unusually long untranslated 5′ region, which is exceedingly GC-rich. Deletion mutagenesis studies *in vitro* suggest that a stable stem-and-loop structure at the 5′ end of this 2.5-kb message is involved in reducing translational efficiency in some cell types but not in others. Since TGFβ2 and β3 also have very long untranslated 5′ leader sequences, it is likely that similar control mechanisms may operate on these mRNAs.

C. TGFβ Latency and Mechanisms of Activation and Inactivation

Activation of the TGFβ latent complex is a possible control point for regulation of bioactivity *in vivo*. This is demonstrated by the fact that cocultures of endothelial cells together with smooth muscle cells or pericytes produce activated TGFβ, whereas cultures of each cell type alone are unable to activate the endogenously synthesized latent form [Antonelli-Orlidge et al., 1989; Sato and Rifkin, 1989].

Recombinant latent TGFβ, secreted from mammalian cells, is synthesized as a relatively simple "small latent complex" [Gentry et al., 1987] composed of a dimeric βLAP, noncovalently associated with the dimeric bioactive dimer. In most other cell types, the latent complex contains an additional component, the latent TGFβ1-binding protein (LTBP), which is covalently attached to βLAP to form the "large latent complex" [Miyazono et al., 1988]. LTBP, is a 150–160-kD glycoprotein, containing several epithelial growth factor–like (EGF-like) repeats [Kanazaki et al., 1990; Tsuji et al. 1990]. Its *in vivo* function has not been fully elucidated but is obviously not necessary for latency. Studies on human erythroleukeumia cells have suggested that the LTBP is necessary for accurate and efficient protein folding and secretion from the cell [Miyazono et al., 1991]. It may also be involved in activation of latent TGFβ at the cell surface [Flaumenhaft et al., 1993].

In vitro, latent TGFβ can be activated by acidification, alkalinization, heating, or the use of chaotropic agents such as SDS and urea, though these mechanisms are clearly not of physiological relevance. The mechanism of activation *in vivo* is still a subject of much study. Miyazono and Heldin [1989] suggested that glycosylation might be important in TGFβ latency, because enzymatic removal of carbohydrate residues results in activation. Lyons et al. [1990] suggested that plasmin was involved in the activation process, and presented data to suggest that proteolytic nicking of the amino terminus of βLAP leads to release of the bioactive dimer. Sato and Rifkin [1989] demonstrated that, in the endothelial/pericyte coculture system, plasmin and urokinase-type plasminogen activator (uPA) are responsible for TGFβ activation. Furthermore, in endothelial cells, basic fibroblast growth factor (bFGF) activates latent TGFβ via elevation of uPA activity [Flaumenhaft et al., 1992]. Thus, the plasmin/uPA system appears to be a physiological regulator of TGFβ activation. Sato and Rifkin [1989] also reported dependency on cell–cell contact for TGFβ activation, and suggested the importance of cell surface association of proteases. More recently, they showed that binding of the latent complex to the mannose-6-phosphate receptor is essential to activation [Dennis and Rifkin, 1991]. This is presumably mediated via mannose residues located in βLAP, and might be a mechanism for concentrating latent TGFβ at the cell surface, in proximity to the cell-associated activating proteases. LTBP might also be important in localization of TGFβ to the cell surface, and its subsequent activation [Flaumenhaft et al., 1993].

Mature TGFβ can also bind a number of other proteins, including soluble β-glycan [Andres et al., 1989], decorin [Yamaguchi et al., 1990], and α_2 macroglobulin [O'Connor-McCourt and Wakefield, 1987], any or all of which might be involved in downregulation of bioactivity. Its ability to bind to specific components of the extracellular matrix (ECM), such as thrombospondin, fibronectin, collagen IV, and heparin [Murphy-Ullrich et al., 1992; Fava and McClure, 1987; Paralkar et al., 1991; McCaffrey et al., 1989], might also be a means to provide a pool of potentially bioactive growth factor, or to localize TGFβ bioactivity to specific regions of the embryo.

D. TGFβ Receptors

Eight or nine distinct membrane proteins that have the capacity to bind to TGFβ with high affinity have been detected by ligand-directed radioactive labelling studies [Massagué, 1992; O'Grady et al., 1991; MacKay and Danielpour, 1991; Mitchell et al., 1992]. The type I, II, and III receptors are found on virtually all cell types in culture, whereas the other receptor types are far more restricted in expression. The type I and II receptors bind with higher affinity to TGFβ1 and TGFβ3 than they do to TGFβ2 [Lin et al. 1992], although there is a subset of these receptors that has high affinity for TGFβ2 [Cheifetz et al., 1990]. The molecular distinction between the high- and low-affinity TGFβ2-binding receptors is not known. The type III receptor binds all TGFβs with equal affinity. However, the lack of type III receptors on some TGFβ-responsive cell lines [Cheifetz et al., 1990; Merwin et al., 1991] suggested that this receptor is not signal transducing.

By studying ethylnitrosourea (ENU)-mutated epithelial cell lines which are refractile to the action of TGFβ, Massagué and colleagues [Laiho et al., 1990a] provided further evidence that the type I and II receptors are the major signal transducers. These receptors loose their capacity to bind TGFβ in the nonresponsive cell lines, whereas the type III receptor does not [Laiho et al., 1990a]. Molecular cloning of the type III receptor has now confirmed the hypothesis that this "receptor" has very little potential to be signal transducing itself, as it is a large proteoglycan molecule with a small cytoplasmic domain [Wang et al., 1991; Lopez-Casillas et al., 1991].

The type I and II receptors have molecular weights of 53 kD and 75 kD, respectively, as assessed by radiolabelled ligand-binding studies and electrophoretic analysis [Laiho et al., 1990a]. Studies on TGFβ-insensitive ENU-mutated cell lines suggested that the type I receptor depends on the presence of functional type II receptors for TGFβ binding, because mutant cells of only two classes have been isolated; those that lack functional type I receptors (D type) and those that lack *both* the type I and type II receptors (DR type) [Laiho et al., 1990a]. Furthermore, the D and DR mutants show genetic complementarity [Laiho et al., 1991]. This hypothesis has more recently been validated by studies with the recombinant type II receptor.

The type II receptor was cloned by Lin et al. [1992], and is a member of a novel family of transmembrane serine/threonine kinases, which includes the *C. elegans daf-1* gene and two activin receptor genes. The type II TGFβ receptor polypeptide has 565 amino acids. It contains a short, cysteine-rich extracellular ligand-binding domain (136 amino acids), which possesses three potential N-linked glycosylation sites, and a larger cytoplasmic domain (377 amino acids) with a serine/threonine kinase moiety. Recombinant type II receptor has serine/threonine autophosphorylating activity [Lin et al., 1992], and a mutation in the putative ATP-binding site of the kinase domain abolishes signal transduction, without interfering with ligand binding [Wrana et al., 1992]. Transfection of a type II receptor expression vector into the DR class of TGFβ-insensitive cells restores TGFβ binding to both the type I and type II receptor, suggesting that DR mutants are defective in the type II receptor alone, and that the type I receptor is dependent on interaction with type II for ligand binding activity [Wrana et al., 1992]. Thus, in the absence of molecular clones for the type I receptor, it has been deduced that initiation of signal transduction requires TGFβ interaction with a type I/type II heteromeric complex.

It has been disputed whether receptor heterodimerization is necessary for TGFβ responses in all cell types, because some responses have been reported in cell lines which appear to lack functional type II receptors [Ohta et al., 1987; Brandes et al., 1991; Geiser et al., 1992]. Indeed, Geiser et al. [1992] showed that, in certain tumor cell lines, the growth inhibitory response to TGFβ is mediated via a different route to that of other gene transcriptional responses, the former requiring both type I and type II receptors, but the latter occurring in the absence of type II receptors. The possibility that structural diversity in the type II receptor may be generated by differential splicing must also be taken into consideration, because this is known to be the case for the activin IIB receptor [Attisano et al., 1992].

The type III receptor has been renamed β-glycan. It is a heparin-chondroitin sulfate proteoglycan of >300 kD, and it can exist in both soluble and cell surface–associated forms. TGFβs bind to β-glycan via the core protein [Andres et al., 1989; Wang et al., 1991; Lopez-Casillas et al., 1991]. The membrane-associated form of β-glycan may be important in targeting TGFβ to the type I and II receptors, or in activation of the latent complex. Interestingly, bFGF can also bind to β-glycan via glycosaminoglycans (GAGs), and can alter the GAG composition of β-glycan [Andres et al., 1992]. β-Glycan might thus have functions related to multiple growth factors. On endothelial cells, which lack β-glycan, the related proteoglycan, endoglin, may perform the same (unknown) function of this "receptor" [Cheifetz et al., 1992].

E. Signal Transduction

The mechanisms of TGFβ signal transduction are still largely unknown. The novel molecular structure of the type II receptor (previous section) might implicate novel signal transduction pathways, and the diverse biological responses of cells to TGFβ may be a warning of the potential complexities of these pathways, and the difficulties yet to be encountered in trying to unravel them.

In different cell types, TGFβ can rapidly transcriptionally activate a number of cellular protooncogenes, including *c-fos* [Kerr et al., 1990], *c-sis* [Leof et al., 1986], *c-jun, junB* [Pertovaara et al., 1989] and *ras* [Mulder and Morris, 1992]. The c-Fos, c-Jun, and JunB proteins might positively or negatively modulate downstream gene transcriptional responses [see Kerr et al., 1990]. The c-Sis or Ras proteins may mediate a positive growth response to TGFβ via their own signal transduction pathways [Leof et al., 1986; Mulder and Morris, 1992]. TGFβ can also downregulate gene transcription as an early event, including that of the growth-related genes, *c-myc, KC,* and *JE* [Coffey et al., 1988].

Phosphorylation of nuclear proteins is known to be an early cellular response to TGFβ [Kramer et al., 1991a], but there are several conflicting reports on the involvement of different early response biochemical pathways for TGFβ signalling [reviewed in Roberts and Sporn, 1990]. Ohtsuki and Massagué [1992], using specific kinase inhibitors, excluded the involvement of protein kinases A, C, and G, and cyclic nucleotide–dependent kinases in the response of Mv1Lu epithelial cells to TGFβ. They did however, implicate serine/threonine kinases as important in mediating the induction of JunB and plasminogen activator inhibitor-I (PAI-I) by TGFβ in these cells.

Kramer et al. [1991b] demonstrated a rapid phosphorylation of the cAMP-responsive element binding protein (CREB) in Mv1Lu cells, in response to TGFβ. Interestingly, this did not enhance binding of the CREB to cAMP-response elements (CREs), but did enhance binding of CREB to the TPA-response element (TRE) of the collagenase gene, which required some interaction with a Fos-related protein. Because many of the transcriptional effects of TGFβ are mediated via TREs [Roberts and Sporn, 1990], this CREB-phosphorylating activity might be of major significance in mediating such effects.

The intracellular signalling mechanisms for negative growth regulation of epithelial cells have been most thoroughly studied, but there is considerable controversy in this area, possibly because signalling pathways may vary depending on the exact cell type. It is well established that the negative growth effects of TGFβ are exerted distal to the early signalling responses of mitogens. TGFβ acts at the G_1/S boundary rather than at the G_0/G_1 transition, which is characteristic of EGF growth stimulation [Laiho et al., 1990b; Pietenpol et al., 1990b; Howe et al., 1991].

It was proposed that negative growth regulation of both keratinocytes and bronchial epithelial cells by TGFβ is mediated via the tumor suppressor gene, Rb [Laiho et al., 1990b; Pietenpol et al., 1990a]. Hypophosphorylated Rb is active as a growth suppressor, and TGFβ serves to keep Rb in this hypophosphorylated state, thus blocking transit through the cell cycle. This effect might be elicited by inhibition of $p34^{cdc2}$ kinase activity, which is hypophosphorylated and inactivated in response to TGFβ [Howe et al., 1991; Ito et al., 1992; Diaz-Meco et al., 1992]. In keratinocytes, it has been proposed that downregulation of *c-myc* transcription is necessary for TGFβ inhibition of growth [Pietenpol et al., 1990a, 1990b, 1991], and that this is mediated by hypophosphorylated Rb interacting with the *c-myc* gene promoter via an Rb-response element [Pietenpol et al., 1990a, 1991]. More recently, experimental evidence suggests that hypophosphorylated Rb prevents the binding of p106, a transcriptional activator of *c-myc,* to the *c-myc* gene promoter [Moses, 1992].

Diaz-Meco et al. [1992] suggested that, in keratinocytes, TGFβ inhibits phospholipase C (PLC)-mediated hydrolysis of phosphotidylcholine, acting distal to mitogenic activation of $p21^{ras}$. They also showed that exogenous PLC activity could override inhibition of H1 kinase ($p34^{cdc2}$ kinase) activity by TGFβ, and thus block downregulation of *c-myc* in response to TGFβ. Thus, in keratinocytes, TGFβ may act as a negative growth regulator by interfering with PLC, upstream of effects on $p34^{cdc2}$ kinase, Rb, and c-Myc.

Using bronchial epithelial cells, Zentrella et al. [1991], dismissed the involvement of Rb in negative growth regulation of bronchial epithelial cells by TGFβ, because carcinoma lines with their Rb genes deleted are still responsive to TGFβ. However, in bronchial epithelia, other proteins functionally related to Rb may be involved, because expression of viral oncoproteins that bind such molecules abrogates growth inhibition by TGFβ [Laiho et al., 1990b]. Zentrella et al. [1991] also disputed the involvement of hypophosphorylated Rb in the downregulation of *c-myc* by TGFβ, because *c-myc* transcription is downregulated in response to TGFβ, regardless of the phosphorylation state of Rb. Furthermore, constitutive expression of *c-myc* does not affect TGFβ-responsiveness of these cells, which brings into question the role of *c-myc* in negative regulation of Mv1Lu-1 cell growth [Longstreet et al., 1992]. Certainly, there appear to be fundamental differences between keratinocytes and bronchial epithelial cells with respect to TGFβ intracellular signalling mechanisms.

Clearly, there is still much work to be done on signal transduction mechanisms for the TGFβs. One area of particular significance to developmental biology, is the possibility of (reciprocal?) interactions between TGFβs and homeobox-containing genes. The existence of such interactions has been suggested by the provocative reciprocal patterns of gene expression seen during mammalian embryogenesis [Robert et al., 1989; Gatherer et al., 1990], and the fact that similar cellular and molecular interactions have been reported during *Drosophila* development [Immergluck et al., 1990; Panganiban et al., 1990].

F. Functional Redundancy of the TGFβs?

Most of the early work on TGFβ bioactivity in cultured cells was performed using TGFβ1, or did not differentiate between the three isoforms. In this review, therefore, the term TGFβ will be used where no specification of TGFβ isoform was made in the original work. Isoform definition will only be made where this is pertinent to interpretation of the work. It should be borne in mind, however, that there are indeed three isoforms, and they probably do have different *in vivo* functions.

Because the three mammalian TGFβs are so similar in primary structure, the question arose as to whether they could be totally interchangeable with respect to biological function *in vivo.* From nucleotide sequence comparisons, it may be argued that the three isoforms have very distinct biological roles, because the interspecies conservation of isoforms is much greater than the interisoform conservation within one species. Similarly, there is evolutionary conservation of developmental gene expression patterns between mouse and human [Gatherer et al., 1990].

Comparative studies on the bioactivities of the three isoforms *in vitro* have suggested that, in general terms, they are similar. Nevertheless, in a few biological systems there are clearly qualitative differences in biological action between TGFβ1 and β2 [Rosa et al., 1988; Ohta et al., 1987], the most definitive example being that of endothelial cells, whose growth is inhibited by TGFβ1 but not by TGFβ2 [Jennings et al., 1988]. Moreover, in many additional cell culture systems, although bioactivities are qualitatively similar, there can be considerable quantitative differences in the response [Graycar et al., 1989; ten Dijke et al., 1990; Cheifetz et al., 1990], which could be critical to bioaction *in vivo*. The differences in biological potency between the three isoforms can be correlated with differential binding to TGFβ receptors and other binding proteins, such as β-glycan and endoglin [Ohta et al., 1987; Cheifetz et al., 1990; Cheifetz et al., 1992; Wrana et al., 1992]. The major type I and II TGFβ receptors present on endothelial cells bind TGFβ1 and β3 with higher affinity than they bind TGFβ2 [Lin et al., 1992]. Furthermore, endoglin, which replaces β-glycan on endothelial cells, binds TGFβ1 and β3 but not TGFβ2 [Cheifetz et al., 1992]. A summary statement regarding differential isoform bioactivities *in vitro* is that TGFβ1 and β3 tend to be similar, whilst TGFβ2 activity can be somewhat distinct.

Recombinant DNA technology has been used to demonstrate that sequence divergence in the region of amino acid residues 40–82 of the mature TGFβ peptide is responsible for the differential activities of TGFβ1 and TGFβ2 on endothelial cells [Qian et al., 1992]. *In vivo*, additional functional specialization might be encoded by the more divergent βLAPs [Roberts and Sporn, 1990], and by other ligand-associated molecule(s), such as LTBP. Recent evidence suggests that each isoform, though capable of binding to TGFβ1-associated LTBP *per se*, might also bind to as-yet-unknown, high molecular weight protein moities [Olofsson et al., 1992], which could influence their biological activity.

Recently, transgenic mice which lack functional *tgf*β*1* genes have been generated [Shull et al., 1992; Kulkarni et al., 1993]. Interestingly, 30%–50% of the homozygous *tgf*β*1*^{null} mice reach parturition, and appear developmentally normal (though they die at 2–3 weeks of age with an inflammatory disorder). The existence of this subset of developmentally normal mice could be accounted for by several explanations. Firstly, placental transfer and utilization of maternal TGFβ1 by the conceptus. Secondly, that there can, indeed, be functional redundancy between the three TGFβ genes. Thirdly that, although TGFβ2 or β3 do not compensate for lack of TGFβ1, there is functional redundancy at a point distal to receptor signalling. The final explanation is that TGFβ1 serves no function in development, but that the widespread expression seen during vertebrate embryogenesis is an evolutionary relic. These various possibilities remain to be investigated. Obviously, in a review of this nature, we would like to dismiss the last possibility! The fact that 50%–70% of the *tgf*β*1*^{null} mice die *in utero*, would suggest that there might be developmental abnormalities in the majority of these mice.

III. TGFβ DISTRIBUTIONS DURING MAMMALIAN EMBRYOGENESIS

A large number of studies have reported the endogenous expression of all three TGFβ isoforms during mammalian embryogenesis. Various approaches have been taken to assess expression including reverse transcription polymerase chain reaction (RT-PCR) [Rappolee et al., 1988], Northern blot analysis [Miller et al., 1989; Denhez et al., 1990], *in situ* hybridization to cellular RNAs [Sandberg et al., 1988; Lehnert and Akhurst, 1988; Wil-

TABLE I. Differential Localization of RNAs Encoding TGFβ1, TGFβ2, TGFβ3 During Murine Embryogenesis

	TGFβ1	TGFβ2	TGFβ3
Haematopoietic tissue	+	−	−
Endothelia (and precursors)	+	−	−
Thyroid	+	−	−
Parathyroid	+	−	−
Thymus	+	−	−
Epithelia			
Whisker follicles	+	+	+
Salivary gland	+	+	−
Tooth bud	+	+	−
Secondary palate	+	−	+
Bronchial epithelium	−	+(s)	+(c)
Otic epithelium	−	+	−
Olfactory epithelium	−	+	−
Lens epithelium	−	+	−
Retina	−	+	−
Hyperplastic nodules	−	+	−
Suprabasal keratinocytes	−	+	−
Cartilage and bone			
Precartilaginous blastemia	−	+(limb)	+(iv)
Growth zone of long bone	−	+	−
Perichondria	−	−	+
Hypertrophic cartilage	−	−	−
Osteoblasts, osteoclasts	+	−	−
Cardiac tissue			
Prevalvular endothelium	+	−	−
Prevalvular myocardium	−	+	−
Neuronal tissue			
Ventral spinal cord	−	−	−
Ventral forebrain	−	+	−
Skeletal muscle	−	−	+
Mesothelia	−	−	+
Mesenchyme	−	+	+

+, denotes that RNA is abundant at some stage between 9.5 days *post coitum*, and birth, although expression may be transient.
−, is below the detection level of *in situ* hybridization.
Table reproduced from Millan et al. [1991], with permission.

cox and Derynck, 1988; Pelton et al., 1990; Akhurst et al., 1990c], and immunolocalization of protein [Heine et al., 1987; Pelton et al., 1991]. Each of these techniques has specific strengths and weaknesses, and the results are variable depending on the sensitivity and specificity of the technique.

Rappolee et al. [1988] demonstrated by RT-PCR that the onset of expression of the TGFβ1 gene occurs during mouse preimplantation development, whereas *in situ* hybridization studies on early postimplantation embryos have been unable to detect transcripts until 7 days *post-coitum* [Akhurst et al., 1990a; Dickson et al., 1993]. This is certainly a technical difference. It is probable that all embryonic cells possess very low levels of TGFβ transcripts, but that in certain cell types expression levels are greatly enhanced. All three RNAs are expressed in large amounts during postimplantation mouse and human development. Each gene has a distinct expression profile (summarized in Table I) which, in some cases, overlaps with those of the other two genes. These

TABLE II. Embryonic Cell Types Expressing TGFβ Proteins

Cell type	TGFβ1	TGFβ2	TGFβ3	Cell type	TGFβ1	TGFβ2	TGFβ3
Cartilage				Gut			
Perichondrium	+++	+	++	Esophageal epithelium	+++	+	+
Chondrocytes	+	++	++	Stomach epithelium	+++	+	+
Bone				Intestinal epithelium	++	+	+
Periosteum	++	−	+	Basement membrane	−	+++	−
Osteocytes	++	++	+	Muscularis	+	+	++
Tooth				Liver			
Ameloblasts	++	−	+	Capsule	−	−	++
Odontoblasts	−	++	−	Parenchyma	−	−	−
Pulp	+	+++	+	Megakaryocytes	+	−	++
Muscle				Eye			
Smooth	+	+	++	Lens epithelium	−	−	−
Cardiac	+	+	+++	Lens fibers	+++	+	+
Skeletal	+	+	++	Ear			
Heart				Cochlear epithelium	−	+	+++
Endothelium	−	−	−	Basement membrane	−	+++	−
Myocytes	+	+	+++	Central Nervous System			
Lung				Meninges	+	+++	+
Bronchi	++	++	++	Glia	−	++	++
Alveoli	−	−	−	Choroid plexus	−	−	++
Blood vessels				Skin			
Endothelium	−	−	++	Periderm	++	+	++
Smooth muscle	+	+	+++	Epidermis	+++	+++	+++
Kidney				Dermis	+	+++	+
Tubules	++	++	++	Hair follicles	++	++	+
Basement membrane	−	+++	−				
Adrenal							
Cortex	+++	+++	−				
Medulla	−	−	−				

Immunolocalisations of TGF β1, β2 and β3 proteins in the mouse embryo 12.5–18.5 days *post coitum* (dpc). A + indicates that positive staining was seen in this tissue at some stage between 12.5 and 18.5 dpc, while a − means that no staining was detected. The number of + signs denotes relative intensity of staining when compared against the other TGFβ antibodies in that particular tissue and should not be compared against the staining intensity in other tissues. Reproduced, with kind permission, from Pelton et al. [1991].

expression patterns are conserved between mouse and man [Gatherer et al., 1990].

TGFβ immunolocalization studies can be difficult to interpret, because the translation products can occur in so many different forms (unprocessed, latent, bioactive, ECM-bound, and degraded), and it is impossible to differentiate between these different states. Furthermore, the generation of isoform-specific antibodies has been problematic. Nevertheless, a number of detailed immunolocalization studies have been performed (see Table II for summary). In some cases there appear to be reasonable correlations between RNA and protein localizations [Akhurst et al., 1990a; Dickson et al., 1993]. In other cases these localizations have been surprisingly disparate, implicating a large amount of posttranscriptional control of protein production [Pelton et al., 1991].

IV. THE EFFECTS OF TRANSFORMING GROWTH FACTORS β ON SPECIFIC BIOLOGICAL SYSTEMS

A. General

TGFβ was originally identified as "sarcoma growth factor" [DeLarco and Todaro, 1978], which was a mixture of the mitogen, transforming growth factor α (TGFα), and of TGFβ, secreted from fibroblasts transformed by murine sarcoma virus. One of its first reported biological activities was the induction of a transformed phenotype in normal rat kidney fibroblasts; this activity is also dependent on the presence of TGFα, or the related mitogen, EGF [DeLarco and Todaro, 1978]. This first assay for TGFβ thus demonstrates one of the features of the bioactivity of this family of molecules, namely, that the biological response of the cell may be modulated by the presence of other growth factors. In a variety of systems, TGFα, EGF, and bFGF have each been reported to act either synergystically, or antagonistically, with TGFβ in modulating cellular phenotype. TGFβ can enhance the mitogenic effects of bFGF on osteoblasts [Globus et al., 1988] and growth plate chondroblasts [Hiraki et al., 1988], and can act in synergy with bFGF in a *Xenopus* mesoderm induction assay [Kimelman and Kirschner, 1987]. In endothelial cells, these growth factors tend to have antagonistic effects. TGFβ1 inhibits bFGF-induced proliferation of endothelial cells [Baird and Durkin, 1986] and tends to downregulate plasminogen activation, in opposition to bFGF effects [Pepper et al., 1990].

It is now widely accepted that the biological responses to TGFβ depend, not only on cell type but, on the proliferative or differentiative state of the cell, the density of cell culture, the three-dimensional organization of the cultured cells and other physiological parameters. Interpretation of the significance of biological effects *in vitro* can therefore be very difficult, though studies *in vivo* are confounded by the complexity of the organism. Nevertheless, this has not prevented scientists from attempting to address the function that these molecules might play, if any, in embryonic processes.

B. TGFβ Effects and the Extracellular Matrix

TGFβ is a major modulator of cell growth and differentiation, and it can affect these processes in either a positive or negative manner. However, one of its most important bioactivities is in the modulation of ECM deposition, composition, and cell–substratum interaction. The ECM is not an inert support for tissue structure, but a molecularly complex, and dynamic substratum for cells, the composition of which is dependent on body site. It can be rapidly modified by cell and tissue interactions, mediated by soluble proteins and proteins attached to the cell surface, such as growth factors, proteases, and enzymatic modifiers of complex carbohydrates. The ECM has defined bioactivities influencing cell proliferation, differentiation, cell shape, adhesion, migration, and morphogenetic interactions. As examples, endothelial cell phenotype *in vitro* is determined by the substratum onto which the cells are seeded [Madri et al., 1983]. Similarly, the differentiated state of cultured primary mammary epithelial cells depends on the composition of the underlying ECM [Li et al., 1987]. Thus, modulation of ECM by TGFβ might be one of its most important dynamic functions during embryogenesis, and several of this growth factor's biological effects might be mediated via changes in the ECM.

Modification of ECM is particularly important in angiogenesis, chondrogenesis, and epithelial-mesenchymal interactions, the details of which will be dealt with in separate sections. In this section we will review specific effects of TGFβ on ECM components and, conversely, the potential role of ECM in modifying TGFβ bioactivity.

In general, TGFβ is thought to increase deposition of ECM components, including the major matrix molecules, such as fibronectin and type I collagen, as well as more specialized ECM proteins, such as thrombospondin, tenascin, osteopontin, and osteonectin, to name a few [reviewed by Roberts and Sporn, 1990]. Increased synthesis of ECM proteins is initiated at the level of transcriptional control of the encoding genes, and can be potentiated posttranscriptionally [Roberts and Sporn, 1990]. Indeed, TGFβ can modulate the production of different fibronectin isoforms by acting at the level of control of differential splicing of the encoding RNA [Balza et al., 1988; Kocher et al., 1990]. TGFβ also increases the synthesis of extracellular and membrane-associated proteoglycans, such as syndecan, and extracellular GAGs [Chen et al., 1987; Rapraeger, 1989; Bassols and Massagué, 1988; Nugent and Edelman, 1992]. The laying down of specialized ECM components is of central importance to the genesis of many tissue types, the obvious examples being cartilage and bone.

This increased production of ECM proteins has also been directly implicated in mediating some of the biological effects of TGFβ on cells in culture. Thus, elevated fibronectin synthesis can mimic the differentiation-inhibitory activity of TGFβ on myoblasts in culture [Heino and Massagué, 1990], and may be involved in mediating TGFβ growth inhibition of endothelial cells [Madri et al., 1988]. Furthermore, enhanced synthesis of heparin sulfate proteoglycans associated with the cell surface may serve to locally concentrate heparin-binding growth factors to the cell surface. This may explain the synergistic activity of TGFβ and bFGF in inducing mitogenesis of Balb/c3T3 cells [Nugent and Edelman, 1992]. Elevated synthesis of certain ECM components, such as collagen IV and thrombospondin, which themselves bind TGFβ [Paralkar et al., 1991; Murphy-Ullrich et al., 1992], might be a regulatory circuit to provide the cell with a localized pool of potentially bioactive TGFβ. Finally, TGFβ-1-induction of the proteoglycan decorin, which is known to bind to and inactivate this growth factor, may be an autoregulatory mechanism for control of TGFβ bioactivity [Yamaguchi et al., 1990].

TGFβ also elevates ECM deposition by its action on genes encoding proteases and protease inhibitors. It is known to decrease synthesis and secretion of several proteases, including tissue plaminogen activator, uPA, thiol protease, collagenase, and transin (stromelysin); and to increase synthesis and secretion of the protease inhibitors PAI, and of the tissue inhibitor of metalloproteases (TIMP) [reviewed in Roberts and Sporn, 1990]. These complementary biological activities, by enhancing synthesis of ECM components, have been proposed to play a role in stabilizing developing structures, such as during branching morphogenesis of lung and mammary gland [Heine et al., 1990; Matrisian et al., 1992; Robinson et al., 1991]. The balanced antagonistic activities of bFGF in increasing protease secretion from endothelial cells (promoting invasiveness), and of TGFβ in supporting ECM elaboration (stabilizing structures), is also thought to be of central importance in precise regulation of angiogenesis [Saksela et al., 1987].

Another effect of TGFβ is on cell–substratum interaction, mediated by modulation of the expression of integrins. This diverse set of molecules are integral membrane heterodimeric protein complexes composed of one α and one β chain, which are receptors for specific ECM components. TGFβ selectively modulates the expression of integrins which bind fibronectin, collagen, laminin, and other ECM proteins [Roberts and Sporn, 1990]; thus the profile of integrin expression at the cell surface is altered by TGFβ, which might have profound effects on interaction of the cell with its substratum.

Modulation of ECM and integrin expression by TGFβ might be important in the regulation of cell migration in a number of systems. Dulannet and Duband [1992] recently demonstrated that premigratory neural crest cells could be stimulated to migrate prematurely in culture, when treated with TGFβ. They showed that this was not due to differences in ECM composition surrounding premigratory and migratory cells, but rather due to the differential ability of the cells to respond to ECM proteins. Thus the *de novo* expression or activation of (a) specific integrin(s) (under the control of TGFβ) may be necessary for initiation of neural crest cell migration. Additionally, smooth muscle and endothelial cell migration, in response to wounding, might be regulated, in part, by changes in the profile of integrin expression elicited by TGFβ [Basson et al., 1992].

In addition to control of ECM by TGFβ, the composition of the ECM may also transcriptionally regulate the production of TGFβ1. Strueli et al. [1993] demonstrated transcriptional induction of TGFβ1, but not TGFβ2 or -β3, by perturbation of cell–substratum interaction. This might be an autoregulatory mechanism of ensuring homeostatic equilibrium between cells and substrate, which would be of essential importance both in embryonic development and wounding [see Martin et al., 1993].

C. TGFβ and Epithelial–Mesenchymal Interactions

Interactions between epithelia and mesenchyma are central to the induction and morphogenesis of a number of organ systems, including lung, kidney, mammary gland, tooth, and hair. The molecular nature of the earliest inductive signals which emanate from the mesenchyme to act, often instructively, on the overlying epithelium, remains to be elucidated. But after this initial induction, there follows a cascade of reciprocal interactions between mesenchyme and epithelium, which regulate the ensuing morphogenesis of the forming organ. Much experimental evidence would implicate the TGFβs and other growth factors in these later processes.

TGFβ1 was the first isoform reported to be expressed at sites of epithelial-mesenchymal interaction [Heine et al., 1987; Lehnert and Akhurst, 1988]. In general, the observation is that TGFβ1 mRNA is induced in the epithelial component of the developing structure, and that TGFβ1 protein is predominantly localized in the subjacent mesenchyme. This leads to the conclusion that there is paracrine TGFβ activity on mesenchymal cells. More extensive and detailed studies have now confirmed the widespread conclusion that each of the *tgf*β genes is transiently expressed within the epithelial component of structures in the process of undergoing morphogenesis, including salivary glands, mammary glands, lung, tooth, hair, and secondary palate. In many cases, the isoforms have overlapping patterns of expression, as seen for TGFβ1 and β3 during palatogenesis [Fitzpatrick et al., 1990]. In other cases, there are subtle differences in isoform expression between different cells of the developing organ [Pelton et al., 1990; Millan et al., 1991; Robinson et al., 1991].

The precise functions of the TGFβs in epithelial-mesenchymal interactions have been debated by a number of laboratories working on different organ systems. The first paradox in all these studies is that rapidly proliferating epithelial cells apparently synthesize a very potent epithelial growth inhibitor. In the early stages of tooth development there is a direct correlation between sites of DNA synthesis and TGFβ1 gene transcription [Vaahtokari et al.,1991]. Some individuals have suggested that TGFβ might support growth and inhibit differentiation of embryonic epithelial [Heine et al., 1990], although this would seem somewhat unprecedented. In contrast, Silberstein and Daniel [1987] demonstrated that the growing end buds of the developing mammary gland, which also synthesize all three TGFβ mRNAs [Robinson et al., 1991], are potently growth-inhibited by localized exogenous application of this growth factor. They postulated that there must be some mechanism, possibly at the level of control of translation, secretion, or activation, whereby the epithelial cells evade the negative growth effects of the endogenous TGFβ. In support of this hypothesis, they demonstrated that immunohistochemically stainable TGFβ never accumulates in front of the growing end buds, but only surrounds the subtending ducts of the mammary gland [Robinson et al., 1991].

In most cases, paracrine regulation of mesenchymal cell activity by TGFβ has been concluded to be a general principle in epithelial-mesenchymal interactions, but the precise molecular mechanisms and cellular response are still open to question. Furthermore, although it is generally concluded that TGFβ acts on the mesenchyme, it is clear, at least in the tooth, that the initiation of epithelial TGFβ expression depends on a signal coming from the mesenchyme. Vaahtokari et al. [1991], using tissue recombinations in organ culture, demonstrated that epithelial TGFβ1 expression is only induced when epithelium (either dental or oral) is cultured in combination with dental mesenchyme, but not with nondental jaw mesenchyme, which is consistent with the inability of the latter to support tooth development.

Modulation of ECM deposition must be one of the most important functions of TGFβ derived from the epithelium. Whether this serves to enhance epithelial invasiveness and tissue remodeling, or to stabilize the forming organ structure, has been debated, but not directly addressed. Some workers [Akhurst et al., 1990a; Fitzpatrick et al., 1990] have suggested that induction of specific ECM components, such as tenascin, would enhance the plasticity of the local tissue environment. Tenascin is a TGFβ1-inducible ECM protein which shows a restricted developmental expression profile, similar to that of TGFβ1 [Akhurst et al., 1990b]. It is known to promote the growth of epithelial cells in culture, and can mediate the disruption of cell–cell interactions and cell–substratum interactions, presumably by competing for binding to other ECM components such as fibronectin [Chiquet-Ehrismann et al., 1988, 1989]. Because of these activities, TGFβ control of tenascin expression has been implicated in tumor invasiveness [Chiquet-Ehrismann et al., 1989], and could equally well promote the invasion of embryonic epithelia into mesenchyma.

An alternative view, is that TGFβ1 stimulation of ECM synthesis is a process required to stabilize the structure of the growing organ [Heine et al., 1990; Robinson et al., 1991]. Robinson et al. [1991] suggested that, in mammary gland development, TGFβ synthesized in the cells of the growing end buds, is not secreted until the epithelial cells of the growing end bud are incorporated into the subtending duct, and that its function there is to elaborate the fibrous supportive ECM ensheathing the duct. Heine et al. [1989] pointed out the importance of the ECM in cleft formation and in the stabilization of organ structure during the formation of branching organs such as the lung, salivary gland, and mammary gland. The accumulation of collagens at the clefting site, in particular, is important in cleft formation, and Heine et al. [1990] noted a strong correlation between sites of deposition of extracellular TGFβ protein and of fibronectin, collagens I and III, and GAGs, within the forming clefts of the fetal mouse lung. They suggested that the role of TGFβ synthesized in the epithelium is to enhance cleft formation by localized induction of ECM, and to stabilize the resultant structure. It is known that TGFα and EGF can inhibit lung-branching morphogenesis in culture by stimulating the synthesis of proteases which degrade the key ECM proteins accumulating at the cleft site [Matrisian et al., 1992]. Conversely, TGFβ might be a promoter of branching morphogenesis.

In contrast to the assumption that TGFβ induces ECM deposition, Weinberg et al. [1990] showed that TGFβ, when added to cultured hair follicle organoids, although it has no effect on protease activity alone, acts in synergy with exogenous TGFα to elevate protease secretion by the follicular cells. Under certain conditions, therefore, it would appear that TGFβ might have an unpredicted effect, i.e., degradation of ECM, which might enhance invasiveness of the follicle during the growth phase of the hair cycle.

Because the development of all these organ systems takes several days, an all-encompassing interpretation of the results is that TGFβ might have a biphasic effect which depends on the expression of other growth factors, their receptors, and of ECM receptors, such as the integrins. At the early stages of organ formation, TGFβ might enhance the invasive capacity of the epithelium. As development proceeds, other factors might influence the secondary responses to TGFβ activity, providing a local environment that leads to accumulation of ECM proteins, and stabilization of organ structure. Additionally, TGFβ1 may act directly on epithelial cells to control epithelial integrity during morphogenesis [Heino et al., 1989].

One major function of TGFβ in epithelial-mesenchymal interactions is probably in the condensation of mesenchymal cells beneath the anlage of the developing organ—an essential step in the initiation of tooth, hair follicle, and mammary gland development. Certainly, epithelial expression of TGFβ1 during tooth and hair follicle development commences early enough for this function to be realized. Mesenchymal condensation is also important in the initiation of chondrogenesis, another process involving interactions between epithelium and mesenchyme [Frenz et al., 1992]. Mesenchymal condensation might be mediated both by chemotactic attraction of cells to the local source of TGFβ [Yang and Moses, 1990] and by increased cell division. TGFβ is known to be mitogenic to many mesenchymal cell types, including embryonic mesenchyme [Choy et al., 1990; Linask et al., 1991]. By stimulating the biosynthesis of proteoglycans and GAGs, TGFβ will also modulate the ability of the mesenchyme to act as a reservoir for other growth factors, such as bFGF, and may well potentiate their effects by presentation of the growth factor to its receptor. In this manner, TGFβ may stimulate mesenchymal cell division both directly and indirectly.

D. Endothelial Cells, Vasculogenesis, and Angiogenesis

Vasculogenesis, the *de novo* differentiation of endothelial cells from primitive mesoderm to form primitive vessels, and angiogenesis, the sprouting of new blood vessels from existing ones, probably share some features in common. Certainly, expression of TGFβ1 mRNA in the early murine embryo has been associated with both these processes. Transcripts are seen at high levels in the forming yolk-sac blood islands, in the cardiac mesoderm prior to the differentiation of endothelial cells, and in endothelial cells—both those lining the primitive heart tube and those undergoing angiogenesis at many other embryonic body sites [Akhurst et al., 1990a].

As with many of the biological systems discussed in this chapter, data on the biological effects of TGFβ on endothelial cells can be highly confusing, because of the exact cell types studied (e.g., bovine aortic versus rat microvascular endothelial cells), because of the variable nature of the culture system (e.g., 2-d versus 3-d), and due to the influence of other factors, particularly bFGF. TGFβ is certainly angiogenic *in vivo* [Roberts and Sporn, 1990], although its activity on cells in culture would suggest the opposite. It is potently inhibitory to many of the processes considered essential for the early events of angiogenesis, including cell proliferation [Jennings et al., 1988], secretion of protease activity [Saksela et al., 1987], migration [Heimark et al., 1986; Sato and Rifkin, 1989], and invasion [Muller et al., 1987; Mignatti et al., 1989]. In general, TGFβ effects oppose, and are antagonistic to, the effects of bFGF, the other potent inducer of angiogenesis. It has been suggested that the angiogenic capacity of TGFβ, unlike that of bFGF, is a consequence of secondary effects from other cells, rather than of its action on the endothelial cell *per se* [Roberts and Sporn, 1990; Yang and Moses, 1990].

Gene expression studies, which suggest that TGFβ1 is involved in vasculogenesis and angiogenesis in the early mammalian embryo [Akhurst et al., 1990a], are paradoxical in the light of its known *in vitro* activities on endothelial cells. However, a number of studies have suggested that TGFβ might be involved in later stages of angiogenesis rather than in early events, i.e., inducing and maintaining the differentiated state of the endothelial cells, and modulating adjacent ECM to allow remodelling or stabilization of the newly formed vessel. Its mode of action might therefore be very similar to that proposed for TGFβ in epithelial-mesenchymal interactions (see Section IV.C).

Madri et al. [1988] demonstrated that microvascular endothelial cells, seeded within a three-dimensional collagen gel, are induced, by TGFβ, to form structures that resemble capillary tubes. This is considered to be induction of a differentiated phenotype, and was documented as being a direct effect on the endothelial cells, because very few differences in the composition of the secreted ECM were seen between TGFβ-treated and untreated cultures. Later reports suggested that there is, in fact, modulation of ECM secretion during this process [Merwin et al., 1990; Madri et al., 1992], and that the altered matrix, and altered cell–substratum interactions, might contribute to the differentiative changes seen within the endothelial cells. In this system, TGFβ modulates expression of integrins, cell adhesion molecules, and tight junctional complex formation [Merwin et al., 1990; Madri et al., 1992].

The idea that the major function of TGFβ derived from the endothelium might be to modulate ECM composition and affect cell phenotypes in an indirect manner would also explain the observation of differential TGFβ1 RNA and protein distribution during cardiac valve morphogenesis [Akhurst et al., 1990a]. In this situation, TGFβ synthesized in the endothelium was proposed to affect cell growth, cell movement, and tissue integrity of the underlying valve mesenchyme, by modulating production of ECM.

It is surprising that embryonic endothelial cells synthesize the highest level of TGFβ1 transcripts at a time when they are most actively dividing [Akhurst et al., 1990a]. However, it is possible that these cells are refractile to growth inhibition at early developmental stages. Indeed, in the three-dimensional culture system of Madri et al. [1988], TGFβ has no effect on cell division. Furthermore, Yang and Moses [1990] showed a *stimulation* of endothelial cell division by very low concentrations of TGFβ. They suggested that there may be threshold effects, and that at certain concentrations, TGFβ might stimulate rather than inhibit the growth of endothelial cells. It is also possible that an important function of TGFβ in the early events of vasculogenesis and angiogenesis is the augmentation of bFGF activity, either by increasing synthesis of this growth factor [Plouet and Gospodarowicz, 1989], or by the induction of GAGs which are essential for the binding of bFGF to its receptor [Nugent and Edelman, 1992].

E. TGFβ and Myogenesis

Until recently, it has been widely accepted that TGFβ is a potent inhibitor of myogenic differentiation *in vitro*. A number of independent laboratories reported that primary myoblasts, and established cell lines characteristic of both smooth and skeletal muscle, were reversibly prevented from terminal differentiation when grown under low-mitogen, differentiation-inducing conditions, in the presence of TGFβ [Olson et al., 1986; Massagué et al., 1986]. This differentiation-inhibitory effect was later shown to be mediated directly by downregulation of myogenic factors, such as myogenin [Heino and Massagué, 1990], MyoD1 [Vaidya et al., 1989], and *Myf 4* [Salminen et al., 1991], and indirectly via accumulation of the ECM proteins [Heino and Massagué, 1990]. Studies on cardiomyocytes have been more difficult, because of the unavailability of appropriate cell lines. However, Parker et al. [1991] showed that cultured neonatal rat cardiomyocytes were not inhibited from differentiating by TGFβ, but were induced to alter their profile of contractile protein gene expression from a neonatal to a more fetal type.

Recent reports on the bioactivity of TGFβ on cultured cells, have suggested that TGFβ, rather than being a myogenic inhibitor, can be an inducer of cardiac, skeletal, and smooth muscle cell differentiation. Zentrella and Massagué [1992] demonstrated that the response of cultured myoblasts to TGFβ can be modulated by the culture conditions. The rat cell line, L_6E_9, which is inhibited from differentiating by TGFβ1 in low-serum medium, can be induced to differentiate by TGFβ if the cells are actively cycling in mitogen-rich medium. This is a consequence of the downregulation of *c-myc* expression, with consequent withdrawal from the cell cycle, and concomitant downregulation of the myogenic inhibitory factor gene, *Id*. Similarly, cultures of Axolotl precardiac mesoderm can be induced to differentiate into cardiac muscle either by TGFβ or PDGF, but not by other growth factors [Muslin and Williams, 1991]. Finally, Slager et al. [1993], demonstrated that TGFβ1 and TGFβ2 are both capable of stimulating cultured embryonic stem cells to differentiate into skeletal and cardiac myocytes. Additionally, TGFβ has been reported to be capable of inducing myocyte characteristics in nonmyogenic cells [Eghbali et al., 1991].

There have been variable reports on the expression of TGFβ mRNAs and proteins during cardiac and skeletal myogenesis. Immunohistochemical staining has suggested high levels of all three proteins in fetal cardiac and skeletal muscle [Pelton et al., 1991; Dickson et al., 1993], though levels of their mRNAs are not high, and appear to be localized to nonmyogenic cells [Akhurst et al., 1990a; Millan et al., 1991; Dickson et al., 1993]. A recent detailed study on the expression of the three TGFβ isoforms during early cardiogenesis, however, points to a role for TGFβ2 in regulating cardiomyogenesis [Dickson et al., 1993]. TGFβ2 transcripts are present, at high levels, in all cells of the early embryo which have the potential to differentiate into cardiomyocytes. As cardiomyogenesis proceeds, transcript levels are downregulated, but staining for TGFβ2 protein increases as the cardiomyocyte differentiates. Additionally, the foregut endoderm, a putative source of cardiomyogenic factors, expresses high levels of this mRNA. It would appear, therefore, that TGFβ2, derived from paracrine and/or autocrine sources, might be an early regulator of mammalian cardiomyogenesis.

Whether or not TGFβ2 is a myogenic inducer or inhibitor *in vivo,* remains to be seen. It is highly probable that it could have a biphasic effect: inducing myogenesis during early development, when cell division is rapid, but inhibiting the process as mitogens become limiting [Zentrella and Massagué, 1992].

F. The Role of TGFβs in Septation of the Heart

Over the last 3 to 4 years there has been mounting evidence for the involvement of TGFβs in inductive tissue interactions that result in the formation of mesenchymal cardiac cushion tissue, which contributes to valve and septa formation.

Prior to cushion tissue formation, the heart can be considered as a simple bilayered tube, possessing an inner, endothelial component, and an outer, myocardial component, separated by an extensive basement membrane, the cardiac jelly. As cardiogenesis proceeds, the tube becomes regionalized and starts to bend. Thickened cardiac jelly persists only in the region of the outflow tract and atrioventricular junction, both areas of future morphogenesis into which mesenchymal cushion cells will migrate. In the atrioventricular region, the majority of mesenchymal cushion cells arise by transformation of endocardial cells. This is a biphasic process; the endothelial cells become activated (become hypertrophic and lose cell–cell contacts) and then they actively invade the underlying cardiac jelly [reviewed by Runyan et al., 1992].

Using a cultured chick cardiac explant system, it has been extensively demonstrated that this process is controlled by inductive interactions between the myocardium and the endothelium. Tissue recombination experiments have shown that an inductive signal secreted from atrioventricular, but not ventricular myocardium, acts on the overlying endocardium to initiate this event. There are both spatial and temporal restrictions on the ability of the myocardium to provide the inductive signal and of the endothelium to respond [Runyan et al., 1992]. Potts and Runyan [1989] provided evidence that one of the TGFβs might contribute to this inductive signal, because anti-TGFβ antibodies could block the endothelial-mesenchymal transition in culture. Furthermore, recombinant TGFβ in combination with ventricular myocardium could support this process in culture, whereas neither alone was capable of eliciting the effect. More recently, this group have implicated TGFβ3 as the active isoform, because antisense oligonucleotides directed against this RNA, but not those directed against RNAs of the other TGFβ isoforms, could block the endothelial-mesenchymal transformation [Potts et al., 1991].

No similar functional studies on cardiac cushion tissue formation have been performed in mammals. However, the provocative transitory localization of TGFβ2 transcripts within myocardial cells of the atrioventricular canal and outflow tract, suggest that TGFβ2 might be a component of this induction signal in mice [Millan et al. 1991; Dickson et al., 1993]. An alternative explanation for the localized myocardial expression of TGFβ2 RNA in the atrioventricular/outflow region of the mouse heart, is that these cells have distinct differentiative properties, which are maintained by high-level expression of this putative cardiomyogenic regulator [discussed in Dickson et al., 1993]. Interestingly, TGFβ3 RNA expression is not seen during mouse cardiac development until well after cardiac cushion tissue has formed. Thus, there appear to be somewhat surprising differences in data relating to TGFβ isoform expression and utilization between avian and mammalian heart development [Millan et al., 1991; Runyan et al., 1992; Dickson et al., 1993].

G. TGFβ in Chondrogenesis and Osteogenesis

Endochondral bone formation follows a sequence of well-characterized events. The first is the elaboration of ECM by prechondrogenic mesoderm cells, which are otherwise indistinguishable from their neighboring cells. This step is essential to the condensation of mesoderm to form the precartilaginous blastemae [Frenz et al., 1992]. Cytodifferentiation of cartilage takes place within the precartilaginous blastemae, followed by cartilage maturation, which involves proliferation, hypertrophy, and calcification. Osteoblasts invade the cartilage model and differentiate to form bone.

Adult bone is one of the richest sources of TGFβ [Seyedin et al., 1985], and *in situ* hybridization studies on mammalian embryos have shown that all three isoforms are expressed in different subsets of cells during both bone and cartilage formation [Heine et al., 1987; Sandberg et al., 1988; Lehnert and Akhurst, 1988; Gatherer et al., 1990; Pelton et al., 1990; Millan et al., 1991]. One of the first-described *in vitro* bioactivities for TGFβ1 and β2 was the induction of cartilage condensates from rat mesenchymal cells [Seyedin et al., 1985], however, the exact sequence of molecular and cellular events modulated by the TGFβs during cartilage and bone formation remains to be resolved, because many reports on their bioactivity have been contradictory. This is most probably because of the complexity of the *in vitro* systems studied. These often involve many cell types, taken from variable body sites and at different developmental stages.

The TGFβs are known to have mainly positive effects on proliferation and differentiation of cells from the chondrocyte and osteocyte lineages [Rosen et al., 1988; Kulyk et al., 1989; Carrington and Reddi, 1990]; this would be compatible with their function as inducers of bone and cartilage formation. Additionally, they can affect the osteoclast lineage by influencing the developmental fate of the progenitor cells [Chenu et al., 1988]. Some of these effects might act directly on the cell and others might be mediated by modulation of ECM. Certainly some of the effects of TGFβ on bone and cartilage cells are modified by the presence of other growth factors, such as bFGF [Centrella et al., 1987a]. Related bone morphogenetic molecules [Wang et al., 1988] might also influence their action.

In examining the effect of TGFβs on micromass cultures of chick limb-bud mesenchyme, Kulyk et al. [1989] demonstrated a striking increase in production of ECM components, including sulfated GAGs, in response to this growth factor. Leonard et al. [1991] suggested that it is this action on ECM accumulation which stimulates precartilage condensation. Interestingly, TGFβ2 RNA is highly expressed within the forming precartilage blastemae of mouse and human limb buds at this stage of development [Gatherer et al., 1990; Millan et al., 1991]. Frenz et al. [1992] found that TGFβ could elicit cartilage formation within the otic capsule, and that this activity mimicked the effect of otic epithelial cells on otic mesenchyme. They thus concluded that TGFβ synthesized in the otic epithelium had a paracrine cartilage-inducing activity on otic mesenchyme.

Many studies on the effects of TGFβ on cartilage and bone formation *in vitro* and *in vivo* have emphasized the importance of concentration dependence (threshold effects) and developmental "windows" of TGFβ responsiveness. The mitogenic effect of TGFβ on cultured osteoblasts is dose-dependent [Centrella et al., 1987b], as is the initiation of chondrogenesis and osteogenesis by subperiosteal injection of TGFβ *in vivo* [Joyce et al., 1990]. Carrington and Reddi [1990] demonstrated variable differentiative responses of chick limb-bud mesodermal cells to TGFβ1, depending on the developmental time of exposure. This suggests that the ability of the mesodermal cell to respond to TGFβ (which is dependent on receptor binding, activation, and signal transduction), varies with the extent of differentiation of the responding cell. A similar observation was made on the differentiative response of otic mesenchymal cells to TGFβ [Frenz et al., 1992].

TGFβ, implanted locally into the chick limb bud *in vivo,* inhibits chondrogenesis, in a temporally and spatially restricted fashion; the TGFβ-sensitive window correlates with the period of mesenchymal cell condensation, i.e., stages 22–25 [Hayamizu et al., 1991]. Once again, the magnitude of the TGFβ response is correlated with the differentiative stage of the cell. Furthermore, the qualitative response to TGFβ can be completely reversed if the implant is performed at a later stage of limb development. Thus, proximal TGFβ1 implants performed *after* the early TGFβ-sensitive window (at stages 25–26), can lead to *induction* of exogenous cartilage elements.

Clearly, there is still much work to be done. Better-defined *in vitro* culture systems must be used, more *in vivo* experiments need to be performed, and the question of how these molecules exert their effects must be addressed before a full understanding of the exact function of TGFβs in bone formation can be achieved.

V. FUTURE DIRECTIONS

Until now, most of the information pertaining to the role of TGFβs in vertebrate embryogenesis has focused on the effects of these growth factors on cells in culture, and on determination of the sites of expression of TGFβ mRNAs and proteins during embryogenesis. Both approaches have been informative, but are also limiting. TGFβ bioactivity is regulated at many levels. Ideally, one would endeavor to determine sites of TGFβ bioactivity within the embryo, currently a technically difficult objective.

A concept which needs to be emphasized for studies *in vivo* is that of threshold effects. This point is illustrated by the effects of the TGFβ-related molecule, activin, in *Xenopus* mesoderm induction. The dorsoventral character of the induced mesoderm is determined by activin concentration in this assay [Green et al., 1992]. There are now several examples of TGFβ showing threshold effects, both *in vitro* and *in vivo* [Yang and Moses, 1990; Centrella et al., 1987b; Joyce et al., 1990]. Thus gradients of TGFβ produced from a localized source might have different effects on the same cell type, depending on the distance of the cell from the source. Additionally, the qualitative and quantitative responses of the same cell type *in vivo* might be regulated in a spatially and temporally restricted manner. One example is the different *in vivo* responses of limb-bud mesoderm cells to locally applied TGFβ [Carrington and Reddi, 1990]. These effects are presumably due to spatial and temporal regulation of expression of receptors and intracellular signalling molecules.

Molecular analysis of the TGFβ receptors is in its infancy, and will certainly be an exponential growth area over the next few years. Preliminary studies suggest that the type II receptor is differentially expressed in murine fetal tissues (personal observations), but the possibility of generation of type II receptor diversity by differential splicing and/or post-translational modification has not been addressed. There may be a related family of receptor genes, and ligand binding or signal transduction might be regulated by interaction with other receptors and/or surface–associated proteins.

The understanding of the TGFβ signal transduction pathway(s) will be another major objective. Intracellular signalling mechanisms may vary considerably between cell types, as is seen in negative growth regulation of keratinocytes versus bronchial epithelial cells by TGFβ. Additionally, Geiser et al. [1992] suggested that different phenotypic effects of TGFβ on the same cell type, might be regulated via different intracellular signalling pathways. Finally, the interaction of TGFβ with other transcriptional regulatory genes, such as the homeobox (HOX) or myogenic regulatory genes, needs to be addressed, as do the molecular mechanisms by which other growth factors modify TGFβ effects.

Developmentally viable mice, homozygous for a defective *tgf*β*1* gene have now been generated. Whether gene "knock-outs" for TGFβ2 and -β3 and the type I and II receptors, will be similarly viable remains to be seen. This transgenic approach will provide excellent tools with which to address some of the above questions.

ACKNOWLEDGMENTS

Apologies to all those individuals whose work is not quoted; only selected papers were chosen in the interests of brevity. Thanks to my colleagues within the lab for comments on the manuscript. Work in the author's laboratory is supported by the Medical Research Council, Wellcome Trust, and Cancer Research Campaign.

REFERENCES

Akhurst RJ, Lehnert SA, Faissner AJ, Duffie E (1990a): TGFβ in murine morphogenetic processes: the early embryo and cardiogenesis. Development 108:645–656.

Akhurst RJ, Lehnert SA, Gatherer D, Duffie E (1990b): The role of TGFβ in mouse development. Ann N Y Acad Sci 593:259–271.

Akhurst RJ, Fitzpatrick DR, Gatherer D, Lehnert SA, Millan FA (1990c): The role of TGFβs in mammalian embryogenesis. Progr Growth Factor Res 2:153–168.

Andres JL, Stanley K, Cheifetz S, Massagué J (1989): Membrane-anchored and soluble forms of βglycan, a polymorphic proteoglycan that binds transforming growth factor-β. J Cell Biol 109:3137–3145.

Andres JL, DeFalcis D, Noda M, Massagué J (1992): Binding of two growth factor families to separate domains of the proteoglycan βglycan. J Biol Chem 267:5927–5930.

Antonelli-Orlidge A, Saunders KB, Smith SR, D'Amore PA (1989): An activated form of transforming growth factor β is produced by cocultures of endothelial cells and pericytes. Proc Natl Acad Sci USA 86:4544–4548.

Assoian RK, Komoriya A, Meyers CA, Miller DM, Sporn MB (1983): Transforming growth factor-β in human platelets. Identification of a major storage site, purification, and characterization. J Biol Chem 258:7155–7160.

Assoian RK, Fleurdelys BE, Stevenson HC, Miller PJ, Madtes DK, Raines EW, Ross R, Sporn MB (1987): Expression and secretion of type β transforming growth factor by activated human macrophages. Proc Natl Acad Sci USA 84:6020–6024.

Attisano L, Wrana JL, Cheifetz S, Massagué J (1992): Novel activin receptors: Distinct genes and alternative splicing generate a repertoire of serine/threonine kinase receptors. Cell 68:97–108.

Baird A, Durkin T (1986): Inhibition of endothelial cell proliferation by β-type transforming growth factor: interactions with acidic and basic fibroblast growth factors. Biochem Biophys Res Commun 138:476–482.

Balza E, Borsi L, Allemanni G, Zardi L (1988): Transforming growth factor β regulates the levels of different fibronectin isoforms in normal human fibroblasts. FEBS Lett 228:42–44.

Bassols A, Massagué J (1988): Transforming growth factor β regulates the expression and structure of extracellular matrix chondroitin/dermatan sulfate proteoglycans. J Biol Chem 263:3039–3045.

Basson CT, Kocher O, Basson MD, Asis A, Madri JA (1992): Differential modulation of vascular cell integrin and extracellular matrix expression in vitro by TGF-β1 correlates with reciprocal effects on cell migration. J Cell Physiol 153:118–128.

Brandes ME, Wakefield L, Wahl SM (1991): Modulation of monocyte type I transforming growth factor-β receptors by inflammatory stimuli. J Biol Chem 266:19697–19703.

Burt DW, Jakowlew SB (1992): A new interpretation of a chick TGFβ4 cDNA. Mol Endocrinol 6:989–992.

Burt DW, Paton IR (1992): Evolutionary origins of the TGFβ gene family. DNA Cell Biol 11:497–510.

Carrington JL, Reddi AH (1990): Temporal changes in the response of chick limb bud mesodermal cells to transforming growth factor β-type I. Exp Cell Res 186:368–373.

Centrella M, McCarthy TL, Canalis E (1987a): Mitogenesis of fetal rat bone cells simultaneously exposed to type β transforming growth factor and other growth regulators. FASEB J 1:312–317.

Centrella M, McCarthy TL, Canalis E (1987b): Transforming growth factor β is a bifunctional regulator of replication and collagen synthesis in osteoblast-enriched cell cultures from fetal rat bone. J Biol Chem 262:2869–2874.

Cheifetz S, Hernandez H, Laiho M, ten Dijke P, Iwata KK, Massagué J (1990): Distinct transforming growth factor-β (TGF-β) receptor subsets as determinants of cellular responsiveness to three TGF-β isoforms. J Biol Chem 265:20533–20538.

Cheifetz S, Bellon T, Cales C, Vera S, Bernabeu C, Massagué J, Letarte M (1992): Endoglin is a component of the transforming growth factor-β receptor system in human endothelial cells. J Biol Chem 267:19027–19030.

Chen JK, Hoshi H, McKeehan WL (1987): Transforming growth factor type β specifically stimulates synthesis of proteoglycan in human adult arterial smooth muscle cells. Proc Natl Acad Sci USA 84:5287–5291.

Chenu C, Pfeilschifter J, Mundy GR, Roodman GD (1988): Transforming growth factor β inhibits formation of osteoclast-like cells in long-term human marrow cultures. Proc Natl Acad Sci USA 85:5683–5687.

Chiquet-Ehrismann R, Kalla P, Pearson CA, Beck K, Chiquet M (1988): Tenascin interferes with fibronectin action. Cell 53:383–390.

Chiquet-Ehrismann R, Kalla P, Pearson CA (1989): Participation of tenascin and transforming growth factor β in reciprocal epithelial-mesenchymal interactions of MCF7 cells and fibroblasts. Cancer Res 49:4322–4325.

Choy M, Armstrong MT, Armstrong PB (1990): Regulation of proliferation of embryonic heart mesenchyme: role of transforming growth factor β1 and the interstitial matrix. Biol 141:421–425.

Coffey RJ Jr, Bascom CC, Sipes NJ, Graves Deal R, Weissman BE, Moses HL (1988): Selective inhibition of growth-related gene expression in murine keratinocytes by transforming growth factor β. Mol Cell Biol 8:3088–3093.

Coletta AA, Wakefield LM, Howell FV, van Roozendaal KE, Danielpour D, Ebbs SR, Sporn MB, Baum M (1990): Anti-oestrogens induce secretion of active transforming growth factor β in human fetal fibroblasts. Br J Cancer 62:405–409.

Daopin S, Piez KA, Ogawa Y, Davie DR (1992): Crystal structure of transforming growth factor β2, an unusual fold for the superfamily. Science 257:369–373.

DeLarco J, Todaro GJ (1978): Growth factors from murine sarcoma virus-transformed cells. Proc Natl Acad Sci USA 75:4001–4005.

Denhez F, Lafayatis R, Kondaiah P, Roberts AB, Sporn MB (1990): Cloning by polymerase chain reaction of a new mouse TGFβ, mTGF-β3. Growth Factors 3:139–146.

Dennis PA, Rifkin DB (1991): Cellular activation of latent transforming growth factor β requires binding to the cation-independent mannose-6-phosphate/insulin-like growth factor receptor. Proc Natl Acad Sci USA 88:580–584.

Derynck R, Jarrett JA, Chen EY, Eaton DH, Bell JR, Assoian RK, Roberts AB, Sporn MB, Goeddel DV (1985): Human transforming growth factor-β cDNA sequence and expression in tumor cell lines. Nature 316:701–705.

Diaz-Meco MT, Dominguez I, Sanz L, Munico MM, Berra E, Cornet ME, Garcia de Herreros A, Johansen T, Moscat J (1992): Phospholipase C-mediated hydrolysis of phosphatidylcholine is a target of transforming growth factor β1 inhibitory signals. Mol Cell Biol 12:302–308.

Dickson M, Slager HG, Duffie E, Akhurst RJ (1993): TGFβ2 RNA and protein localisations in the early embryo suggest a role in cardiac development and myogenesis. Development 117:625–639.

Dulannet M, Duband J-L (1992): Transforming growth factor-β control of cell-substratum adhesion during avian neural crest emigration in vitro. Development 116:275–287.

Eghlbali M, Tomek R, Woods C, Bhambi B (1991): Cardiac fibroblasts are predisposed to convert into myocyte phenotype: specific effects of transforming growth factor β. Proc Natl Acad Sci USA 88:795–799.

Fava RA, McClure DB (1987): Fibronectin-associated transforming growth factor. J Cell Physiol 131:184–189.

Fitzpatrick DR, Denhez F, Kondaiah P, Akhurst RJ (1990): Differential expression of TGFβ isoforms in murine palatogenesis. Development 109:585–595.

Flaumenhaft R, Abe M, Rifkin DB (1992): Basic fibroblast growth factor-induced activation of latent transforming growth factor β in endothelial cells: regulation of plasminogen activator activity. J Cell Biol 118:901–909.

Flaumenhaft R, Abe M, Sato Y, Miyazono K, Harpel J, Heldin C-H, Rifkin DB (1993): Role of latent TGF-β binding protein in the activation of latent TGF-β by co-cultures of endothelial and smooth muscle cells. J Cell Biol 120:995–1002.

Frenz DA, Galinovic-Schwartz V, Liu W, Flanders KC, Van der Water TR (1992): Transforming growth factor β1 is an epithelial-derived signal peptide that influences otic capsule formation. Dev Biol 153:324–336.

Frolik CA, Dart LL, Meyers CA, Smith DM, Sporn MB (1983): Purification and initial characterization of a type β transforming growth factor from human placenta. Proc Natl Acad Sci USA 80:3676–3680.

Gatherer D, ten Dijke P, Baird DT, Akhurst RJ (1990): Expression of TGFβ isoforms during first trimester human embryogenesis. Development 110:445–460.

Geiser AG, Kim S-J, Roberts AB, Sporn MB (1991): Characterisation of the mouse transforming growth factor-β1 promoter and activation by the Ha-ras oncogene. Mol Cell Biol 11:84–92.

Geiser AG, Burmester JK, Webbink R, Roberts AB, Sporn MB (1992): Inhibition of growth by transforming growth factor-β following fusion of two non-responsive human carcinoma cell lines. J Biol Chem 267:2588–2593.

Gentry LE, Webb NR, Lim GJ, Brunner AM, Ranchalis JE, Twardzik DR, Lioubin MN, Marquardt H, Purchio AF (1987): Type 1 transforming growth factor β: amplified expression and secretion of mature and precursor polypeptides in Chinese hamster ovary cells. Mol Cell Biol 7:3418–3427.

Glick AB, Flanders KC, Danielpour D, Yuspa SH, Sporn MB (1989): Retinoic acid induces transforming growth factor-β2 in cultured keratinocytes and mouse epidermis. Cell Regul 1:87–97.

Globus RK, Patterson-Buckendahl P, Gospodarowicz D (1988): Regulation of bovine bone cell proliferation by fibroblast growth factor and transforming growth factor β. Endocrinology 123:98–105.

Graycar JL, Miller DA, Arrick BA, Lyons RM, Moses HL, Derynck R (1989): Human transforming growth factor-β3: recombinant expression, purification, and biological activities in comparison with transforming growth factors-β1 and -β2. Mol Endocrinol 3:1977–1986.

Green JBA, New HV, Smith JC (1992): Responses of embryonic Xenopus cells to activin and FGF are separated by multiple dose thresholds and correspond to distinct axes of the mesoderm. Cell 71:731–739.

Hayamizu TF, Sessions SK, Wanek N, Bryant SV (1991): Effects of localised application of transforming growth factor β1 on developing chick limbs. Dev Biol 145:164–173.

Heimark RL, Twardzik DR, Schwartz SM (1986): Inhibition of endothelial regeneration by type-β transforming growth factor from platelets. Science 233:1078–1080.

Heine U, Munoz EF, Flanders KC, Ellingsworth LR, Lam HY, Thompson NL, Roberts AB, Sporn MB (1987): Role of transforming growth factor-β in the development of the mouse embryo. J Cell Biol 105:2861–2876.

Heine UI, Munoz EF, Flanders KC, Roberts AB, Sporn MB (1990): Colocalisation of TGFβ-1 and collagen I and III, fibronectin and glycosaminoglycans during lung branching morphogenesis. Development 109:29–36.

Heino J, Ignotz RA, Hemler ME, Crouse C, Massagué J (1989): Regulation of cell adhesion receptors by transforming growth factor-β. Concomitant regulation of integrins that share a common β 1 subunit. J Biol Chem 264:380–388.

Heino J, Massagué J (1990): Cell adhesion and decreased myogenic gene expression implicated in the control of myogenesis by transforming growth factor β. J Biol Chem 265:10181–10184.

Hiraki Y, Inoue H, Hirai R, Kato Y, Suzuki F (1988): Effect of transforming growth factor β on cell proliferation and glycosaminoglycan synthesis by rabbit growth-plate chondrocytes in culture. Biochim Biophys Acta 969:91–99.

Howe PH, Draetta G, Leof EB (1991): Transforming growth factor β1 inhibition of $p34^{cdc2}$ phosphorylation and histone H1 kinase activity is associated with G1/S-phase growth arrest. Mol Cell Biol 11:1185–1194.

Ikeda T, Lioubin MN, Marquardt H (1987): Human transforming growth factor β2 production by a prostatic adenocarcinoma cell line, purification, and initial characterization. Biochemistry 26:2406–2410.

Immergluck K, Lawrence PA, Bienz M (1990): Induction across germ layers in Drosophila mediated by a genetic cascade. Cell 62:261–268.

Ito M, Yasui W, Kyo E, Yokozaki H, Makayama H, Ito H, Tahara E (1992): Growth inhibition of transforming growth factor β on human gastric carcinoma cells: receptor and postreceptor signaling. Cancer Res 52:295–300.

Jennings JC, Mohan S, Linkhart TA, Widstrom R, Baylink DJ (1988): Comparison of the biological actions of TGFβ-1 and TGFβ-2: differential activity in endothelial cells. J Cell Physiol 137:167–172.

Joyce ME, Roberts AB, Sporn MB, Nolander ME (1990): Transforming growth factor-β and the initiation of chondrogenesis and osteogenesis in the rat femur. J Cell Biol 110:2195–2207.

Kanazaki T, Olofsson A, Moren A, Wernstedt C, Hellman U, Miyazono K, Claesson-Welsh L, Heldin C-H (1990): TGFβ1 binding protein: a component of the large latent complex of TGFβ with multiple repeat sequences Cell 61:1051–1061.

Kerr LD, Miller DB, Matrisian LM (1990): TGF-β1 inhibition of transin/stromelysin gene expression is mediated through a fos binding sequence. Cell 61:267–278.

Kim S-J, Glick A, Sporn MB, Roberts AB (1989): Characterization of the promoter region of the human transforming growth factor-β 1 gene. J Biol Chem 264:402–408.

Kim S-J, Park K, Koeller D, Kim KY, Wakefield LM, Sporn MB, Roberts AB (1992): Posttranscriptional regulation of the transforming growth factor-β1 gene. J Biol Chem 267:13702–13707.

Kimelman D, Kirschner M (1987): Synergistic induction of mesoderm by FGF and TGF-β and the identification of an mRNA coding for FGF in the early Xenopus embryo. Cell 51:869–877.

Kocher O, Kennedy SP, Madri JA (1990): Alternative splicing of endothelial cell fibronectin mRNA in the IIICS region. Functional significance. Am J Pathol 137:1509–1524.

Kramer IM, Koorneef I, de Vries C, de Groot RP, De Laat SW, van den Eijnden-van Raaij AJM, Kruijer W (1991a): Phophorylation of nuclear protein is an early event in TGFβ action. Biochem Biophys Res Commun 175:816–822.

Kramer IM, Koorneef I, DeLaat SW, van den Eijnden-van Raaij AJM (1991b): Transforming growth factor β1 induces phosphorylation of the cyclic AMP responsive element binding protein in ML-CC164 cells. EMBO J 10:1083–1089.

Kulkarni AB, Chang-Goo H, Becker D, Geiser A, Lyght M, Flanders KC, Roberts AB, Sporn MB, Ward JM, Karlsson S (1993): Transforming growth factor-β1 null mutation in mice causes excessive inflammatory response and early death. Proc Natl Acad Sci USA 90:770–774.

Kulyk WM, Rodgers BJ, Greer K, Robert RK (1989): Promotion of embryonic chick limb cartilage differentiation by transforming growth factor-β. Dev Biol 135:424–430.

Lafyatis R, Lechleider R, Kim S-J, Jakowlew S, Roberts AB, Sporn MB (1990): Structural and functional characterisation of the transforming growth factor β3 promoter. J Biol Chem 265:19128–19136.

Laiho M, Weis FMB, Massagué J (1990a): Concomitant loss of transforming growth factor (TGF)-β receptor types I and II in TGF-β-resistant cell mutants implicates both receptor types in signal transduction. J Biol Chem 265:18518–18524.

Laiho M, DeCaprio JA, Ludlow JW, Livingston DM, Massagué (1990b): Growth regulation by TGF-β linked to suppression of retinoblastoma protein phosphorylation. Cell 62:175–185.

Laiho M, Weis FMB, Boyd FT, Ignotz RA, Massagué J (1991): Responsiveness to transforming growth factor-β (TGF-β) restored by genetic complementation between cells defective in TGF-β receptors I and II. J Biol Chem 266:9108–9112.

Lehnert SA, Akhurst RJ (1988): Embryonic expression patterns of TGF β type-1 RNA suggests both paracrine and autocrine mechanisms of action. Development 104:263–273.

Leof EB, Proper JA, Goustin AS, Shipley GD, DiCorleto PE, Moses HL (1986): Induction of c-sis mRNA and activity similar to platelet-derived growth factor by transforming growth factor β: a proposed model for indirect mitogenesis involving autocrine activity. Proc Natl Acad Sci USA 83:2453–2457.

Leonard CM, Fuld HM, Frenz DA, Downie SA, Massagué J, Newman SA (1991): Role of transforming growth factor-β in chondrogenic pattern formation in the embryonic limb: stimulation of mesenchymal condensation and fibronectin gene expression by exogenous TGF-β and evidence for endogenous TGF-β-like activity. Dev Biol 145:99–109.

Li ML, Aggeler J, Farson DA, Hatier C, Hassell J, Bissell MJ (1987): Influence of a reconstituted basement membrane and its components on casein gene expression and secretion in mouse mammary epithelial cells. Proc Natl Acad Sci USA 84:136–140.

Lin HY, Wang XF, Ng EE, Weinberg RA, Lodish HF (1992): Expression cloning of the TGF-β type II receptor, a functional transmembrane serine/threonine kinase. Cell 68:775–785.

Linask KK, D'Angelo M, Gehris AL, Greene RM (1991): Transforming growth factor receptor profiles of human and murine embryonic palate mesenchymal cells. Exp Cell Res 192:1–9.

Longstreet M, Miller B, Howe PH (1992): Loss of transforming growth factor β1 (TGF-β1)-induced growth arrest and $p34^{cdc2}$ regulation in ras transfected epithelial cells. Oncogene 7:1549–1556.

Lopez-Casillas F, Cheifetz S, Doody J, Andres JL, Lane WS, Massagué J (1991): Structure and expression of the membrane proteoglycan βglycan, a component of the TGFβ receptor system. Cell 67:785–795.

Lyons RM, Gentry LE, Purchio AF, Moses HL (1990): Mechanism of activation of latent recombinant transforming growth factor β1 by plasmin. J Cell Biol 110:1361–1367.

MacKay K, Danielpour D (1991): Novel 150- and 180-kDa glycoproteins that bind transforming growth factor (TGF)-β1 but not TGF-β2 are present in several cell lines. J Biol Chem 266:9907–9911.

Madri JA, Williams SK, Wyatt T, Mezzio C (1983): Capillary endothelial cell cultures: phenotypic modulation by matrix components. J Cell Biol 97:153–165.

Madri JA, Pratt BM, Tucker AM (1988): Phenotypic modulation of endothelial cells by transforming growth factor-β depends upon the composition and organization of the extracellular matrix. J Cell Biol 106:1375–1384.

Madri JA, Bell L, Merwin JR (1992): Modulation of vascular cell behaviour by transforming growth factors β. Mol Reprod Dev 32:121–126.

Martin P, Dickson MC, Millan FA, Akhurst RJ (1993): Rapid induction and clearance of TGFβ is an early response to wounding in the mouse embryo. Dev Genet 14:225–238.

Massagué J (1992): Receptors for the TGFβ family. Cell 69:1067–1070.

Massagué J, Cheifetz S, Endo T, Nadal-Ginard B (1986): Type β transforming growth factor is an inhibitor of myogenic differentiation. Proc Natl Acad Sci USA 83:8206–8210.

Matrisian LM, Ganser GL, Kerr LD, Pelton RW, Wood LD (1992): Negative regulation of gene expression by TGFβ. Mol Reprod Dev 32:111–120.

McCaffrey TA, Falcone DJ, Brayton CF, Agarwal LA, Welt FG, Weksler BB (1989): Transforming growth factor-β activity is potentiated by heparin via dissociation of the transforming growth factor-β/alpha 2-macroglobulin inactive complex. J Cell Biol 109:441–448.

Merwin JR, Anderson J, Kocher O, van Itallie C, Madri JA (1990): Transforming growth factor β modulates extracellular matrix organisation and cell-cell junctional complex formation during in vitro angiogenesis. J Cell Physiol 142:117–128.

Merwin JR, Newman W, Beall D, Tucker A, Madri JA (1991): Vascular cells respond differentially to transforming growth factors-β1 and β2. Am J Pathol 138:37–51.

Mignatti P, Tsuboi R, Robbins E, Rifkin DB (1989): In vitro angiogenesis on the human amniotic membrane: requirement for basic fibroblast growth factor-induced proteinases. J Cell Biol 108:671–682.

Millan FA, Kondaiah P, Denhez F, Akhurst RJ (1991): Embryonic gene expression patterns of TGFβs 1, 2 and 3 suggest different developmental functions in vivo. Development 111:131–144.

Miller DA, Lee A, Matsui Y, Chen EY, Moses HL, Derynck R (1989): Complementary DNA cloning of murine transforming growth factor β3 (TGFβ3) precursor and the comparative expression of TGFβ1 and TGFβ3 and TGFβ1 in murine embryos and adult tissues. Mol Endocrinol 3:1926–1934.

Miller DM, Ogawa Y, Iwata KK, ten Dijke P, Purchio AF, Soloff MS, Gentry LE (1992): Characterisation of the binding of transforming growth factor-β1, -β2, and -β3 to recombinant β1-latency-associated peptide. Mol Endocrinol 6:694–702.

Mitchell EJ, Fitz-Gibbon L, O'Connor-McCourt MD (1992): Subtypes of βglycan and of type I and type II transforming growth factor-β (TGF-β) receptors, with different affinities for TGF-β1 and TGF-β2, are exhibited by human placental trophoblast cells. J Cell Physiol 150:334–343.

Miyazono K, Heldin CH (1989): Role for carbohydrate structures in TGF-β 1 latency. Nature 338:158–160.

Miyazono K, Hellman U, Wernstedt C, Heldin CH (1988): Latent high molecular weight complex of transforming growth factor β 1. Purification from human platelets and structural characterizations. J Biol Chem 263:6407–6415.

Miyazono K, Olofsson A, Colosetti P, Heldin C-H (1991): A role of the latent TGFβ-1 binding protein in the assembly and secretion of TGFβ1. EMBO J 10:1091–1101.

Moses HL (1992): TGFβ regulation of epithelial cell proliferation. Mol Reprod Dev 32:179–184.

Mulder KM, Morris SL (1992): Activation of p21-ras by transforming growth factor β in epithelial cells. J Biol Chem 267:5029–5031.

Muller G, Behrens J, Nussbaumer U, Böhlen P, Birchmeier W (1987): Inhibitory action of transforming growth factor β on endothelial cells. Proc Natl Acad Sci USA 84:5600–5604.

Murphy-Ullrich JE, Schultz-Cherry S, Hook M (1992): Transforming growth factor β complexes with thrombospondin. Mol Biol Cell 3:181–188.

Muslin AJ, Williams LT (1991): Well-defined growth factors promote cardiac development in axolotl mesodermal explants. Development 112:1095–1101.

Noma T, Glick AB, O'Reilly MA, Miller J, Roberts AB, Sporn MB (1991): Molecular cloning and structure of the human transforming growth factor-β2 gene promoter. Growth Factors 4:247–255.

Nugent MA, Edelman ER (1992): Transforming growth factor β1 stimulates the production of basic fibroblast growth factor binding proteoglycans in Balb/c3T3 cells. J Biol Chem 267:21256–21264.

O'Connor-McCourt MD, Wakefield LM (1987): Latent transforming growth factor-β in serum. A specific complex with alpha 2-macroglobulin. J Biol Chem 262:14090–14099.

O'Grady P, Kuo M-D, Baldassare JJ, Huang SS, Huang JS (1991): Purification of a new type high molecular weight receptor (type V) of transforming growth factor β from bovine liver: Identification of type V receptor in cultured cells. J Biol Chem 266:8583–8589.

Ohta M, Greenberger JS, Anklesaria P, Bassols A, Massagué J (1987): Two forms of transforming growth factor-β distinguished by multipotential haematopoietic progenitor cells. Nature 329:539–541.

Ohtsuki M, Massagué J (1992): Evidence for the involvement of protein kinase activity in transforming growth factor β signal transduction. Mol Cell Biol 12:261–265.

Olofsson A, Miyazono K, Kanzaki T, Colosetti P, Engstrom U, Heldin C-H (1992): Transforming growth factor-β1, -β2 and -β3 secreted by a human glioblastoma cell line. J Biol Chem 267:19482–19488.

Olson EN, Sternberg E, Hu JS, Spizz G, Wilcox C (1986): Regulation of myogenic differentiation by type β transforming growth factor. J Cell Biol 103:1799–1805.

Panganiban GEF, Reuter R, Scott MP, Hoffman FM (1990): A *Drosophila* growth factor homolog, *decapentaplegic*, regulates homeotic gene expression within and across germ layers during midgut morphogenesis. Development 110:1041–1050.

Paralkar V, Vukicevic S, Reddi AH (1991): Transforming growth factor β type 1 binds to collagen IV of basement membrane matrix: implications for development. Dev Biol 143:303–308.

Parker TG, Chow K-L, Schwartz RJ, Schneider MD (1991): Differential regulation of skeletal α-actin transcription in cardiac muscle by two fibroblast growth factors. Proc Natl Acad Sci USA 87:7066–7070.

Pelton RW, Dickinson ME, Moses HL, Hogan BLM (1990): In situ hybridisation analysis of TGFβ3 RNA expression during mouse development: comparative study with TGFβ1 and TGFβ2. Development 110: 609–620.

Pelton RW, Saxena B, Jones M, Moses HL, Gold LI (1991): Immunohistochemical localisation of TGFβ1, TGFβ2, and TGFβ3 in the mouse embryo: expression patterns suggest multiple roles during embryonic development. J Cell Biol 115:1091–1105.

Pepper MS, Belin D, Montesano R, Orci L, Vassalli J-D (1990): Transforming growth factor-β1 modulates basic fibroblast growth factor-induced proteolytic and angiogenic properties of endothelial cells. J Cell Biol 111:743–755.

Pertovaara L, Sistonen L, Bos TJ, Vogt PK, Keski-Oja J, Alitalo K (1989): Enhanced *jun* gene expression is an early genomic response to transforming growth factor β stimulation. Mol Cell Biol 9:1255–1262.

Pietenpol JA, Stein RW, Moran E, Yaciuk P, Schlegel R, Lyons RM, Pittelkow MR, Munger K, Howley PM, Moses HL (1990a): TGF-β1 inhibition of *c-myc* transcription and growth in keratinocytes is abrogated by viral transforming proteins with pRB binding domains. Cell 61:777–785.

Pietenpol JA, Holt JT, Stein RW, Moses HL (1990b): Transforming growth factor β1 suppression of *c-myc* gene transcription: role in inhibition of keratinocyte proliferation. Proc Natl Acad Sci USA 87:3758–3762.

Pietenpol JA, Munger K, Howley PM, Stein RW, Moses HL (1991): Factor-binding element in the human *c-myc* promoter involved in transcriptional regulation by transforming growth factor β1 and by the retinoblastoma gene product. Proc Natl Acad Sci USA 88:10227–10231.

Plouet J, Gospodarowicz D (1989): Transforming growth factor β-1 positively modulates the bioactivity of fibroblast growth factor on corneal endothelial cells. J Cell Physiol 141:392–399.

Potts JD, Runyan RB (1989): Epithelial-mesenchymal cell transformation in the embryonic heart can be mediated, in part, by transforming growth factor β. Dev Biol 134:392–401.

Potts JD, Dagle JM, Walder JA, Weeks DL, Runyan RB (1991): Epithelial-mesenchymal transformation of embryonic cardiac endothelial cells is inhibited by a modified antisense oligodeoxynucleotide to transforming growth factor β3. Proc Natl Acad Sci USA 88:1516–1520.

Qian SW, Burmester JK, Merwin JR, Madri JA, Sporn MB, Roberts AB (1992): Identification of a structural domain that distinguishes the actions of the type 1 and 2 isoforms of transforming growth factor β on endothelial cells. Proc Natl Acad Sci USA 89:6290–6294.

Rappolee DA, Brenner CA, Schultz R, Mark D, Werb Z (1988): Developmental expression of PDGF, TGF-α, and TGF-β genes in preimplantation mouse embryos. Science 241:1823–1825.

Rapraeger A (1989): Transforming growth factor (type β) promotes the addition of chondroitin sulfate chains to the cell surface proteoglycan (syndecan) of mouse mammary epithelia. J Cell Biol 109:2509–2518.

Robert B, Sassoon D, Jacq B, Gehring W, Buckingham M (1989): Hox-7, a mouse homeobox gene with a novel pattern of expression during embryogenesis. EMBO J 8:91–100.

Roberts AB, Anzano MA, Sporn MB (1983): Purification and properties of a type β transforming growth factor from bovine kidney. Biochemistry 22:5692–5698.

Roberts AB, Sporn MB (1990): The transforming growth factor βs. In Sporn MB & Roberts AB, (eds.): Peptide Growth Factors and Their Receptors—Handbook of Experimental Pharmacology. Heidelberg: Springer-Verlag, pp 419–472.

Robinson SD, Silberstein GB, Roberts AB, Flanders KC, Daniel CW (1991): Regulated expression and growth inhibitory effects of transforming growth factor-β isoforms in mouse mammary gland development. Development 113:867–878.

Rosa F, Roberts AB, Danielpour D, Dart LL, Sporn MB, Dawid IB (1988): Mesoderm induction in amphibians: the role of TGF-β2-like factors. Science 239:783–785.

Rosen DM, Stempien SA, Thompson AY, Seyedin SM (1988): Transforming growth factor-β modulates the expression of osteoblast and chondroblast phenotypes in vitro. J Cell Physiol 134:337–346.

Runyan RB, Potts JD, Weeks DL (1992): TGF-β3-mediated tissue interaction during embryonic heart development. Mol Reprod Dev 32:152–159.

Saksela O, Moscatelli D, Rifkin DB (1987): The opposing effects of basic fibroblast growth factor and transforming growth factor β on the regulation of plasminogen activator activity in capillary endothelial cells. J Cell Biol 105:957–963.

Salminen A, Braun T, Buchberger A, Jurs S, Winter B, Arnold H-H (1991): Transcription of the muscle regulatory gene MYF4 is regulated by serum components, peptide growth factors and signaling pathways involving G proteins. J Cell Biol 115:905–917.

Sandberg M, Vuorio T, Hirvonen H, Alitalo K, Vuorio E (1988): Enhanced expression of TGF-β and *c-fos* mRNA in the growth plates of developing human long bones. Development 102:461–470.

Sato Y, Rifkin DB (1989): Inhibition of endothelial cell movement by pericytes and smooth muscle cells: activation of a latent transforming growth factor-β1-like molecule by plasmin during co-culture. J Cell Biol 109:309–315.

Schlunegger MP, Grutter MG (1992): An unusual feature revealed by the crystal structure at 2.2Å resolution of human transforming growth factor-β2. Nature 358:430–434.

Seyedin SM, Thomas TC, Thompson AY, Rosen DM, Piez KA (1985): Purification and characterization of two cartilage-inducing factors from bovine demineralized bone. Proc Natl Acad Sci USA 82:2267–2271.

Shull MM, Ormsby I, Kier AB, Pawlowski S, Diebold RJ, Yin M, Allen R, Sidman C, Proetzel G, Calvin D, Annunziata N, Doetschman T (1992): Targeted disruption of the mouse transforming growth factor β-1 gene results in multifocal inflammatory disease. Nature 359:693–699.

Silberstein GB, Daniel CW (1987): Reversible inhibition of mammary gland growth by transforming growth factor-β. Science 237:291–293.

Slager HG, van Inzen W, van den Eijnden-van Raaij AJM, Mummery CL (1993): Retinoic acid concentration-dependent differentiation of aggregated embryonic stem cells: effects of TGF-β isoforms on muscle formation. Dev Genet (in press).

Strueli CH, Schmidhauser C, Kobrin M, Bissell MJ, Derynck R (1993): Extracellular matrix regulates expression of the TGF-β1 gene. J Cell Biol 120:253–269.

ten Dijke P, Iwata KK, Goddard C, Pieler C, Canalis E, McCarthy TL, Centrella M (1990): Recombinant TGF type β3: biological activities and receptor binding properties in isolated bone cells. Mol Cell Biol 10:4473–4479.

Tsuji T, Okada F, Yamaguchi K, Nakamura T (1990): Molecular cloning of the large subunit of transforming growth factor type β masking protein and expression of the mRNA in various tissues. Proc Natl Acad Sci USA 87:8835–8839.

Vaahtokari A, Vainio S, Thesleff I (1991): Associations between transforming growth factor β1 RNA expression and epithelial-mesenchymal interactions during tooth development. Development 113:985–994.

Vaidya TB, Rhodes SJ, Taparowsky EJ, Konieczny SF (1989): Fibroblast growth factor and transforming growth factor β repress transcription of the myogenic regulatory gene MyoD1. Mol Cell Biol 9:3576–3579.

Wang EA, Rosen V, Cordes P, Hewick RM, Kriz MJ, Luxenberg DP, Sibley BS, Wozney JM (1988): Purification and characterization of other distinct bone-inducing factors. Proc Natl Acad Sci USA 85:9484–9488.

Wang XF, Lin HY, Ng EE, Downward J, Lodish HF, Weinberg RA (1991): Expression cloning and characterisation of the TGF-β type III receptor. Cell 67:797–805.

Weinberg WC, Brown PD, Stetler-Stevenson WG, Yuspa SH (1990): Growth factors specifically alter hair follicle proliferation and collagenolytic activity alone or in combination. Differentiation 45:168–178.

Wilcox JN, Derynck R (1988): Developmental expression of transforming growth factors α and β in the mouse fetus. Mol Cell Biol 8:3415–3422.

Wrana JL, Attisano L, Carcamo J, Zentrella A, Doody J, Laiho M, Wang X-F, Massagué J (1992): TGFβ signals through a heteromeric protein kinase receptor complex. Cell 71:1003–1014.

Yamaguchi Y, Mann DM, Ruoslahti E (1990): Negative regulation of transforming growth factor-β by the proteoglycan decorin. Nature 346:281–284.

Yang EY, Moses HL (1990): Transforming growth factor β1-induced changes in cell migration, proliferation, and angiogenesis in the chicken chorioallantoic membrane. J Cell Biol 111:731–741.

Zentrella A, Weis FMB, Ralph DA, Laiho M, Massagué J (1991): Early gene responses to transforming growth factor-β in cells lacking growth-suppressive RB function. Mol Cell Biol 11:4952–4958.

Zentrella A, Massagué J (1992): Transforming growth factor β induces myoblast differentiation in the presence of mitogens. Proc Natl Acad Sci USA 89:5176–5180.

ABOUT THE AUTHOR

ROSEMARY J. AKHURST is a Reader in the Department of Medical Genetics at Glasgow University, Scotland, where she organizes an undergraduate course in medical genetics and teaches some aspects of developmental molecular biology to undergraduate students. After receiving a B.Sc. in biochemistry from Imperial College of Science and Technology, London University, in 1978, she obtained a Ph.D. in 1981 for her work at the Beatson Institute for Cancer Research, in Scotland, on the molecular biology of hepatic growth and neoplasia. This was followed by postdoctoral research at the California Institute of Technology where she studied actin gene structure and regulation during early sea-urchin development under the supervision of Dr. Eric Davidson. She was a lecturer in the Department of Biochemistry and Molecular Genetics in St. Mary's Hospital Medical School for four years, before moving to Glasgow in 1988. Dr. Akhurst's research has involved investigations of the molecular and cellular regulation of development and differentiation. In particular she has interests in control of homeostasis of the epidermis and in myogenesis, in particular cardiomyogenesis. More recently her laboratory has focused on the family of molecules encoding transforming growth factors beta and their receptors, and the role that they play in developmental processes.

Growth Factors and Signal Transduction in Development: 123–138

The Ras Signal Transduction Pathway in Development

Richard T. Hamilton

I. INTRODUCTION

Ras is the prototypical member of a superfamily of small G-proteins that have mainly been identified by virtue of their sequence similarity to the protein encoded by the *ras* protooncogene. So far, about 30 proteins related to Ras have been identified.

There are four major groups of monomeric, eukaryotic G-proteins: Ras, Rho, Rab, and Arf. They mediate regulation of three different aspects of cellular function:

1. The Ras proteins mediate signals for cell growth and differentiation [Downward, 1990; Hall, 1990; Haubruck and McCormick, 1991]
2. The Rho proteins regulate organization of the cytoskeletal network [Ridley and Hall, 1992; Ridley et al., 1992].
3. The Rab and Arf proteins mediate intracellular trafficking (movement of vesicles from the endoplasmic reticulum to the Golgi apparatus) and secretion [Balch, 1990; Serafini et al., 1991] as well as endocytosis [Bucci et al., 1992; van der Sluijs et al., 1992]

The small G-proteins are activated by binding of guanosine triphosphate (GTP). They are inactive when bound to guanosine diphosphate (GDP). Therefore, the G-protein is involved in a cycle of activation (GTP-bound) and inactivation (GDP-bound). Regulation of the kinetic properties of this cycle determines the ratio of active to inactive G-protein. The enzymatic activity associated with G-proteins is the hydrolysis of GTP. The guanosine triphosphatase (GTPase) activity of G-proteins determines the ratio of GTP/GDP-bound protein and therefore determines the proportion of active G-protein. The intrinsic GTPase activity of the small G-proteins is stimulated by the GTPase activating protein (GAP). GAP decreases the activity of G-proteins by stimulating GTP hydrolysis, therefore promoting their conversion to the GDP-bound inactive state. However, GAP may also interact directly with GTP-bound G-proteins to stimulate a conformational change that is necessary for their activation. Upon cell stimulation, GAP is postulated either to be inhibited, and so allow activated Ras-GTP to accumulate, or to interact with Ras at its effector domain and convey a downstream signal from Ras. Guanine nucleotide releasing proteins (GNRP) or guanine nucleotide dissociation stimulators (GDS) release GDP from G-proteins allowing them to bind GTP and become activated. Guanine nucleotide dissociation inhibitors (GDI) inhibit exchange of guanine nucleotides on G-proteins, therefore having the opposite effect of the GNRPs or GDSs on G-protein function.

The use of GTP and its hydrolysis as a molecular switch for enzyme activity is evolutionarily ancient. The best-known bacterial G-protein is EF-Tu, which drives incorporation of new amino acyl tRNAs onto the ribosome-mRNA complex for peptide elongation. The small G-proteins are highly conserved in sequence throughout evolution.

Their ability to hydrolyze GTP is also a conserved function. However, the signal transduction pathways with which they are associated vary with the organism and even between cell types in the same organism. Therefore, although the physical and enzymatic functions of these proteins are conserved through evolution, the biological functions are not conserved.

The GDP/GTP cycle, as it is known, is the primary control point in the inductive pathway [Satoh et al., 1992a; Lowy and Willumsen, 1993]. In this review, I describe what is known about the use of this fundamental molecular switch in the various developmental pathways where its use has been investigated. In particular, I review the prototypical *ras* protooncogene and its signal transduction pathway in development, with some reference to other small G-proteins. The greatest progress in tracing along the Ras signal transduction pathway in development has been made in those systems where genetics can be used.

II. INDUCTION IN THE DEVELOPING COMPOUND EYE OF *DROSOPHILA*

All multicellular organisms generate different cell types and arrange them into complex arrays during development. The molecular mechanisms of induction of cell fate have been intensely studied in the *Drosophila* retina because of the relative ease of scoring mutations in the photoreceptor cells [for reviews see, Ready, 1989; Rubin, 1989; Cagan and Zipursky, 1992]. The adult *Drosophila* retina is composed of about 800 facets or ommatidia; each ommatidium contains 8 photoreceptor neurons (R1-R8 cells) and is surrounded by 4 lens-secreting cone cells and a sheath of pigment cells. Genetic analyses have revealed that the fate of the cells within each ommatidium depends upon cell–cell interactions [Ready et al., 1976; Lawrence and Green, 1979; Reinke and Zipursky, 1988].

Of the eight cells in the ommatidium, the development of the R7 cell has been studied the most. The central R8 cell and a single neighboring uncommitted precursor cell communicate to specify the developmental fate of R7 [Van Vactor et al., 1991]. This process is mediated through Bride of sevenless (Boss) and Sevenless (Sev). *Sevenless* is a gene essential for the normal development of the R7 photoreceptor in particular. Boss is expressed on the surface of the R8 cell. It is the ligand for Sevenless [Krämer et al., 1991], which is a tyrosine kinase receptor expressed on the surface of five cells making contact with the developing R8 cell and on other cells not making contact with R8 [Rubin, 1989]. Upon binding to Sev, Boss is taken up by the R7 precursor cell, and the R7 inductive pathway is activated [Cagan et al., 1992]. Hart et al. [1993] recently found that Boss activates tyrosine phosphorylation of the Sev receptor; that the transmembrane domain of Boss, which contains seven transmembrane segments, is necessary for Boss's function; and that a soluble form of the extracellular domain of Boss acts as an antagonist of the Sev receptor *in vivo* and *in vitro*. The soluble form of the extracellular domain of Boss inhibited the interaction between Boss and Sev *in vivo* and *in vitro,* indicating that this domain of Boss interacts with Sev. This soluble form of Boss also inhibited aggregation of Sev expressing and Boss expressing cells and tyrosine phosphorylation of the Sev receptor. To inhibit these interactions, however, large concentrations of the soluble, partially purified receptor in the range of 40 to 100 $\mu g\ ml^{-1}$ were used. The need to use these high concentrations of impure receptor raises the possibility that the antagonistic effects were caused by an impurity in the preparations of soluble Boss, or that soluble Boss is a weak antagonist and that concentrations high enough to have regulatory effects are not likely to be reached during development. On the other hand, the ability of soluble Boss to inhibit the Boss-Sev interaction was tested *in vivo* in transgenic flies carrying one or two copies of soluble Boss expressed under the control of the hsp70 promoter. Two copies of the soluble Boss construct reduced the rescue of the *boss* phenotype in a *boss* rescue line (expressing reduced amounts of wild-type Boss) from about 76% to about 23% ommatidia containing R7 cells. Therefore, in contrast to soluble forms of other transmembrane ligands—for example, transforming growth factor type α (TGFα) [Massagué et al., 1990; Ankelesaria et al., 1990] and Steel [Copeland et al., 1990; Huang et al., 1990; Zsebo et al., 1990; Brannan, 1991] that bind with high affinity to receptor tyrosine kinases and stimulate tyrosine-kinase activity—the soluble form of Boss antagonizes binding to the Sev receptor and its stimulation of tyrosine kinase activity. If such a soluble form of Boss is released during development, it could act to counter the effects of Boss when Boss is no longer needed.

To determine the steps in the inductive pathway initiated by the Boss-Sev interaction, Simon et al. [1991] used flies expressing a temperature-sensitive mutant of Sev. They identified mutations at seven distinct loci named *Enhancers of sevenless (E(sev)),* which decrease Sev-dependent signaling. One of the *E(sev)* loci was the *Drosophila ras*1 gene. Another locus mapped to a gene, *son of sevenless (sos),* that had been shown before to affect Sev signalling [Rogge et al., 1991]. Simon et al. [1991] found that *sos* encodes a polypeptide with a centrally located domain related in sequence to the CDC25 gene product of *Saccharomyces cerevisiae.* CDC25 catalyzes the exchange of GDP for GTP on Ras, thereby leading to Ras activation. Genetic analysis suggests that the Sos protein acts downstream of Sev as a GNRP for Ras1 during *Drosophila* eye development. Fortini et al. [1992] found that by expressing an activated Ras1^{Val12} mutant protein only in those cells that normally express Sev, they were able to induce R7 differentiation in the absence of functional Sev or of Boss. In addition, by expressing Ras1^{Val12} in these cells, they induced nonneuronal cell precursors to form extra R7 cells.

Recently, two groups [Olivier et al., 1993; Simon et al., 1993b] have provided strong evidence for filling in one of the gaps in the Ras1 signal transduction pathway outlined above. That is, the coupling of the Sev tyrosine kinase to an activator of Ras guanine nucleotide exchange, Sos. In both studies, it was found that the product of the *E(sev)2B* gene is an SH3-SH2-SH3 protein with 63%–64% amino acid identity to the growth factor receptor bound 2 protein (Grb2) of mammalian cells [Lowenstein et al., 1992] and with 57%–61% identity to the Sem-5 protein of *C. elegans* [Clark et al., 1992; Chapter 9, Section IV, this volume]. On the basis of these findings, the *E(Sev)2B* gene was renamed *drk (downstream of receptor kinases).* Having identified *drk* as a gene required for Sev signalling, the two research groups then attempted to put Drk in its proper position in the Sev signalling pathway. As discussed above, other loci, such as *sos* and *ras1,* have been identified in genetic screens as being required for signalling downstream of *sev.* Both groups showed that Drk can bind to Sev and to Sos, *in vitro.* Olivier et al. [1993] showed that Drk bound to activated, autophosphorylated Sev and epidermal growth factor (EGF) receptors, and that it did so through its SH2 domain. They also showed that Drk binds to the carboxy-(C)-terminal tail of Sos. The C-terminal tail of Sos contains a proline-rich region, similar in sequence to the proline-rich region of the protein 3BP-1. Recent evidence has suggested that the SH3 domains of the tyrosine protein kinases Abl and Src can bind to a proline-rich motif in 3BP-1 [Cicchetti et al., 1992; Ren et al., 1993]. Therefore, Olivier et al. [1993] suggest that the SH3 regions of Drk bind to the proline-rich motif in the C-terminal tail of Sos and through this interaction activate the GDP-GTP exchange activity of Sos on Ras1. As these authors point out, "given the evolutionary conservation of tyrosine kinases, the Drk/Sem-5/Grb2 family, Sos, and Ras, it seems probable that the pathway proposed from the above studies will prove to be a general mechanism by which receptor tyrosine kinases communicate with Ras proteins." In a collection of papers, biochemical experiments are described in which the various interactions between receptor tyrosine kinases, Drk/Sem-5/Grb2, Sos, and Ras are further elaborated and corroborated [Egan et al., 1993; Li et al., 1993; Gale et al.,

1993; Rozakis-Adcock, 1993; Buday and Downward, 1993; Chardin et al., 1993]. These experiments done with mammalian cells are described in Section V.

Two other proteins related to Ras seem to be involved in *Drosophila* eye development. The first, Ras2, is 50% homologous to Ras1. The second is a *Drosophila* homologue of the human Rap1 protein [Hariharan et al., 1991]. Fortini et al. [1992] reported that an activated *ras*2 gene induces a different, abnormal phenotype of the eye than *ras*1 and that it is unable to replace *sev.* The dominant eye mutation *roughened* maps to a point mutation in the Drosophila *rap*1 gene [Hariharan et al., 1991]. This mutation, which prevents R7 development in roughly half of the ommatidia, may counteract the function of Ras1 in the developing R7 cell.

Genetic studies have revealed another member of the Sevenless signal transduction pathway further downstream of Ras1 [Dickson et al., 1992]. This is the *Drosophila* homologue of the serine/threonine protein kinase, Raf-1. Several studies in nondevelopmental systems had placed Raf-1 downstream of Ras [Morrison et al., 1988; Kolch et al., 1991; Wood et al., 1992; Williams et al., 1992; discussed further in Section V.D). In the torso system, it was clear that Raf-1 was essential for signalling [Ambrosio et al., 1989], and in the sevenless system it is clear that Ras is similarly essential. Now Dickson et al. [1992] and Doyle and Bishop [1993] have shown that Ras lies in the signal pathway to both systems.

Two other genes that participate in the intracellular transmission of the sevenless signal specifying R7 cell fate in *Drosophila* have been identified and molecularly characterized. They are *gap1* [Gaul et al., 1992] and *sina* [Carthew and Rubin, 1990]. *Gap1* encodes a protein similar to mammalian RasGAP. Loss-of-function mutations in *gap1* mimic constitutive activation of the Sev receptor tyrosine kinase and eliminate the requirement for a functional Sev protein in the R7 cell [Gaul et al., 1992]. Whether Gap1 acts as a negative regulator of Ras activity or as an effector of Ras activity in this pathway is unresolved. Dickson et al. [1992] provide genetic evidence that Raf acts downstream of Ras1 and upstream of Sina in this transduction cascade. Sina is located in the nucleus, which raises the possibility that it is the nuclear target of the sevenless pathway.

III. THE PARTICIPATION OF RAS IN DIFFERENTIATION OF THE TERMINAL, NONSEGMENTED REGIONS OF THE *DROSOPHILA* EMBRYO

The signal transduction pathway regulated by *torso,* which encodes another *Drosophila* receptor tyrosine kinase that is required for embryonic development, is virtually identical to the sevenless pathway.

The initial specification of cell fate in the *Drosophila* embryo depends on four separate systems of maternally expressed genes. Three of these systems, anterior, posterior, and terminal, are required for the anteroposterior axis, and the other is necessary for the dorsoventral axis [Nüsslein-Volhard et al., 1987; Stein and Stevens, 1991; St. Johnston and Nüsslein-Volhard, 1992]. Within each system are controlling genes specifying the differentiation of specific regions of the embryo. The controlling gene in the terminal system is *torso,* which encodes a transmembrane, tyrosine kinase receptor located around the circumference of the embryo. The terminal system specifies the nonsegmented termini of the embryo, called the acron and telson. Loss of function in *torso* or any of six other maternal genes results in deletions of structures from both ends of the embryo. Three genes of the terminal class have been cloned. They are *tor, l(1)ph,* and *csw;* they are homologous to other genes encoding signalling proteins. The *tor* gene encodes a transmembrane protein with molecular weight of about 105,000. The Tor protein has a tyrosine kinase domain similar to that of the platelet-derived growth factor receptor (PDGFR) tyrosine kinase domain [Casanova and Struhl, 1989; Sprenger et al., 1989]. The *l(1)ph* gene encodes the *Drosophi-*

la Raf-1 homologue, a serine/threonine kinase [Mark et al., 1987; Nishida et al., 1988]. The *csw* gene encodes a nonreceptor tyrosine phosphatase homologous to the mammalian PTP1C protein [Perkins et al., 1992; discussed further by Ingebritsen et al., Chapter 9, this volume]. D-Raf-1 and the product of the *csw* gene seem to act downstream of Tor in that, to be expressed, gain-of-function mutations in the *tor* gene require D-Raf-1 and Csw activities [Ambrosio et al., 1989; Perkins et al., 1992]. Other products in the Tor pathway have since been identified. Doyle and Bishop [1993] carried out a mutagenesis screen for dominant suppressors of a *tor* gain-of-function allele (*Su(tor)*). They characterized genetically more than 40 mutations of this type. Two of the 40 correspond to mutations in *ras*-1 and *sos*. Lu et al. [1993] used a different approach to demonstrate that Ras and Sos were members of the Tor signal transduction pathway leading to a terminal-class phenotype. They injected mammalian $p21^{v\text{-}ras}$ variants into early *Drosophila* embryos and found that the maternal-effect phenotypes of *tor* and *csw* null mutations were rescued. Moreover, they found that embryos derived from germ cells lacking Sos activity have a terminal-class phenotype. In addition to demonstrating that the SH3-SH2-SH3 protein, Drk, is a member of the Sevenless pathway [Simon et al., 1993b], Simon et al. [1993a] identified the product of the *crkle* gene in the Tor pathway as being equivalent to Drk. Thus, the identified members of both Sev and Tor pathways are virtually identical and presumably act in the same sequence.

A direct connection between the cytoplasm and the nucleus in the Torso receptor–mediated pathway was suggested by the results of Ronchi et al. [1993]. The anterior body pattern in *Drosophila* is specified by the graded distribution of the bicoid protein, which activates subordinate genes in distinct anterior domains. Ronchi et al. [1993] constructed an artificial promoter consisting of three high-affinity Bicoid binding sites and a naive transcription site that could be transcriptionally activated under *bicoid (bcd)* gene control and then transcriptionally repressed under *tor* gene control. They showed that repression of transcription depends on *D-raf* and is associated with phosphorylation of Bcd protein. Neither *tailless* nor *huckelbein* were required for repression; both genes were previously thought to constitute the only gene targets of the *tor* signalling system. These results suggest that Bcd is phosphorylated by a protein kinase in the Torso pathway to alter its transcriptional regulation of several subordinate regulatory genes in the zygote. Thus, a direct linkage between signal transduction machinery and nuclear activity is suggested; but, as with the sevenless pathway, the components linking the pathway to the nuclear switch have yet to be identified.

IV. THE PARTICIPATION OF RAS IN GONADAL DEVELOPMENT OF *C. ELEGANS*

Another popular cell induction system in which one set of cells specifies the fate of another set of cells is the vulval development of *Caenorrhabditis elegans* hermaphrodites [Horvitz and Sternberg, 1991; Tuck and Greenwald, Chapter 9, this volume]. Normally, in these nematodes, the vulva forms from a subset of six tripotential cells called the vulval precursor cells (VPCs) that are located within the ventral epidermis. Each VPC can have an epidermal fate or either of two vulval fates, in each of which it produces seven or eight vulval progeny. Normally, the three VPCs closer to the anchor cell of the somatic gonad have vulval fates and the three VPCs farther from the anchor cell have epidermal fates. Destroying the anchor cell early in development leads to all six VPCs having an epidermal fate, a result which indicates that the vulval fates are induced by a signal from the anchor cell. Hill and Sternberg [1992] have provided genetic evidence that the *lin-3* gene of the anchor cell encodes the vulval-inducing signal, which is a molecule whose sequence is similar to EGF and TGFα. The *lin-3* gene product acts through the EGF receptor (EGFR) homolog Let-23.

Genetic studies of gonadal development in the nematode *C. elegans* have further revealed an important connection between receptor tyrosine kinases and Ras molecules [Horvitz and Sternberg, 1991]. The *let-23* gene, which encodes a receptor tyrosine kinase of the EGFR family, is required for induction of vulval cells from precursor cells. Another gene essential for vulval induction is *let-60,* a nematode *ras* homolog that has gain-of-function mutations similar to mutations that activate mammalian *ras* genes [Horvitz and Sternberg, 1991]. A potential intermediary between the products of the *let-23* and *let-60* genes is the product of the *sem-5* gene: worms lacking an inhibitory signal normally generated by cells surrounding the vulval precursor cells have been found to exhibit a multivulval phenotype that is suppressed by loss-of-function mutations in *let-23, let-60,* and *sem-5* [Clark et al., 1992]. The *sem-5* gene encodes a protein that has two SH3 domains and one SH2 domain, in the order SH3-SH2-SH3 [Clark et al., 1992]. The Sem-5 protein is 228 amino acids long, which leaves little room for other domains; therefore, it seems that Sem-5 is an intermediary protein consisting entirely of domains thought to mediate protein–protein interactions. Clark et al. [1992] speculated that Sem-5 binds to a phosphotyrosine-containing protein, possibly the Let-23 receptor tyrosine kinase, and regulates the function of the Let-60 Ras, perhaps by acting through a guanine nucleotide exchange protein. Evidence for that speculated sequence of reactions has since been obtained in mammalian cells and in *Drosophila* [Lowenstein et al., 1992; Simon et al., 1993b; Olivier et al., 1993]. Lowenstein et al. [1992] have cloned a mammalian homologue of Sem-5, which they have named Grb2. Simon et al. [1993b] and Olivier et al. [1993] have each cloned and expressed a *Drosophila* homolog of Sem-5, which they have designated Drk (downstream of receptor kinases). I further discuss Grb2 and Drk below, in Section V.

Han et al. [1993] have recently presented evidence for a step further along the vulval differentiation pathway. They have shown that the C. elegans *lin-45 (raf)* gene participates in the differentiation pathway that is stimulated by Ras, the product of the *let-60* gene. The evidence that they present suggests that, as in other tyrosine kinase signal transduction pathways, Raf acts downstream of Ras, or it may act in a separate pathway that is required for Ras function to specify vulval fates.

V. RAS SIGNAL TRANSDUCTION PATHWAYS IN VERTEBRATE CELLS

A. General

Much knowledge about the Ras GDP-GTP cycle and its signal transduction pathway has been obtained from studies done with mammalian cells; however, because of the relative difficulty of doing genetic studies in mammalian developmental systems compared with, for example *Drosophila,* much less is known of the Ras signal transduction pathways in mammalian developmental systems than is known in *Drosophila,* yeast, or *C. elegans.* Nevertheless, an almost complete Ras-initiated signal transduction pathway from cell membranes to nucleus has been traced out using mammalian cells in culture [reviewed by Marx, 1993]. The steps in the pathway have been remarkably conserved and coincide with those determined for invertebrates. As evidence of conservation, some of the components of the mammalian pathway function interchangeably in the invertebrate pathway and vice versa.

In fibroblastic cells in culture, evidence has been obtained that Ras acts downstream of various growth factors. For example, DNA synthesis induced by PDGF and EGF was inhibited by microinjection of an anti-Ras antibody [Mulcahy et al., 1985]; DNA synthesis or gene expression induced by EGF and insulin was blocked by dominant negative mutants of Ras [Cai et al., 1990; Medema et al., 1991]. When Ras-bound GDP–GTP ratios were determined in fibroblasts, it was found that at least four growth factors, PDGF, EGF, insulin, and macrophage colony stimulating factor (MCSF) can within 1 minute increase the amount of Ras-GTP, when the receptor is expressed on the cell surface [Satoh et al., 1990a,b; Gibbs et al., 1990; Burgering et al., 1991]. Although none of these studies were done in developmental systems, we can guess that, because growth factors are a central component regulating development, where growth factors are involved in development, so too will Ras.

A collection of recent papers describing studies done with mammalian cells and with the *Drosophila* system clarifies the picture of how receptors turn Ras on [Egan et al., 1993; Li et al., 1993; Gale et al., 1993; Rozakis-Adcock et al., 1993; Buday and Downward, 1993; Chardin et al., 1993; Olivier et al., 1993; Simon et al., 1993b; Baltensperger et al., 1993; Skolnik et al., 1993]. I discuss the studies done with mammalian cells in this section; those on *Drosophila* development are discussed above in Section II. The results of these studies lead to a scheme where receptor activation results in autophosphorylation, creating a binding site for Grb2, as described previously [Lowenstein et al., 1992]. Grb2 then associates with Sos1, recruiting it to the activated receptor in the plasma membrane where Ras activation is presumed to take place. As with the studies done in *Drosophila,* the sites of interaction between Grb2 and Sos1 were mapped. The *Drosophila* Sos and the two mouse Sos, mSos-1 and -2, each contain a 20-kD stretch of proline-rich sequence at their carboxyl termini through which they bind Drk and Sem5 *in vitro,* as described above [Simon et al., 1993b; Olivier et al., 1993; and *in vivo* [Egan et al., 1993; Li et al., 1993; Rozakis-Adcock et al., 1993; Buday and Downward, 1993]. In addition, the SH3 domains of human Sos (hSos) are required for interaction with Grb2 [Chardin et al., 1993]. In fibroblasts, EGF stimulates the formation of a ternary complex containing SOS-Sem5/Grb2-EGF tyrosine kinase receptor [Egan et al., 1993; Li et al., 1993; Rozakis-Adcock et al., 1993; Buday and Downward, 1993]: that is, Sos is brought into the vicinity of its target Ras at the plasma membrane by the adapter Sem5/Grb2. Cell activation does not alter the intrinsic GNRP activity of Sos, so that regulation of nucleotide exchange on Ras probably results from an increased amount of Sos brought into proximity with Ras [Buday and Downward, 1993]. Another adapter protein, SH2-containing transforming protein (Shc), has been implicated in this signalling pathway, because it associates with Sem5/Grb2 in stimulated cells [Rozakis-Adcock et al., 1993]. Shc is a protein containing an SH2 domain that is a target for many tyrosine kinases [Sun et al., 1991, Pelicci et al., 1992].

This scheme of interactions and reactions has been demonstrated for both the insulin receptor [Baltensperger et al., 1993; Skolnik et al., 1993] and the EGF receptor [Egan et al., 1993; Li et al., 1993; Gale et al., 1993; Rozakis-Adcock et al., 1993; Buday and Downward, 1993; Chardin et al., 1993]. Baltensperger et al. [1993] found, in COS cells, that complexes of the p85 regulatory subunit of phosphatidylinositol-3 kinase, or Grb2, with the insulin receptor substrate-1 (IRS-1), would bind to heterologously expressed Drosophila Sos. These results suggest that Grb2, p86, or other proteins with SH2-SH3 adapter sequences, could link Sos to IRS-1 signalling complexes as part of the mechanism by which insulin activates Ras. Skolnik et al. [1993] demonstrated in L6 myoblast cell lines that activation of mitogen-activated protein (MAP) kinases by insulin was enhanced by stable overexpression of Grb2 but not by overexpression of mutant Grb2 proteins with single amino acid changes in the SH2 and SH3 domains. In addition, a dominant negative form of Ras blocked activation of MAP kinases by insulin in cells overexpressing GRB2. A complex of Sos and GRB2 formed in cells overexpressing GRB2. Upon stimulation of cells by insulin, this complex bound to tyrosine-phosphorylated IRS-1 and Shc. Whereas the activated EGFR binds to the complex of Grb2 and Sos directly, the activated insulin receptor does not bind directly to the complex of Grb2 and Sos, but by phosphorylating IRS-1 causes IRS-1 and Shc to bind to the complex.

Most of these events occur at the membrane. Recent studies have provided evidence for extending the Ras pathway further along towards the nucleus [Moodie et al., 1993; Van Aelst et al., 1993]. Moodie et al. [1993] coupled recombinant c-Ha-Ras to silica beads activated by an N-hydroxyl succinimide group. The coupled Ras maintained its ability to bind guanine nucleotides and was used to probe rat brain cytosol for signalling proteins that specifically interacted with Ras-GTP. The coupled, activated Ras formed complexes with Raf-1 and mitogen-activated protein kinase kinase (MAPKK). Van Aelst et al. [1993] used a *Saccharomyces cerevisiae* genetic screen to detect interactions between RAS and RAF, and between RAS and Byr2. Byr2 is a protein kinase believed to be in the Ras1 pathway of *Schizosaccharomyces pombe*. They also found that the catalytic domain of RAF bound to MAPKK. RAS and MAPKK interacted, but only when RAF was overexpressed. Based on these results and results from many other different researchers, Raf-1 kinase is placed downstream of Ras and MAPKK directly downstream of Raf.

Recent strong evidence has been provided that c-Raf-1 is the immediate downstream target of Ras [Zhang et al., 1993; Hughes et al., 1993; Warne et al., 1993; Vojtek et al., 1993]. Zhang et al. [1993] found that Ras binds in a dose-dependent fashion to the amino-terminal regulatory domain of Raf-1 (residues 1–257). The Raf regulatory domain [Raf-1(1–257)] contains the c-Raf zinc-cysteine finger, and is a domain present in other signal transduction proteins, which the authors point out, "may also enable these proteins to interact directly with Ras or other Ras-like proteins." Interaction of Ras and the Raf regulatory domain [Raf-1(1–257)] took place *in vitro* and in a yeast expression system. Raf-1(1–257)] bound GTP-Ras in preference to GDP-Ras. Raf-1(1–257) competed with Ras-GAP for binding to the Ras effector domain. Consistent with this last finding, the binding of Raf(1–257) to c-Ha-Ras did not alter the kinetics of Ras GTPase activity, but inhibited the stimulation of Ras GTPase activity by Ras-GAP. Mutations in and around the Ras effector domain impaired Ras binding to Raf-1(1–257) and Ras transforming activity in parallel. The Ras carboxy-terminal domain, including the S-farnesyl moiety, did not seem to be essential for the Ras-Raf interaction. As further evidence of a Ras effector domain–Raf regulatory domain interaction, Rap1b was demonstrated to interact strongly with the Raf amino-terminal domain. Rap1b is a low molecular weight GTP-binding protein identical to Ras in the effector domain [Pizon et al., 1988].

Warne et al. [1993] also demonstrated a direct interaction of Ras and the amino-terminal region of Raf-1 *in vitro*. They found that the amino-terminal, cysteine-rich region of c-Raf-1, expressed as a glutathione-S-transferase fusion protein, binds to Ras with a dissociation constant of 50 nM. Only activated GTP-bound Ras bound Raf, and this interaction was inhibited by a peptide having the sequence of the effector domain of Ras. In agreement with Zhang et al. [1993], it was found that Raf competitively inhibited GAP and neurofibromin binding to Ras. In addition, Raf weakly stimulated the GTPase activity of Ras.

A fourth group, Vojtek et al. [1993], used a two-hybrid system screen of a mouse cDNA library to identify a conserved region of 81 residues at the N-terminus of Raf that interacts with Ras. They showed that Raf interacts with wild-type and activated Ras but not with an effector domain mutant of Ras or with a dominant-interfering Ras mutant. They also showed that Ras and the N-terminal region of Raf bind to each other *in vitro* and that this direct interaction depended on GTPs being bound to Ras.

In determining whether the MAPK pathways were conserved between vertebrates and *Schizosaccharomyces pombe,* Hughes et al. [1993] found that complementation of Byr1 in fission yeast by mammalian MAPKK required the coexpression of Raf kinase. Byr1 is 55% identical in amino acid sequence with mammalian MAPKK in the catalytic domain and 38% identical overall. Again, these results suggest that the pathways are similar in function and that Raf kinase directly phosphorylates and activates MAPKK.

B. Proliferation and Differentiation of Hemopoietic Cells

When various kinds of T-cell, B-cell, and mast cell lines were stimulated by lymphokines such as interleukin-2 (IL-2), IL-3, IL-5, granulocyte macrophage colony stimulating factor (GMCSF), and Steel factor, the amount of Ras-GTP increased [Satoh et al., 1991; Duronio et al., 1992; Graves et al., 1992; Heideran et al., 1992]. Erythropoietin also activates Ras in human erythroleukemic cells [Torti et al., 1992]. The role of tyrosine kinases in the formation of Ras-GTP was suggested by the finding in a mast cell line that herbimycin, an inhibitor of tyrosine kinases, blocks the formation of Ras-GTP by IL-3 and GMCSF [Satoh et al., 1992b].

C. Activation of T Lymphocytes

Downward et al. [1990] found an increase in the amount of Ras-GTP when they stimulated T lymphocytes with phytohemagglutinin or an anti-CD3 antibody that mimics activation by the antigen. Stimulation of CD2 also raised the level of Ras-GTP [Graves et al., 1991]. In human T cells, an inhibitor of tyrosine kinases blocked the increase of Ras-GTP by an anti-CD3 antibody. This finding suggests that tyrosine phosphorylation contributes to the activation of Ras in T cells as well as in mast cells [Izquierdo et al., 1992].

A new GNRP-containing pathway to Ras that is specific to the hematopoietic system was recently outlined by Gulbins et al. [1993]. Hematopoietic cells contain a protooncogene called *vav* [Katzav et al., 1989, 1991]. Vav is a 95-kD protein with two SH2 domains, two SH3 domains, a cysteine-rich region with similarity to protein kinase C, and a region highly similar to proteins with guanine nucleotide exchange activity on Ras-like GTPases [reviewed in Puil and Pawson, 1992; Hu et al., 1993]. Gulbins et al. [1993] found that human T-cell lysates or Vav immunoprecipitates contained GNRP activity that was higher after T-cell antigen receptor (TCR)-CD3 triggering. They also found that a Vav fragment containing the suspected GNRP domain was active. Recent work has shown that Vav is tyrosine phosphorylated in response to stimulation of surface membrane receptors in a variety of hematopoietic cell lines [reviewed in Hu et al., 1993]. Vav may function in hematopoietic cell signalling by coupling tyrosine kinase pathways to Ras-like GTPases through the regulation of guanine nucleotide exchange.

D. Neuronal Differentiation of Pheochromocytoma Cells

The pheochromocytoma cell line, PC12, differentiates to a neuronlike cell after nerve growth factor (NGF) is added to the culture medium. Several experiments have shown that Ras is involved in the signalling pathways that lead to neurite outgrowth in these cells. Constitutively active Ras can mimic the action of NGF [Noda et al., 1985]. Expression of oncogenic Ras in PC12 cells leads to activation of MAPKK and MAPK, hyperphosphorylation of Raf-1, and ultimately neurite outgrowth [Thomas et al., 1992; Wood et al., 1992]. An anti-Ras antibody, when injected into the cells, inhibits the neuronal differentiation triggered by NGF [Hagag et al., 1986]. A dominant negative mutant of Ras prevents NGF from inducing the differentiation of PC12 cells [Szeberenyi et al., 1990]. Furthermore, several groups have found that NGF can cause an increase in the amount of Ras-GTP in PC-12 cells [Qui and Green, 1991; Li et al., 1992; Nakafuku et al., 1992]. Two other differentiation factors of PC12 cells, FGF and IL-6, also stimulate Ras [Nakafuku et al., 1992]. As in so many other types of cells, tyrosine phosphorylation seems to be important in the activation of Ras; for example, several tyrosine kinase inhibitors block formation of Ras-GTP in PC12 cells, possibly by preventing tyrosine phosphorylation of the NGF receptor [Qui and Green, 1991; Nakafuku et al., 1992].

Another member of the Ras pathway in mammalian neurons has been identified. This is the brain-specific p140 Ras-GNRP (also termed $CDC25^{Mm}$). It was cloned from cDNA libraries by exploiting either its sequence [Shou et al., 1992; Wei et al., 1992] or its functional similarity to CDC25 [Martegani et al., 1992], the yeast GNRP, which was the first GNRP to be identified [Broek et al., 1987; Robinson et al., 1987]. The ligand, receptor, and adaptor protein for p140 Ras-GRF have not yet been identified, so the members of this Ras pathway are still in the early stages of being resolved.

E. Downstream Targets of Ras

The functioning of Ras signal transduction pathways that control differentiation requires GAP, which interacts with Ras and stimulates the GTPase activity of Ras. Therefore, it was thought that GAP acts to downregulate Ras. But recent evidence is in favor of GAP being an effector of Ras, at least in *Xenopus*. In the *Xenopus* oocyte, germinal vesicle breakdown (GVBD) caused by insulin-like growth factor or insulin requires the participation of Ras; GVBD caused by progesterone, however, does not need Ras. Duchesne et al. [1993] have mapped regions of GAP necessary for Ras-GAP-mediated signal transduction and find that its SH3 domain is essential. A monoclonal antibody to this region blocked GVBD induced by the oncogenic protein $Ha\text{-}Ras^{Lys12}$ in *Xenopus* oocytes. The monoclonal antibody, which recognizes the peptide corresponding to amino acids 275 to 351 within the amino-terminal domain of GAP, did not inhibit its Ha-Ras GTPase activity. They found that when they injected into the oocytes peptides corresponding to amino acids 275 to 351, and 317 to 326 of GAP, GVBD induced by insulin or by $Ha\text{-}Ras^{Lys12}$ was blocked, but not that induced by progesterone. These results provide the strongest evidence so far for GAP acting as an effector of Ras. They also provide evidence that SH3 domains can regulate cell signal transduction. These results are an important step in understanding the signal transduction pathway or pathways through Ras and beyond. Because the SH3 domain of p120-GAP is believed to be involved in cytoskeletal interaction, one Ras-initiated pathway may lead directly to the cytoskeleton and to cytoskeleton changes. On the other hand, p120-GAP contains an SH2 domain, which is responsible for interaction with tyrosine phosphoproteins, and it interacts with several proteins containing phosphotyrosines [Pawson, 1988]. Two of these tyrosine phosphoproteins are of unknown function and are referred to as p190 and p62 [Ellis et al., 1990]. Wong et al. [1992] have cloned the GAP-associated tyrosine phosphoprotein, p62. It has extensive sequence similarity to a putative hnRNP protein, GRP33. Recombinant human p62 purified from insect Sf9 cells binds to DNA and to mRNA. Like many proteins involved in mRNA processing, recombinant p62 is modified by methylation on arginine residues. Also, p62 binds tightly to $p21^{ras}$GAP *in vitro*. The binding to GAP depends on phosphorylation of tyrosine residues on p62 and occurs through SH2 groups of GAP. These findings suggest a pathway from $p21^{ras}$ through GAP to the nucleus. An even more direct route to the nucleus is suggested by the finding that the other tyrosine phosphoprotein interacting with the SH2 domain of GAP, p190, has a central region whose sequence corresponds to that of a transcriptional repressor of the glucocorticoid receptor gene [Settleman et al., 1992]. Thus, GAP might serve as a bifurcation point in the ras signal transduction pathway; the route taken could depend on the ancillary proteins, present and active, that interact with either the SH2 or SH3 domains of GAP. The interactions of GAP with p62 and with p190 have not yet been established in a developmental pathway. The arguments for and against GAP and/or neurofibromin being a $p21^{ras}$ effector that would lead to the activation of Raf have been summarized in a recent review by Marshall [1993].

Recently, MacNicol et al. [1993] reported that Raf-1 kinase is essential for early *Xeno-*

pus development and mediates the induction of mesoderm by FGF. As reported elsewhere in this review, Raf-1 acts downstream of Ras in mammalian tissue culture cells, in *Drosophila* eye development, and in nematode gonad development. Therefore, it is most likely that Raf-1 acts downstream of Ras in *Xenopus* mesoderm induction by FGF. MacNicol et al. [1993] found that when they injected a dominant negative *raf*-1 mutant into an animal cap explants from *Xenopus* embryos there was a complete blockage of mesoderm induction stimulated by basic FGF. Injection of the mutant mRNA into embryos at the 2-cell stage blocked normal development during neurula stages and caused severe posterior truncations in tadpoles. This mutant phenotype specifically blocked the activation of endogenous Raf kinase activity. The posterior defect caused by the mutant *raf*-1 is similar to the defect observed in *Drosophila raf*-1 mutants. In *Drosophila,* the absence of Raf-1 leads to normal blastula development, but the posterior three abdominal segments are truncated and the telson is missing (see Section II above). As discussed above, Raf-1 mediates the action of the Torso tyrosine kinase receptor, which is related to the PDGF and FGF receptors [Sprenger et al., 1989]. The *Xenopus* and *Drosophila* FGF receptor signal transduction pathways, which result in similar phenotypes, are most likely very similar. This remarkable conservation will no doubt guide the studies done with mammalian developmental systems.

F. The Superfamily of Ras Proteins

Two of the members of the ras superfamily deserve special mention; they are Rac and Rho. Recent findings published in two important papers show that Rho regulates the assembly of focal adhesions and actin stress fibers in response to growth factors and that Rac regulates growth factor–induced membrane ruffling [Ridley and Hall, 1992; Ridley et al., 1992]. These studies were done with Swiss 3T3 cells, so they do not directly relate to development. Nevertheless, a role for the actin cytoskeleton has been implicated in many cellular functions, such as motility, chemotaxis, cell division, endocytosis, and secretion, all of which are involved in development. Therefore, it is likely that the Rac and Rho pathways, activated by growth factors and leading to changes in structure and activity of the actin cytoskeleton, will be found to be essential elements in developmental pathways.

VI. CONCLUSIONS AND FUTURE PROSPECTS

From a combination of genetic and biochemical studies, a complete route from receptor tyrosine kinases through Ras to the activation of genes in the nucleus has almost been established (Fig. 1). This pathway consists of the following steps: the appropriate ligand external to the cell activates its specific tyrosine protein kinase receptor so that the receptor autophosphorylates on tyrosine. The SH2 group of the adaptor protein, Sem5/Drk/Grb2, binds specifically to the tyrosine-phosphate domain of the receptor. The Shc protein interacts with Sem5/Drk/Grb2 to possibly modify its interaction with the receptor or with the GNRP, Sos. Sem5/Drk/Grb2 binds through its two SH3 domains to the proline-rich C-terminal region of Sos. This last interaction brings Sos into proximity with GDP-Ras at the inner surface of the plasma membrane. Sos activates Ras by promoting the exchange of GTP for GDP. Activated Ras may then use GAP as an effector of its action on Raf kinase. Raf kinase is postulated to be a MAPKKK, although that is still open to question. If it is, then once MAPKK becomes phosphorylated and activated by Raf, MAPKK phosphorylates and activates MAPK. What happens from there has not been resolved, at least at the time of the writing of this review.

This Ras-initiated pathway has been amazingly conserved: almost any member of the pathway from any species in which it has been studied will substitute for its counterpart in another species. Those studying developmental systems in which genetics can be readily

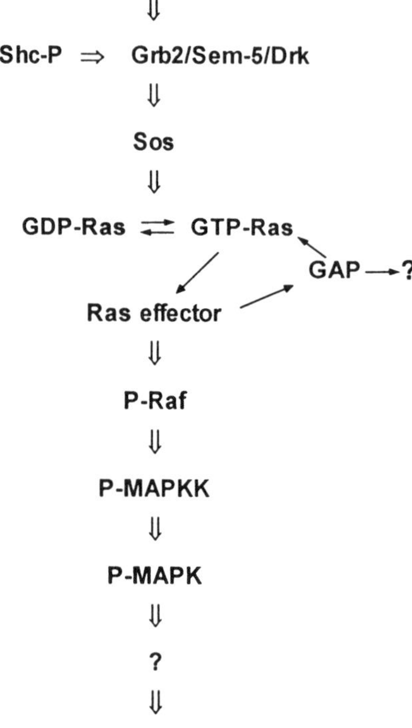

Fig. 1. *See text for the details of this pathway. Arrows point in the direction of the flow of information or signify an interaction between two molecules. It has not been resolved whether GAP downregulates Ras or whether GAP acts as an effector of Ras. P = phosphoryl group.*

applied have collaborated with those studying systems in which a biochemical approach has been used. Out of these collaborations has come a burst of papers outlining what appear to be most of the steps in the Ras signal transduction pathway. Future studies will rapidly fill in the gaps and consolidate the pathway already established. They will also provide information concerning the basis for specificity, which is not yet clear.

ACKNOWLEDGMENTS

An NIH grant (HD 29087) supports the work done in the author's laboratory. I thank Paul Sternberg for generously providing information about work done in his laboratory.

REFERENCES

Ambrosio L, Mahowald AP, Perrimon N (1989): Requirement of the *Drosophila* raf homologue for torso function. Nature 342:288–291.

Ankelesaria P, Teixidó J, Laiho M, Pierce JH, Greenberger JS, Massagué J (1990): Cell-cell adhesian mediated by binding of membrane-anchored transforming growth factor α to epidermal growth factor receptors promotes cell proliferation. Proc Natl Acad Sci USA 87:3289–3293.

Balch WE (1990): Small GTP-binding proteins in vesicular transport. Trends Biochem Sci 15:473–477.

Baltensperger K, Kozma LM, Cherniak AD, Klarlund JK, Chawla A, Banerjee U, Czech MP (1993): Binding of the Ras activator Son of Sevenless to insulin receptor substrate-1 signaling complexes. Science 260:1950–1952.

Brannan CI, Lyman SD, Williams DE, Eisenman J, Anderson DM, Cosman D, Bedell MA, Jenkins NA, Copeland NG (1991): Steel-Dickie mutation encodes a c-Kit ligand lacking transmembrane and cytoplasmic domains. Proc Natl Acad Sci USA 88:4671–4674.

Broek D, Toda T, Michaeli T, Levin L, Birchmeier C, Zoller M, Powers S, Wigler M (1987): The S. cerevisiae *CDC25* gene product regulates the *RAS*/adenylate cyclase pathway. Cell 48:789–799.

Bucci C, Parton RG, Mather IH, Stunnenberg H, Simons K, Hoflack B, and Zerial M (1992): The small GTPase rab5 functions as a regulatory factor in the early endocytic pathway. Cell 70:715–728.

Buday L, Downward J (1993): Epidermal growth factor regulates p21ras through the formation of a complex of receptor, Grb2 adapter protein, and Sos nucleotide exchange factor. Cell 73:611–620.

Burgering BMT, Medema RH, Maassen JA, van de Wetering M, van der Eb AJ, McCormick F, Bos JL (1991): Insulin stimulation of gene expression mediated by p21ras activation. EMBO J 10:1103–1109.

Cagan RL, Krämer H, Hart AC, Zipursky SL (1992): The Bride of Sevenless and Sevenless interaction: internalization of a transmembrane ligand. Cell 69:393–399.

Cagan RL, Zipursky SL (1992): Cell choice and patterning in the Drosophila retina. In Shankland M, Macagno E (eds): Determinants of Neuronal Identity. San Diego: Academic Press, pp 101–135.

Cai H, Szeberenyi J, Cooper GM (1990): Effect of a dominant inhibitory Ha-ras mutation on mitogenic signal transduction in NIH 3T3 cells. Mol Cell Biol 10:5314–5323.

Carthew RW, Rubin GM (1990): *seven in absentia,* a gene required for specification of R_7 cell fate in the Drosophila eye. Cell 63:561–577.

Casanova J, Struhl G (1989): Localized surface activity of torso, a receptor tyrosine kinase, specifies terminal body pattern in *Drosophila.* Genes Dev 3:2025–2038.

Chardin P, Camonis JH, Gale NW, Van Aelst L, Schlessinger J, Wigler MH, Bar-Sagi D (1993): Human Sos 1: a guanine nucleotide exchange factor for Ras that binds to Grb2. Science 260:1338–1343.

Cicchetti P, Mayer BJ, Thiel G, Baltimore D (1992): Identification of a protein that binds to the SH3 region of ABI and is similar to Bcr and GAP-rho. Science 257:803–806.

Clark SG, Stern MJ, Horvitz HR (1992): C. Elegans cell-signalling gene sem-5 encodes a protein with SH2 and SH3 domains. Nature 356:340–344.

Copeland NG, Gilbert DJ, Cho BC, Donovan PJ, Jenkins NA, Cosman D, Anderson D, Lyman SD, Williams DE (1990): Mast cell growth factor maps near the *steel* locus on mouse chromosome 10 and is deleted in a number of *steel* alleles. Cell 63:175–183.

Dickson B, Sprenger F, Morrison D, Hafen E (1992): Raf functions downstream of Ras1 in the sevenless signal transduction pathway. Nature 360:600–602.

Downward J, Graves JD, Warne PH, Rayter S, Cantrell DA (1990): Stimulation of $p21^{ras}$ upon T-cell activation. Nature 346:719–723.

Doyle HJ, Bishop JM (1993): Torso, a receptor kinase required for embryonic pattern formation, shares substrates with the Sevenless and EGF-R pathways in *Drosophila.* Genes Dev 7:633–646.

Duchesne M, Schweighoffer F, Parker F, Clerc F, Frobert Y, Thang MN, Tocqué B (1993): Identification of the SH3 domain of GAP as an essential sequence for Ras-GAP-mediated signaling. Science 259:525–528.

Duronio V, Welham MJ, Abraham S, Dryden P, Schrader JW (1992): $p21^{ras}$ activation via hemopoietin receptors and c-kit requires tyrosine kinase activity but not tyrosine phosphorylation of $p21^{ras}$ GTPase-activating protein. Proc Natl Acad Sci USA 89:1587–1591.

Egan SE, Giddings BW, Brooks MW, Buday L, Sizeland AM, Weinberg, RA (1993): Association of Sos Ras exchange protein with Grb2 is implicated in tyrosine kinase signal transduction and transformation. Nature 363:45–51.

Ellis C, Moran M, McCormick F, Pawson T (1990): Phosphorylation of GAP and GAP-associated proteins by transforming and mitogenic tyrosine kinases. Nature 343:377–381.

Fortini ME, Simon MA, Rubin GM (1992): Signaling by the *sevenless* protein tyrosine kinase is mimicked by Ras1 activation. Nature 355:559–561.

Gale NW, Kaplan S, Lowenstein EJ, Schlessinger J, Bar-Sagi D (1993): Grb2 mediates the EGF-dependent activation of guanine nucleotide exchange on Ras. Nature 363:88–92.

Gaul U, Mardon G, Rubin GM (1992): A putative Ras GTPase activating protein acts as a negative regulator of signaling by the sevenless receptor tyrosine kinase. Cell 68:1007–1019.

Gibbs JB, Marshall MS, Scolnick EM, Dixon RAF, Vogel US (1990): Modulation of guanine nucleotides bound to ras in NIH 3T3 cells by oncogene, growth factors, and the GTPase activating protein (GAP). J Biol Chem 265:20437–20442.

Graves JD, Downward J, Rayter S, Warne P, Tutt AL, Glennie M, Cantrell DA (1991): CD2 antigen mediated activation of the guanine nucleotide binding proteins $p21^{ras}$ in human T lymphocytes. J Immunol 146:3709–3712.

Graves JD, Downward J, Izquierdo-Pastor M, Rayter S, Warne PH, Cantrell DA (1992): Interleukin 2 activates $p21^{ras}$ in normal human T lymphocytes. J Immunol 148:2417–2422.

Gulbins E, Coggeshall M, Baier G, Katzav S, Burn P, Altman A (1993): Tyrosine kinase-stimulated guanine nucleotide exchange activity of Vav in T cell activation. Science 260:822–825.

Hagag N, Halegoua S, Viola M (1986): Inhibition of growth factor-induced differentiation of PC12 cells by microinjection of antibody to ras p21. Nature 319:680–682.

Hall A (1990): The cellular function of small GTP-binding proteins. Science 249:635–640.

Hall A (1992): Signal transduction through small GTPases—a tale of two GAPs. Cell 69:389–391.

Han M, Golden A, Han Y, Sternberg PW (1993): C elegans lin-45 raf gene participates in let-60 ras-stimulated vulval differentiation. Nature 363:133–140.

Han M, Sternberg PW (1990): *let-60,* a gene that specifies cell fates during *C. elegans* vulval induction, encodes a *ras* protein. Cell 63:921–931.

Hariharan IK, Carthew RW, Rubin GM (1991): The Drosophila roughened mutation: activation of a rap homolog disrupts eye development and interferes with cell determination. Cell 67:717–722.

Hart AC, Krämer H, Zipursky SL (1993): Extracellular domain of the boss transmembrane ligand acts as an antagonist of the sevenless receptor. Nature 361:732–736.

Haubruck H, McCormick F (1991): Ras p21: effects and regulation. Biochim Biophys Acta 1072:215–229.

Heideran MA, Molloy CJ, Pangelinan M, Choudhury GG, Wang L, Fleming TP, Sakaguchi AY, Pierce JH (1992): Oncogene 7:147–152.

Hill RJ, Sternberg PW (1992): The gene lin-3 encodes an inductive signal for vulval development in *C. elegans.* Nature 358:470–476.

Horvitz HR, Sternberg PW (1991): Multiple intercellular signalling systems control the development of the *Caenorhabditis elegans* vulva. Nature 351:535–541.

Hu P, Margolis B, Schlessinger J (1993): Vav: A potential

link between tyrosine kinases and Ras-like GTPases in hematopoietic cell signalling. Bioessays 15:179–183.

Huang E, Nocka K, Beier DR, Chu T-Y, Buck J, Lahm H-W, Wellner D, Leder P, Besmer P (1990): The hematopoietic growth factor KL is encoded at the Sl locus and is the c-*kit* receptor, the gene product of the W locus. Cell 63:225–233.

Hughes DA, Ashworth A, Marshall CJ (1993): Complementation of *byr1* in fission yeast by mammalian MAP kinase kinase requires coexpression of Raf kinase. Nature 364:349–352.

Izquierdo M, Downward J, Graves JD, Cantrell DA (1992): Role of protein kinase C in T-cell antigen receptor regulation of $p21^{ras}$: evidence that two $p21^{ras}$ regulatory pathways coexist in T cells. Mol Cell Biol 12:3305–3312.

Katzav S, Cleveland JL, Heslop HE, Pulido D (1991): Loss of the amino-terminal helix-loop-helix domain of the *vav* proto-oncogene activated its transforming potential. Mol Cell Biol 11:1912–1920.

Katzav S, Martin-Zanca D, Barbacid M (1989): *Vav,* a novel human oncogene derived from a locus ubiquitously expressed in hematopoietic cells. EMBO J 8:2283–2290.

Kolch W, Heidecker G, Lloyd P, Rapp UR (1991): Raf-1 protein kinase is required for growth of induced NIH 3T3 cells. Nature 349:426–428.

Krämer H, Cagan RL, Zipursky SL (1991): Interaction of Bride of Sevenless membrane-bound ligand and the Sevenless tyrosine-kinase receptor. Nature 352:207–212.

Lawrence PA, Green SM (1979): Cell lineage in the developing retina of *Drosophila.* Dev Biol 71:142–152.

Li, B-Q, Kaplan D, Kung H-F, Kamata T (1992): Nerve growth factor stimulation of the Ras-guanine nucleotide exchange factor and GAP activities. Science 256:1456–1459.

Li N, Batzer A, Daly R, Yajnik V, Skolnik E, Chardin P, Bar-Sagi D, Margolis B, Schlessinger J (1993): Guanine-nucleotide-releasing factor hSos1 binds to Grb2 and links receptor tyrosine kinases to Ras signalling. Nature 363:85–88.

Lowenstein EJ, Daly RJ, Batzer AG, Li W, Margolis B, Lammers R, Ullrich A, Skolnik EY, Bar-Sagi D, Schlessinger J (1992): The SH2 and SH3 domain-containing protein GRB2 links receptor tyrosine kinases to ras signaling. Cell 70:431–442.

Lowy DR, Willumsen BM (1993): Function and regulation of Ras. In Richardson CC, Abelson JN, Meister L, Walsh CT (eds): Annual Review of Biochemistry. Palo Alto: Annual Reviews, pp 851–891.

Lu X, Chou T-B, William NG, Roberts T, Perrimon N (1993): Control of cell fate determination by $p21^{ras}$/Ras 1, an essential component of torso signaling in *Drosophila.* Genes Dev 7:621–632.

MacNicol AM, Muslim AJ, Williams LT (1993): Raf-1 kinase is essential for early Xenopus development and mediates the induction of mesoderm by FGF. Cell 73:571–583.

Mark GE, MacIntyre RJ, Digan ME, Ambrosio L, Perrimon N (1987): Drosophila melanogaster homologs of the raf oncogene. Mol Cell Biol 7:2134–2140.

Marshall MS (1993): The effector interactions of $p21^{ras}$. Trends Biol Sci 18:250–254.

Martegani E, Vanoni M, Zippel R, Coccetti P, Brambilla R, Ferrari C, Sturani E, Alberghina L (1992): Cloning by functional complementation of a mouse cDNA encoding a homologue of CDC25, a *Saccharomyces cerevisiae* RAS activator. EMBO J 11:2151–2159.

Marx J (1993): Forging a path to the nucleus. Science 260:1588–1590.

Massagué J (1990): Transforming growth factor-α. A model for membrane-anchored growth factors. J Biol Chem 265:21393–21396.

Medema RH, Wubbolts R, Bos JL (1991): Two dominant inhibitory mutants of $p21^{ras}$ interfere with insulin induced gene expression. Mol Cell Biol 11:5963–5967.

Moodie SA, Willumsen BM, Weber MJ, Wolfman A (1993): Complexes of Ras-GTP with Raf-1 and mitogen-activated protein kinase kinase. Science 260:1658–1660.

Morrison DK, Kaplan DR, Rapp U, Roberts TM (1988): Signal transduction from membrane to cytoplasm: growth factors and membrane-bound oncogene products increase Raf-1 phosphorylation and associated protein kinase activity. Proc Natl Acad Sci USA 85:8855–8859.

Mulcahy LS, Smith MR, Stacey DN (1985): Requirement for *ras* proto-oncogene function during serum-stimulated growth of NIH3T3 cells. Nature 313:241–243.

Nakafuku M, Satoh T, Kaziro Y (1992): Differentiation factors, including nerve growth factor, fibroblast growth factor, and interleukin-6, induce an accumulation of an active Ras-GTP complex in rat pheochromocytoma PC12 cells. J Biol Chem 267: 19448–19454.

Nishida Y, Hata M, Ayaki T, Ryo H, Yamagata M, Shikuzi K, Nishizuka Y (1988): Proliferation of both somatic and germ cells is affected in the Drosophila mutants of raf proto-oncogene. EMBO J 7:775–781.

Noda M, Ko M, Ogura A, Liu D, Amano T, Takano T, Ikawa Y (1985): Sarcoma viruses carrying *ras* oncogenes induce differentiation-associated properties on a neuronal cell line. Nature 318:73–75.

Nüsslein-Volhard C, Frohnhöfer HG, Lehman R (1987): Determination of anterior-posterior polarity in *Drosophila.* Science 238:1675–1681.

Olivier JP, Raabe T, Henkemeyer M, Dickson B, Mbamalu G, Margolis B, Schlessinger J, Hafen E, Pawson T (1993): A Drosophila SH2-SH3 adaptor protein implicated in coupling the sevenless tyrosine kinase to an activator of ras guanine nucleotide exchange, Sos. Cell 73:179–191.

Pawson T (1988): Non-catalytic domains of cytoplasmic protein-tyrosine kinases: regulatory elements in signal transduction. Oncogene 3:491–495.

Pelicci G, Lanfrancone L, Grignani F, McGlade J, Cavallo F, Forni G, Nicoletti I, Grignani F, Pawson T, Pelicci PG (1992): A novel transforming protein (SHC) with an SH2 domain is implicated in mitogenic signal transduction. Cell 70:93–101.

Perkins LA, Larsen I, Perrimon N (1992): Corkscrew encodes a putative protein tyrosine phosphatase that functions to transduce the terminal signal from the receptor tyrosine kinase torso. Cell 70:225–236.

Pizon V, Lerosey I, Chardin P, Tavitian A. (1988): Nucleotide sequence of a human cDNA encoding a *ras*-related protein (*rap* 1B). Nucleic Acids Res 16:7719.

Puil L, Pawson T (1992): Vagaries of Vav. Curr Biol 2:275–277.

Qui M, Green SH (1991): NGF and EGF rapidly activate $p21^{ras}$ in PC12 cells by distinct convergent pathways involving tyrosine phosphorylation. Neuron 7:937–946.

Ready DF (1989): A multifaceted approach to neural development. Trends Neurosci 12:102–110.

Ready DF, Hanson TE, Benzer S (1976): Development of the Drosophila retina, a neurocrystalline lattice. Dev Biol 53:217–240.

Ren R, Mayer B, Cicchetti P, Baltimore D (1993): Identification of a ten-amino acid proline-rich SH3 binding site. Science 259:1157–1161.

Reinke R, Zipursky SL (1988): Cell-cell interaction in the Drosophila retina: the *bride of sevenless* gene is required in cell R8 for R7 cell development. Cell 55:321–330.

Ridley AJ, Hall A (1992): The small GTP-binding protein rho regulates the assembly of focal adhesions and actin stress fibers in response to growth factors. Cell 70:389–399.

Ridley AJ, Paterson HF, Johnston CL, Diekmann D, Hall A (1992): The small GTP-binding protein rac regulates growth factor-induced membrane ruffling Cell 70:401–410.

Robinson LC, Gibbs JB, Marshall MS, Sigal IS, Tatchell K (1987): CDC25: A component of the RAS-adenylate cyclase pathway in *Saccharomyces cerevisiae*. Science 235:1218–1221.

Rogge RD, Karlovich CA, Banerjee U (1991): Genetic dissection of a neurodevelopmental pathway: son of sevenless functions downstream of the sevenless and EGF receptor tyrosine kinases. Cell 64:39–48.

Ronchi E, Treisman J, Dostatni N, Struhl G, Desplan C (1993): Down-regulation of the drosophila morphogen bicoid by the torso receptor-mediated signal transduction cascade. Cell 74:347–355.

Rozakis-Adcock M, Fernley R, Wade J, Pawson T, Bowtell D (1993): The SH2 and SH3 domains of mammalian Grb2 couple the EGF receptor to the Ras activator mSos1. Nature 363:83–85.

Rubin GM (1989): Development of the Drosophila retina: inductive events studied at single cell resolution. Cell 57:519–520.

Satoh T, Endo M, Nakafuku M, Aikyama T, Yamamoto T, Kaziro Y (1990a): Accumulation of $p21^{ras}$ in response to stimulation with epidermal growth factor and oncogene products with tyrosine kinase activity. Proc Natl Acad Sci USA 87:7926–7929.

Satoh T, Endo M, Nakafuku M, Nakamura S, Kaziro Y (1990b): Platelet-derived growth factor stimulates formation of active $p21^{ras}$ GTP complex in Swiss mouse 3T3 cells. Proc Natl Acad Sci USA 87:5993–5997.

Satoh T, Nakafuku M, Miyajima A, Kaziro Y (1991): Involvement of ras p21 protein in signal transduction pathways from interleukin 2, interleukin 3, and granulocyte/macrophage colony-stimulating factor, but not from interleukin 4. Proc Natl Acad Sci USA 88:3314–3318.

Satoh T, Nakafuku M, Kaziro Y (1992a): Function of Ras as a molecular switch in signal transduction. J Biol Chem 267:24149–24152.

Satoh T, Uehara Y, Kaziro Y (1992b): Inhibition of interleukin 3 and granulocyte/macrophage colony stimulating factor stimulated increase of active Ras-GTP by herbimycin A, a specific inhibitor of tyrosine kinases. J Biol Chem 267:2537–2541.

Serafini T, Orci L, Amherdt M, Brunner M, Kahn RA, Rothman JE (1991): ADP-ribosylation factor is a subunit of the coat of Golgi-derived vesicles: a novel role for a GTP-binding protein. Cell 67:239–253.

Settleman J, Narasimhan V, Foster LC, Weinberg RA (1992): Molecular cloning of cDNAs encoding the GAP-associated protein p190: implications for a signaling pathway from ras to the nucleus. Cell 69:539–549.

Shou C, Farnsworth CL, Neel BG, Feig LA (1992): Molecular cloning of cDNAs encoding a guanine-nucleotide-releasing factor for RAS p21. Nature 358:351–354.

Simon MA, Bowtell DDL, Dodson GS, Laverty TR, Rubin GM (1991): Ras 1 and a putative guanine nucleotide exchange factor perform crucial steps in signaling by the sevenless tyrosine kinase. Cell 67:701–716.

Simon MA, Carthews RW, Fortini ME, Gaul U, Mardon G, Rubin GM (1993a): Signal transduction pathway initiated by activation of the sevenless tyrosine kinase receptor. Cold Spring Harbor Symp Quant Biol 57:375–380.

Simon MA, Dodson GS, Rubin GM (1993b): An SH3-SH2-SH3 protein is required for $p21^{Ras1}$ activation and binds to sevenless and Sos proteins in vitro. Cell 73:169–177.

Skolnik EY, Batzer A, Li N, Lee C-H, Lowenstein E,

Mohammadi M, Margolis B, Schlessinger J (1993): The function of GRB2 in linking the insulin receptor to Ras signaling pathways. Science 260:1953–1955.

Sprenger F, Stevens LM, Nüsslein-Volhard C (1989): The Drosophila gene torso encodes a putative receptor tyrosine kinase. Nature 338:478–483.

Stein DS, Stevens LM (1991): Establishment of dorsal-ventral and terminal pattern in the Drosophila embryo. Curr Opin Genet Dev 1:247–254.

St. Johnston D, Nüsslein-Volhard C (1992): The origin of pattern and polarity in the Drosophila embryo. Cell 68:201–219.

Sun XJ, Rothenberg P, Kahn CR, Backer JM, Araki E, Wilden PA, Cahill DA, Goldstein BJ, White MF (1991): Structure of the insulin receptor substrate IRS-1 defines a unique signal transduction protein. Nature 352:73–75.

Szeberenyi J, Cai H, Cooper GM (1990): Effect of a dominant inhibitory Ha-ras mutation on neuronal differentiation of PC12 cells. Mol Cell Biol 10:5324–5332.

Thomas SM, DeMarco M, D'Arcangelo G, Halegoua S, Brugge JS (1992): Ras is essential for nerve growth factor and phorbol ester-induced tyrosine phosphorylation on MAP kinases. Cell 68:1031–1040.

Torti M, Marti KB, Altschuler D, Yamamoto K, Lapetina EG (1992): Erythropoietin induces $p21^{ras}$ activation and $p120^{GAP}$ tyrosine phosphorylation in human erythroleukemia cells. J Biol Chem 267:8293–8298.

Van Aelst L, Barr M, Marcus S, Polverino A, Wigler M (1993): Complex formation between RAS and RAF and other protein kinases. Proc Natl Acad Sci USA 90:6213–6217.

van der Sluijs P, Hull M, Webster P, Mâle P, Goud B, Mellman I (1992): The small GTP-binding protein rab4 controls an early sorting event on the endocytic pathway. Cell 70:729–740.

Van Vactor DL, Jr Cagan RL, Krämer H, Zipursky SL (1991): Induction in the developing compound eye of drosophila: multiple mechanisms restrict R7 induction to a single retinal precursor cell. Cell 67:1145–1155.

Vojtek AB, Hollenberg SM, Cooper JA (1993): Mammalian Ras interacts directly with the serine/threonine kinase Raf. Cell 74:205–214.

Warne PH, Viciana PR, Downward J (1993): Direct interaction of Ras and the amino-terminal region of Raf-1 *in vitro*. Nature 364:352–355.

Wei W, Mosteller RD, Sanyal P, Gonzales E, McKinney D, Dasgupta C, Li P, Liu B-X, Broek D (1992): Identification of a mammalian gene structurally and functionally related to the *CDC25* gene of *Saccharomyces cerevisiae*. Proc Natl Acad Sci USA 89:7100–7104.

Williams NG, Roberts TM, Li P (1992): Both $p21^{ras}$ and $pp60^{src}$ are required, but neither alone is sufficient, to activate the Raf-1 kinase. Proc Natl Acad Sci USA 89:2922–2926.

Wong G, Müller O, Clark R, Conroy L, Moran MF, Polakis P, McCormick F (1992): Molecular cloning and nucleic acid binding properties of GAP-associated tyrosine phosphorylation p62. Cell 69:551–558.

Wood KW, Sarnecki C, Roberts TM, Blenis J (1992): Ras mediates nerve growth factor receptor modulation of three signal-transducing kinases. Cell 68:1041–1050.

Zhang X, Settleman J, Kyriakis JM, Takeuchi-Suzuki E, Elledge SJ, Marshall MS, Bruder JT, Rapp UR (1993): Normal and oncogenic $p21^{ras}$ proteins bind to the amino-terminal regulatory domain of c-Raf-1. Nature 364:308–313.

Zsebo KM, Williams DA, Suggs SV (1990): Stem cell factor is encoded at the Sl locus of the mouse and is the ligand for the c-kit tyrosine kinase receptor. Cell 63:213–224.

ABOUT THE AUTHOR

Richard T. Hamilton is Associate Professor of Zoology and Genetics at Iowa State University in Ames, Iowa. After receiving a B.S. degree from the Royal Melbourne Institute of Technology in Melbourne, Australia, he pursued doctoral research at Cornell University in the laboratory of Ray Wu, where he earned a Ph.D. in 1974 for work on DNA sequencing of lambda bacteriophage DNA. This was followed by postdoctoral research at the Salk Institute in La Jolla, California, where he worked with Robert Holley on growth control of mammalian cells. After two years as a postdoctoral fellow, he continued at the Salk Institute in the Department of Cell Biology. He became chairman of that department in 1980. In 1982, he joined the Department of Zoology at Iowa State University. Since 1974, Dr. Hamilton's research has involved explorations of the control by growth factors and oncogenes of gene expression in mammalian cells.

Growth Factors and Signal Transduction in Development: 139–163

Protein Phosphatases in Cell Proliferation, Differentiation, and Development

Thomas S. Ingebritsen, Kelaginamane T. Hiriyanna, and Keli Hippen

I. INTRODUCTION

Reversible phosphorylation of proteins is a major regulatory mechanism in signal transduction pathways that control cell proliferation, differentiation, and development. The best-understood phosphorylation reactions in eukaryotic cells occur on serine, threonine and tyrosine residues. There are three functionally distinct classes of protein kinases and phosphatases involved in regulating cellular phosphorylation of proteins: (1) those that act only on serine or threonine, (2) those that act only on tyrosine, and (3) those that act on serine, threonine, and tyrosine.

Protein phosphatases have several kinds of roles in these signal transduction pathways. Generally, they act as cell proliferation inhibitors and/or tumor suppressors by opposing the actions of protein kinases that stimulate cell proliferation. However, some protein dephosphorylation reactions seem to be required for cell proliferation, and some growth factors activate protein phosphatases. Protein phosphatases also stimulate and inhibit cell cycle progression in the G_2 and M phases. Finally, protein phosphatases are necessary for differentiation of certain eukaryotic cells and are involved in developmental regulation in metazoans. We review here genetic, cell biological, and biochemical studies that have provided insights concerning the role of protein phosphatases in these cellular processes.

II. PROPERTIES AND CLASSIFICATION OF PROTEIN

A. Primary Structures

There are three families of protein phosphatase genes. Two of the gene families encode protein serine/threonine phosphatases and a third family encodes enzymes that function as protein tyrosine phosphatases or dual-specificity protein phosphatases (PP) [Cohen et al., 1990b; Pot and Dixon, 1992].

One gene family of protein serine/threonine phosphatases consists of two isozymes of PP-2C while the second, termed the PP-1/2A/2B gene family, consists of 11 members [Cohen et al., 1990b; Steele et al., 1992]. Ten members are eukaryotic enzymes and the other member, termed ORF 221, is found in prokaryotes. Other eukaryotic members of the family are RdgC, SIT 4, PP-$2B_w$, PP-V, PP-X, PP-Y, and PP-Z. Several members of this family have two or more isozymes. PP-1 enzymes, isolated from various organisms from yeast to humans, show extreme amino acid sequence conservation; the overall homology is greater than 74%, and homology over a 70-amino-acid stretch, comprising part of the catalytic domain, is greater than 92% [Bollen and Stalmans, 1992]. This level of homology is similar to that observed among members of families of structural proteins such as tubulin. The homology shared between PP-1, PP-2A, and PP-2B is also quite high in the 70-amino-acid region of

TABLE I. PTPs Discussed in this Review

PTP class	PTP	Distinguishing Characteristics	Potential function
Transmembrane	CD45	Extracellular domain is heavily glycosylated, similar to ligand-binding domains	Stimulation of cell proliferation
	LAR	Extracellular domain contains FN III and Ig-like domains	Development
	DPTP99A	Extracellular domain contains FN III and Ig-like domains	Development
	DPTP10D	Extracellular domain contains FN III domains only	Development
	RPTPα	Extracellular domain is very small	Stimulation of cell proliferation
	PTPγ		Inhibition of cell proliferation
Intracellular	PTP-1	Endoplasmic reticulum localization	Inhibition of cell proliferation
	PTP-1B	Endoplasmic reticulum localization	Inhibition of cell proliferation
	TC-PTP	Endoplasmic reticulum localization	Inhibition of cell proliferation
	PTP-MEG	Contains domain with homology to band 4.1 protein and/or ezrin	Unknown
	PTP-H1	Contains domain with homology to band 4.1 protein and/or ezrin	Unknown
	PTP-1C*	Contains two SH2-like domains	Development
	PTP-1D**	Contains two SH2-like domains	Development
	Csw	Contains two SH2-like domains	Development
	Dictyo PTP-1	Spatiotemporal regulated expression	Development
Dual-specificity	VH-1	Vaccinia virus gene	Unknown
	HePTPase	Mitogen stimulated expression	Stimulation of cell proliferation
	PAC-1	Nuclear localization	Stimulation of cell proliferation
	3CH34	Mitogen stimulated expression	Stimulation of cell proliferation
	Cdc25		Cell cycle regulation

*PTP-1C is also known as SH-PTP1, HCP, and HSP.
**PTP-1D is also known as SH-PTP2 and sypPTP.

the catalytic domain (82% and 60%, respectively).

The protein tyrosine phosphatase (PTP) gene family is a large group of enzymes with over 60 members identified to date [Pot and Dixon, 1992]. The PTP catalytic domain, defined by sequence homology, is 230 amino acids in length and contains a primary structural feature called the HC box. This is an 11-amino-acid stretch of high homology with the consensus sequence [I/V]HCXAGXXR[S/T]G, where the amino acid sequence is given using the one-letter code and X stands for an unconserved amino acid. The cysteine residue in this HC box is essential for PTP activity, and seems to function as a nucleophile which attacks the substrate phosphate ester forming a phospho-enzyme intermediate.

The PTP gene family can be subdivided into three classes: transmembrane, intracellular, and dual-specificity (Table 1) [Pot and Dixon, 1992]. The first group are single-pass transmembrane proteins that contain one or two intracellular PTP-like domains and an extracellular domain of variable structure. The intracellular class of PTPs do not have transmembrane or extracellular domains but typically have N- or C-terminal extensions that may have targeting or regulatory functions (Section II.C.2 and II.D.2). The dual-specificity group of protein phosphatases dephosphorylates serine and threonine residues as well as tyrosine. The prototype for this class is the vaccinia virus PTP, VH-1. This class also contains the yeast *S. pombe* cell division cycle gene product Cdc25, as well as Cdc25

TABLE II. Enzymological Properties and Functions of Type 1 and 2 Serine/Threonine Phosphatases

Protein phosphatase	Type	Phosphorylase kinase specificity	I-1, I-2 DARPP-32 sensitivity	Okadaic acid (IC_{50})	Divalent cation requirement	Function
PP-1	1	β Subunit	Yes	20 nM	None	Inhibition of cell proliferation Differentiation Cell cycle control
PP-2A	2	α Subunit	No	0.2 nM	None	Inhibition of cell proliferation Differentiation Cell cycle control
PP-2B	2	α Subunit	No	5 μM	Ca^{2+}	Stimulation of cell proliferation
PP-2C	2	α Subunit	No	Insensitive	Mg^{2+}	Unknown

IC_{50} = concentration which results in 50% inhibition.

homologues in other eukaryotes. The intracellular and dual-specificity PTPs have a single catalytic domain.

B. Biochemical Properties

Protein serine/threonine phosphatases were initially categorized on the basis of their enzymological properties [Ingebritsen and Cohen, 1983; Cohen, 1991]. Two classes, termed type 1 (consisting of PP-1) and type 2 (consisting of PP-2A, PP-2B, and PP-2C), were defined on the basis of substrate specificity and sensitivity to regulatory molecules (Table II). The type 1 enzyme dephosphorylates the β-subunit of phosphorylase kinase and is potently inhibited by three heat-stable protein inhibitors. The type 2 enzymes preferentially dephosphorylate the α-subunit of phosphorylase kinase and are unaffected by the inhibitor proteins. The individual enzymes can be further distinguished by their requirements for divalent cations and their sensitivity to inhibition by okadaic acid [Cohen, 1991]. Other members of the PP-1/2A/2B family were identified by molecular biological techniques and have not been well characterized biochemically [Cohen et al., 1990b].

Studies of a large number of phosphorylated proteins involved in metabolic regulation indicate that several of the protein serine/threonine phosphatases (PP-1, PP-2A, PP-2C) have broad and overlapping substrate specificities [Ingebritsen and Cohen, 1983; Cohen, 1989]. Similarly, studies with synthetic phosphopeptides based on known tyrosine phosphorylation sites from eukaryotic cell proteins indicate that several PTPs (PTP-5, PTP-1, YOP-51, CD45, and LAR) also have broad substrate specificities [Hippen et al., 1994a]. Additionally, it has been found that acidic amino acids on the N-terminal side of phosphotyrosine are positive recognition elements for some PTPs (Hippen et al., 1994b; Hippen et al., 1993; Ruzzene et al., 1993; Cho et al., 1993]. Nevertheless, there are clear differences in substrate preferences among the three protein serine/threonine phosphatases and among the five PTPs that may influence substrate selection *in vivo*. In contrast the protein serine/threonine phosphatase, PP-2B, has a very narrow substrate specificity.

C. Targeting Subunits or Domains

Many protein phosphatases contain targeting domains, or associate with other proteins that direct the catalytic subunit to specific cellular locations and, at least in some cases, alter the substrate specificity of the enzymes.

1. Protein serine/threonine phosphatases. Protein phosphatase-1 interacts with several targeting proteins that direct it to specific subcellular locations, alter its substrate specificity, and regulate its activity [Hubbard and Cohen, 1993]. The best-characterized targeting subunit is a 124-kD protein that is specifically expressed in striated muscle. The protein, termed the G-subunit, targets PP-1 to glycogen and to the sarcoplasmic reticulum. In addition this protein enhances the activity of PP-1 towards glycogen-bound substrates and serves as a site for regulation of PP-1 activity by cAMP, Ca^{2+}, and insulin (Section II.D.1). Other PP-1 targeting subunits include a liver G subunit that affects the catalytic and regulatory properties of PP-1 in a manner distinct from that of the skeletal muscle G subunit, smooth and skeletal muscle myofibril-targeting subunits (M subunits), and the product of the fission yeast *sds*22 gene that may function as a nuclear targeting subunit.

The protein serine/threonine phosphatase, PP-2A, can be isolated as a monomer, heterodimer, or heterotrimer consisting of a 37-kD catalytic (C) subunit alone, the C subunit and a 61-kD A subunit, or the C and A subunits with one of three B subunits (54 kD, 55 kD or 72–74 kD) [Shenolikar and Nairn, 1991]. Unlike the G and M subunits of PP-1, the *in vivo* function of the A and B subunits of PP-2A are not known, although the phosphatase activities of the various forms of the enzyme differ *in vitro*. In addition to the A and B subunits, the small T tumor antigens from SV40, polyoma, and BK viruses as well as the polyoma middle T antigen are able to form stable complexes with the dimeric form (C plus A subunits) of PP-2A via contacts with the C subunit [Mumby and Walter, 1991]. Studies with polyoma middle T-antigen have indicated that this association is necessary, but not sufficient, for transformation [Grussenmeyer et al., 1987]. The exact function of this association and of the association of the C and A subunits with the B subunit, is unknown. However, the middle T-antigen of polyoma virus contains a transmembrane domain, and is localized to the membrane [Carmichael et al., 1982]; this would represent a possible redistribution of normally cytosolic PP-2A. Alternatively, association of the A and C subunits of PP-2A with SV40 small T antigen, like their association with the B subunit, affects substrate selectivity [Yang et al., 1991; Scheidtmann et al., 1991].

Protein phosphatase-2A is also regulated by association of the catalytic subunit with other proteins. Cayla et al. [1990], have shown that the specificity of PP-2A can be changed from serine/threonine residues to serine/threonine/tyrosine by a 40-kD protein abundant in rabbit skeletal muscle (0.25 μM estimated) and *Xenopus* oocytes (0.75 μM estimated).

2. Protein tyrosine phosphatases. Most PTPs are also targeted to specific cellular locations. The transmembrane PTPs are localized to the plasma membrane and they may be activated by association with extracellular ligands, although as yet no ligands have been found for any of the various PTPs of this class. Transmembrane PTP activity may also be controlled by association with intracellular proteins. For example, the cytoskeletal protein fodrin was recently found to be associated with CD45 *in vivo* [Lokeshwar and Bourguignon, 1992]. Additionally, it was shown *in vitro* that fodrin, and the closely related protein spectrin, can bind CD45, and that association causes an increase in the PTP activity of CD45.

Many members of the intracellular class of PTPs have regulatory domains that are responsible for subcellular localization. A C-terminal hydrophobic domain targets PTP-1B to the endoplasmic reticulum [Frangioni et al., 1992], and similar domains are also present near the C-terminus of the T-cell PTP [Cool et al., 1989] and of PTP-1 [Guan et al., 1990]. T-cell PTP [Cool et al., 1989] also has a basic region with homology to nuclear localization signals. PTP-MEG [Gu et al., 1991] and PTP-H1 [Yang and Tonks, 1991] contain N-terminal extensions with homologies to erythrocyte band 4.1, a cytoskeletal protein, in a region of the latter protein thought to mediate inositol phospholipid–dependent association with glycophorin. PTP-1C, PTP-1D and the product of the *Drosophila corkscrew* gene each contain two N-terminal SH2 (*S*rc *H*omology *2*) domains that are probably involved in targeting the PTPs to activated growth factor receptors (Section VII.B).

D. Regulators of Protein Phosphatase Activity

Protein phosphatases are regulated by second messengers (cAMP and Ca^{2+}), protein inhibitors, and by reversible phosphorylation of the catalytic or regulatory subunits.

1. Protein serine/threonine phosphatases. Protein phosphatase-2B is a Ca^{2+}-dependent enzyme that is further activated by calmodulin [Cohen, 1989]. This enzyme is also known as calcineurin because of its previous description as a major calmodulin-binding protein in brain. Protein phosphatase-2B is composed of a 61-kD catalytic (A) subunit and a 19-kD Ca^{2+}-binding (B) subunit. The B subunit associates with the A subunit in the absence of Ca^{2+}, and binding of the divalent cation to the B subunit produces 10% of maximal activity. Calmodulin interacts with the A subunit only in the presence of Ca^{2+} to produce a further 10-fold increase in the activity of PP-2B. Ca^{2+} and calmodulin binding relieves autoinhibition by a domain from the C-terminal portion of PP-2B.

The *Drosophila retinal degeneration C* (*rdgC*) gene [Steele et al., 1992] encodes a novel protein serine/threonine phosphatase with 30% homology to PP-1, PP-2A, and PP-2B. It has a putative Ca^{2+}-binding domain and thus may be regulated by Ca^{2+}.

Inhibitor-1 (I-1) and dopamine and cyclic AMP-regulated phosphoprotein (DARPP-32) are two related heat-stable proteins that are potent and specific inhibitors of PP-1 [Huang and Glinsman, 1976; Nimmo and Cohen, 1978; Williams et al., 1986]. Inhibitor-1 is expressed ubiquitously in mammalian tissues, whereas the expression of DARPP-32 is restricted to certain cells in the kidney and nervous system that express D1 dopamine receptors [Ouimet et al., 1984]. A unique property of the two inhibitor proteins is that their activity is dependent on their prior phosphorylation on a specific threonine residue by cAMP-dependent protein kinase (PKA).

Phosphorylation of I-1 and DARPP-32 by PKA has several functions. First, it is a mechanism for amplifying the cAMP signal. This is because many PKA substrates are also substrates for PP-1. A second role is as a predominant route for transduction of the cAMP signal. A third role is in the integration of Ca^{2+} and cAMP signals. DARPP-32 and I-1 are among the best substrates for the Ca^{2+} and calmodulin-dependent protein serine/threonine phosphatase, PP-2B, which has a very restricted substrate specificity [Ingebritsen and Cohen, 1983; Hemmings et al., 1984]. The extent of phosphorylation of I-1 and DARPP-32 is enhanced by cAMP and decreased by Ca^{2+}.

Three other proteins that inhibit PP-1 have been identified; however, their biological functions are not understood. Inhibitor-2 is a heat-stable protein that is unrelated to I-1 and DARPP-32 and that does not require prior phosphorylation to be an active PP-1 inhibitor [Foulkes and Cohen, 1980]. Two additional PP-1 inhibitors have been purified from bovine thymus nuclei [Beullens et al., 1992]. However, the inhibition caused by these proteins is substrate-dependent, whereas inhibition of PP-1 by I-1, DARPP-32, and inhibitor-2 occurs with all substrates.

The G subunit that targets PP-1 to the glycogen particle and to sarcoplasmic reticulum in skeletal muscle is also a site for integration of signal transduction pathways [Hubbard and Cohen, 1993]. PKA phosphorylates the G subunit at two serine residues, termed sites 1 and 2. This phosphorylation causes the release of the glycogen-bound PP-1 catalytic subunit, thereby decreasing its activity towards glycogen--bound substrates (i.e., glycogen phosphorylase, glycogen synthase, and phosphorylase kinase). Additionally, the soluble PP-1 catalytic subunit interacts with cytoplasmic I-1 leading to further PP-1 inhibition (see above). Site 1 of the G subunit is also phosphorylated by an insulin-stimulated protein serine/threonine kinase that is closely related to the ribosomal S6 kinase II. Phosphorylation of site-1 only does not cause PP-1 dissociation from the G subunit but results in a two- to threefold increase in PP-1 activity towards glycogen synthase and phosphorylase kinase. Thus phosphorylation of the G subunit by PKA and the insulin-stimulated protein serine/threonine kinase integrate signal transduction pathways that regulate glycogen synthesis and breakdown. Conversely site 2 is dephosphorylated by PP-2A and PP-2B, thus allowing PP-1 to rebind to the G subunit. The involvement of PP-2B provides an additional route for Ca^{2+} regulation of PP-1.

Another method for regulation of protein serine/threonine phosphatase activity is by direct phosphorylation of the catalytic subunit. Protein phosphatase-1 has been shown to be a substrate for the protooncogene tyrosine kinase $pp60^{c\text{-}src}$ [Johansen and Ingebritsen, 1986]. Phosphorylation decreases PP-1 activity by increasing the Michaelis constant (K_m) for substrate dephosphorylation. The site of phosphorylation is on the C-terminal 4-kD portion of the molecule. Protein phosphatase-2A is also a substrate for the protein tyrosine kinase $pp60^{v\text{-}src}$, as well as the protein tyrosine kinase, $pp56^{lck}$, and the tyrosine kinase receptors for insulin and epidermal growth factor (EGF) [Chen et al., 1992]. Tyrosine-307 of PP-2A was found to be the phosphorylated residue and its phosphorylation resulted in a 10-fold decrease in PP-2A activity. PP-2B is phosphorylated on serine-197 by protein kinase C (PKC), a Ca^{2+} and phospholipid dependent protein kinase, and this phosphorylation increases the K_m for dephosphorylation of ^{32}P-labelled myosin light chains, while having no effect on V_{max} [Hashimoto and Soderling, 1989]. It is not yet clear whether any of these phosphorylation reactions occur *in vivo*.

2. Protein tyrosine phosphatases. Two heat-stable inhibitors of protein tyrosine phosphatases [Ingebritsen, 1989] have been identified and partially purified from bovine brain. The two proteins potently inhibit two PTPs from bovine brain (PTP-4 and PTP-5) and rat brain PTP-1 [K.T. Hiriyanna and T.S. Ingebritsen, unpublished results] but have only weak activity towards five other PTPs from bovine brain. The inhibitor proteins have no activity against protein serine/threonine phosphatases.

Protein tyrosine phosphatase activity is also influenced by phosphorylation. Protein tyrosine phosphatase-1D, a PTP containing an SH2 domain, is phosphorylated on tyrosine in response to growth factor simulation of cells (Section VII.B). The phosphorylation of PTP-1D increases its activity. The transmembrane PTP, CD45, is phosphorylated on serine residues in resting T cells, and when $[Ca^{+2}]_i$ is increased, the level of CD45 serine phosphorylation decreases. Concommitant with this decrease in phosphorylation is a decrease in membrane-associated PTP activity [Ostergaard and Trowbridge, 1991].

E. Pharmacological Inhibitors

1. Protein serine/threonine phosphatases. The recent discovery that okadaic acid inhibits the protein phosphatases PP-1 and PP-2A has sparked much investigation of these phosphatases [Cohen et al., 1990a]. Okadaic acid is a 38-carbon polyether fatty acid found in the digestive glands of shellfish and sponges, and causes diarrhetic shellfish poisoning and is a tumor promoter. It is a potent and specific inhibitor of PP-1 and PP-2A, but has only very weak activity towards PP-2B, and no activity towards PP-2C or members of the PTP family [Cohen et al., 1990a; Suganuma et al., 1992]. Many other compounds with widely varying structures have recently been reported to inhibit the activity of PP-1 and/or PP-2A, including tautomycin (a structural analogue of okadaic acid), calyculin A [Ishihara et al., 1989], microcystin-LR (a cyclic heptapeptide) [MacKintosh et al., 1990], cantharidin (the toxic constituent of blister beetles), and the herbicide, endothall [Li and Casida, 1992]. One difference noted between these substances is that the concentration that causes 50% inhibition (IC_{50}) is 10-to-100-fold lower for okadaic acid inhibition of PP-2A than for inhibition of PP-1, whereas calyculin A, tautomycin, and microcystin-LR seem to inhibit both phosphatases with similar IC_{50}s.

Recently, the immunosuppressive drugs, cyclosporin A and FK506, were shown to be specific inhibitors of the protein phosphatase PP-2B [Liu et al., 1991]. Inhibition was dependent upon the immunosuppressant-binding proteins cyclophilin and FK-506 binding protein. This discovery has identified PP-2B as a player in the signalling pathways of the T cell receptor and the immunoglobulin E (IgE) receptor, as well as identifying another compound that can be used to investigate protein phosphatase functions (Section III.A.1).

2. Protein tyrosine phosphatases. Several compounds have been found to inhibit, inactivate, or enhance the activity of PTPs *in vitro*. All PTPs have an absolute requirement for reducing conditions, owing to their active-site cysteine residue. In agreement with this is the finding that the cysteine-modifying alkylating reagents, iodoacetate and iodoacetamide, were both able to inactivate LAR and PTP-1B [Pot et al., 1991; Tonks et al., 1991]. In addition, phenylarsine oxide, which covalently reacts with vicinal thiols was found, both *in vivo* and *in vitro,* to inactivate two membrane-associated PTPs (PTP-HA1 and PTP-HA2) that mediate pp15 dephosphorylation in insulin-stimulated 3T3-L1 adipocytes [Liao et al., 1991]. Phenylarsine oxide was also found to inhibit CD45 *in vitro* [Garcia-Morales et al., 1990]. Another compound that has been found to be inhibitory for all PTPs tested thus far is the phosphate analogue orthovanadate (VO_4^{-2}) [Gordon, 1991].

III. INHIBITION OF CELL PROLIFERATION AND TUMORIGENESIS BY PROTEIN PHOSPHATASES

A. Protein Serine/Threonine Phosphatases

Stimulation of animal cells with growth factors or other mitogens, or cell transformation, results in enhanced phosphorylation of cell proteins on serine/threonine residues. Indeed a large number of protein serine/threonine kinases (e.g., PKC, Ca^{2+} and calmodulin-dependent protein kinases, Raf-1, MAP kinases, S6 kinases) are known to be activated under these conditions [Nishida and Gotoh, 1993]. In view of this, one would expect that protein serine/threonine phosphatases would inhibit cell proliferation and transformation by opposing the actions of these mitogen-activated protein kinases. Several lines of evidence indicate that this is indeed the case.

1. Effects of protein serine/threonine phosphatase inhibitors. Okadaic acid, a potent and specific inhibitor of PP-1 and PP-2A, is a tumor promoter [Suganuma et al., 1988; Schonthal, 1992]. This implies that PP-1 and/or PP-2A normally act to oppose tumorigenesis. Cells treated with this inhibitor exhibit enhanced phosphorylation of many cell proteins and increased expression of more than 10 different genes including c-*fos* and c-*jun*.

It has been suggested that tumor promotion by protein serine/threonine phosphatase inhibitors may result from the sustained phosphorylation of proteins that are substrates for PKC [Cohen et al., 1990a]. However, comparative studies of the actions of phorbol esters, which activate PKC, and of okadaic acid indicate that there are differences in the kinetics of gene activation and in the pattern of phosphoproteins induced by the two types of compounds. Both treatments induce AP-1 transcription complex proteins [Kim et al., 1990; Schonthal et al., 1991], nuclear kappa-B transcription factor (NFκB) [Thevenin et al., 1990], and serum response elements. However, the kinetics of expression of these genes are quite different in cells treated with okadaic acid compared to those treated with phorbol esters. Furthermore, okadaic acid does not increase the phosphorylation of an 80-kD protein that is the major PKC substrate in cells treated with phorbol esters [Mahadevan et al., 1991]. Conversely, the phosphorylation of nucleolin is induced in cells treated with okadaic acid but not in those treated with phorbol esters [Issinger et al., 1988; Schneider et al., 1989].

A similar conclusion was reached from comparative studies carried out with reference to epidermal growth factor receptor (EGFR) phosphorylation in response to okadaic acid and phorbol ester [Hernandez-Sotomayor et al., 1991]. Both phorbol esters and okadaic acid caused hyperphosphorylation of the EGFR, but okadaic acid was more potent. Moreover, phorbol esters increased the phosphorylation of threonine-654 of the EGFR, whereas okadaic acid treatment did not. These and the other observed differences, taken together, suggest that tumor promotion by okadaic acid may not be mediated through the phorbol ester pathway [Schonthal, 1992]. This gives credence to the idea that the tumorigenesis could be induced by multiple mechanisms.

The results of studies using okadaic acid and inhibitor-2 indicate that PP-1 is essential for maintaining the tumor suppressor activity of the retinoblastoma susceptibility gene product (pRB). The pRB protein undergoes cell cycle–dependent phosphorylation and/or dephosphorylation. The protein becomes phosphorylated on serine/threonine residues just before cells enter S phase and is dephosphorylated at the end of M phase. pRB is also underphosphorylated in quiescent cells (i.e., those in G_0 and/or G_1, or in terminal differentiation or senescence) [Buchkovich et al., 1989; Mihara et al., 1989; DeCaprio et al., 1992]. Treatment with okadaic acid (specific for PP-1 or PP-2A) or with I-2 (specific for PP-1) inhibits pRB dephosphorylation so that PP-1 may be involved in pRB dephosphorylation [Ludlow et al., 1993].

2. Regulation of protein serine/threonine phosphatases by mitogens or oncogenes. Treatment of cultured keratinocytes with transforming growth factor type β (TGF-β) rapidly activated PP-1. The identity of the protein phosphatase was established on the basis of its sensitivity to I-1 and cross-reactivity with a anti-PP-1 antibody. This suggests that PP-1 may have a role in growth arrest induced by TGF-β [Gruppuso et al., 1991].

Recent studies also indicate that some oncoviruses (SV40 and polyoma virus) target PP-2A catalytic and regulatory subunits to cause transformation [Walter et al., 1990; Ulug et al., 1992] (see also Section II.C.1). Small T antigen of SV40 and the middle T antigen of polyoma virus form stable complexes with PP-2A and inhibit the dephosphorylation of crucial proteins like p53, consequently enhancing transformation [Scheidtmann et al., 1991]. The SV40 small T antigen has also been found to inhibit dephosphorylation of pRB [Fuijiki, 1992].

B. Protein Tyrosine Phosphatases

A role for PTPs is expected in signalling pathways involved in growth inhibition and tumor suppression because many growth factor receptors and oncogene products exhibit protein tyrosine kinase activity.

1. Overexpression of PTPs. Direct evidence for this hypothesis has been obtained through overexpression of several intracellular PTPs in transformed cell lines. Overexpression of PTP-1B or rat brain PTP-1 suppressed the transforming potential of the *neu* and *v-src* oncogenes, respectively, in NIH 3T3 cells [Brown-Shimer et al., 1992; Woodford-Thomas et al., 1992]. Overexpression of a truncated T-cell PTP in baby hamster kidney cells caused a 50% inhibition of growth [Cool et al., 1990] and led to cytokinetic failure and asynchronous nuclear division [Cool et al., 1992]. Overexpression of rat brain PTP-1 in NIH 3T3 cells also led to a substantial increase in the number of multinucleate cells [Woodford-Thomas et al., 1992].

2. Genetic and cytological correlations of PTP gene abnormalities with tumorigenesis. A strong correlation between the chromosomal aberrations in the PTPγ locus and many renal carcinomas has been reported [LaForgia et al., 1991]. Correlations of neoplasticity with chromosomal aberrations in or near many other PTP loci (i.e., CD45, PTP1B, RPTPα, HCP, LAR, TCPTP) have also been made [Charbonneau and Tonks, 1992].

3. Growth inhibitors enhance PTP activity. TGF-β and interleukin-6 (IL-6) stimulate PTP activity and decrease tyrosine phosphorylation of cell proteins in M1 myeloblastic cells [Zafriri et al., 1993]. Somatostatin is a growth inhibitory tetradecapeptide hormone that activates heterotrimeric G protein sensitive to pertussis toxin. It has now been shown that treatment of an undifferentiated human pancreatic cancer cell line (MIA PaCa-2) with somatostatin causes an increase in membrane-associated PTP activity and that this activation of PTP activity is sensitive to pertussis toxin [Pan et al., 1992].

IV. STIMULATION OF CELL PROLIFERATION AND TUMORIGENESIS BY PROTEIN PHOSPHATASES

The initial hypothesis that protein phosphatases function in growth inhibitory and tumor suppressor signalling pathways is made more complicated because it is now clear that protein phosphatases can also have a positive function in the regulation of cell proliferation.

A. Protein Serine/Threonine Phosphatases

1. Lymphocyte activation. Lymphocyte activation is a useful model system for studying complex signal transduction pathways involved in cell proliferation, differentiation, and development. Protein phosphatase-2B is a major cytosolic calmodulin-binding protein in lymphocytes [Kinkaid et al., 1987]. However, the physiological role of calcineurin in T-cell activation only came to light with the recent studies regarding the mechanism of action of the immunosuppressive drugs, cyclosporin A and FK-506. These drugs have been observed to form complexes with cellular proteins termed cyclophilin and FK-506 binding protein (FKBP) respectively [Sigal and Dumont, 1992]. The drug–protein complexes specifically inhibit calcium-mediated events of T-cell activation which includes Ca^{2+}-induced transcription of the interleukin-2 (IL-2) gene. The drug–receptor protein complexes, but not the free drugs or their binding proteins, specifically bind to and inhibit PP-2B [Liu et al., 1991; Fruman et al., 1992; Schreiber, 1992]. From several studies that followed these early observations [Schreiber, 1992; Sigal and Dumont, 1992], the following model has emerged. Elevated intracellular Ca^{2+}, produced in response to the initiation of T-cell activation, activates PP-2B. Subsequent to PP-2B activation, a cytoplasmically localized subunit of nuclear factor of activated T cells (NF-AT) is translocated into the nucleus and becomes associated with a nuclear-localized subunit to generate the active NF-AT transcription factor. NF-AT induces transcription of a subset of genes that includes IL-2. Further support for this model comes from the observation that overexpression of PP-2B stimulates NF-AT-dependent transcription in Jurkat cells [O'Keefe et al., 1992; Clipstone and Crabtree, 1992]. The mechanism by which PP-2B controls NF-AT activation is not yet clear. Dephosphorylation of the cytoplasmic subunit of NF-AT that is dependent on protein phosphatase-2B may be necessary for translocation of the NF-AT subunit into the nucleus; or PP-2B may control an upstream element that is required for nuclear translocation of the NF-AT subunit [Schrieber, 1992].

2. PP-2B opposes mating factor–induced growth arrest in yeast. Two yeast genes (CNA1 and CNA2) coding for the PP-2B catalytic A subunit, and a gene coding for the regulatory B subunit (CNB), have been cloned from *Saccharomyces cerevisiae.* Mutants lacking all three genes were viable, but they were hypersensitive to α-mating factor–induced growth arrest, and they were defective in their ability to recover from this growth arrest [Cyert and Thorner, 1992]. Cyclosporin A and FK-506 inhibit yeast PP-2B after forming complexes with cellular receptor proteins. The effects of these drugs on pheromone-induced growth arrest were similar to those observed in PP-2B null mutants [Foor et al., 1992].

3. Effect of mitogens and cell transformation on the activity or expression of protein serine/threonine phosphatases. Insulin, EGF, or platelet-derived growth factor (PDGF) treatment of Swiss mouse 3T3 cells was found to promote a rapid and transient increase (maximal effect <15 min) in the activity of PP-1, measured with phosphorylase phosphatase as substrate [Chan et al., 1988]. Another group found that insulin (but not EGF) increased the activity of PP-1 in Swiss mouse 3T3 cells using the ribosomal S6 protein as substrate, but the effect was slow to develop, requiring about 2 hr (Olivier et al., 1988). Insulin treatment also increases PP-1 activity in human and rabbit skeletal muscle [Kida et al., 1990; Dent et al., 1990] and in hepatic tissues [Toth et al., 1988], and promotes the dephosphorylation of proteins involved in metabolic regulation in liver, skeletal muscle, and adipose tissue [Cohen, 1989]. In skeletal muscle the increase in PP-1 activity is due, at least in part, to the phosphorylation of the G subunit that targets PP-1 to the glycogen particle (see above). Surprisingly, the level of PP-1 mRNA, as judged by Northern blot analysis and S1 nuclease, a single strand specific nuclease, protection assays, was decreased by 50% in rhabdomyosarcoma cells following insulin treatment [Thompson and Sommercorn, 1992], while the mRNA transcript levels of several other genes (i.e., *c-src, c-fos, c-myc, c*-Ha-*ras*) were increased. The decrease in PP-1 mRNA may serve to limit the PP-1 activation in response to insulin.

Rat hepatocellular carcinomas exhibit elevated expression of a type 1 phosphatase (PP-1α) suggesting a positive involvement of this phosphatase in hepatocarcinogenesis or in hepatic cell proliferation [Sasaki et al., 1990].

B. Protein Tyrosine Phosphatases

1. Role in lymphocyte activation. Interaction of T and B lymphocytes with antigen results in activation leading to proliferation and differentiation into mature, immune-responsive cells. The first step in the activation process is the binding of antigen to a specific cell surface receptor. The T-cell receptor (TCR) complex consists of an antigen-binding TCR component made up of clonotypic α–β heterodimers which interact with a CD3 component (consisting of ϵ–γ and ϵ–δ dimers) and with ζ–ζ or ζ–η dimers [Clevers et al., 1988; Janeway, 1992]. The antigenic signal is then transduced across the membrane into the cell and ultimately to the nucleus. The hallmarks of activation include (a) phospholipase C–mediated inositol 1,4,5-triphosphate (IP3) and diacylglycerol (DAG) generation, (b) elevation of cellular Ca^{2+} and protein kinase C activation, and (c) increase in the tyrosine phosphorylation of cellular proteins.

Activation of a protein tyrosine kinase(s) that phosphorylates CD3 and the ζ chain is one of the earliest steps in this signal transduction pathway [Sefton and Campbell, 1991; Shaw and Thomas, 1991; Klausner and Samelson, 1991]. The tyrosine phosphorylation sites on these two proteins are located on sequences referred to as the antigen recognition activation motifs (ARAM). These sequences are also known as antigen receptor homology 1 motifs (ARH1) or tyrosine-based activation motifs (TAM). They seem to play a key role in coupling the receptor to downstream elements of the signalling pathway [Reth, 1989; Samelson and Klausner, 1992; Weiss, 1993].

CD45, a lympocyte-specific transmembrane PTP [Charbonneau et al., 1988] has been shown to be essential for T- and B-cell activation via their respective antigen receptors [Shaw and Thomas, 1991; Trowbridge, 1991; Cambier, 1992]. Compelling evidence for the importance of CD45 in signalling comes from studies involving T cells that lacked CD45. $CD4^+$ or $CD8^+$ T cells that were $CD45^-$ were unable to proliferate in response to antigen [Pingel and Thomas, 1989], were incapable of generating phosphatidylinositol-derived IP3 [Koretzky et al., 1990], and did not show increased tyrosine phosphorylation of proteins or increased interleukin production in response to antigen [Koretzky et al., 1991]. These results indicate that CD45 is essential for TCR signalling and probably is an early component of the pathway that is required for the activation of a tyrosine kinase.

None of the components of the T-cell receptor complex are themselves kinases; however, two members of the src family of protein tyrosine kinases, $p56^{lck}$ and $p59^{fyn}$, have been implicated in T-cell receptor signalling. The

$p56^{lck}$ kinase is exclusively expressed in T cells and binds with high affinity to the cytoplasmic domains of CD4 and CD8. It has also been shown that *in vitro* dephosphorylation of the regulatory tyr-505 of $p56^{lck}$ by CD45 stimulates protein kinase activity [Mustelin and Altman, 1990]. Also these two proteins interact to form a complex *in vivo* [Schraven et al., 1991].

The $p59^{fyn}$ protein kinase is associated with the TCR/CD3 complex although it is not exclusive to T cells [Samelson et al., 1990]. The activity of $p59^{fyn}$ is probably stimulated by CD45 [Shiroo et al., 1992]. Further, T cells obtained from transgenic mice that overexpress $p59^{fyn}$ showed an increased sensitivity to antigen stimulation [Cooke et al., 1991], although a similar increase in sensitivity was not seen when $p56^{lck}$ was overexpressed in transgenic mice.

A third protein tyrosine kinase implicated in T-cell activation is ZAP-70. It associates with the activated T-cell receptor complex, presumably through its SH2 domain that binds the phosphorylated ζ and CD3 chains of the TCR. The ZAP-70 protein kinase seems to act downstream from $p56^{lck}$ and $p59^{fyn}$ and it does not seem to be modulated by CD45 [Chan et al., 1991; Samelson and Klausner, 1992; Weiss, 1993].

B-lymphocyte activation exhibits many parallels to the T-cell activation process. Membrane-associated IgM and IgD act as primary antigen receptors. Elevation of cellular Ca^{2+}, activation of PKC, and enhanced tyrosine phosphorylation of B-cell proteins has been observed in response to antigen binding. Importantly, B cells lacking CD45 are not activated in response to antigen, which again indicates the essential role of CD45 in B-cell signalling [Justement et al., 1991].

Mechanisms for the regulation of CD45 in T and B cells are not well established. A Ca^{2+}-dependent regulation of CD45 activity that is mediated by a protein serine/threonine phosphatase in T cells has been suggested (Section II.D.2). The extracellular domain of CD45 could function as a receptor and thus regulate the PTP activity of this protein. However, studies have recently been carried out using chimeric CD45 molecules in which the CD45-derived cytoplasmic domain was intact and the extracellular and transmembrane domains were replaced by the corresponding domains of a major histocompatability class I protein [Hovis et al., 1993] or replaced with the short myristylated N-terminal sequence of the $pp60^{src}$ protein kinase [Volarevic et al., 1993]; these studies have shown that the extracellular and transmembrane domains of CD45 are not critical for its function in TCR-mediated signal transduction. The chimeric molecules could fully replace the CD45 requirement for T-cell activation in cells that were $CD45^-$.

2. Overexpression of PTPα. Overexpression of the transmembrane PTP, PTPα, in rat embryo fibroblasts induced transformation of the cells and caused persistent activation of c-Src by dephosphorylating tyrosine-527 on c-Src [Zheng et al., 1992]. The transformed cells exhibited enhanced colony formation and had the ability to form tumors with a short latency period in nude mice. These results, along with evidence for their role in T-cell activation, indicate the positive role played by phosphatases in cell proliferation and transformation.

3. The expression of many PTPs is enhanced by mitogens. An intracellular PTP, HePTPase, is expressed exclusively in the thymus and spleen of the mouse [Zanke et al., 1992]. The expression of HePTase mRNA increased 10- to 15-fold upon stimulation with phytohemagglutinin, concanavalin A, oligopolysaccharides, or anti-CD3 antibody. The expression of several members of the VH1 family of dual-specificity protein phosphatases is also regulated by mitogens. The Pac-1 nuclear-localized PTP is expressed predominantly in hematopoietic tissues. It is transiently expressed during the G_1 phase in mitogen-activated T cells and is constitutively expressed in T cells infected with human T-cell leukemia virus (HTLV-1) [Rohan et al., 1993]. Expression of the *pac*-1 gene is also increased by treating human peripheral blood T cells with lectins and phorbol esters [Rohan et al., 1993]. Another VH1-type PTP, 3CH34, is induced by serum in 3T3 cells [Charles et al., 1992].

V. PROTEIN PHOSPHATASES IN CELL CYCLE REGULATION

Protein phosphatases have multiple roles in regulating cell cycle progression. They stimulate and inhibit progression from G_2 into M phase, they are required for exit from M phase, and they are required for progression from G_1 to S phase.

A. Regulation of the G_2 to M Phase Transition

1. PTPs are required for M-phase entry. The Cdc25 protein is a member of the dual-specificity subgroup of the PTP family [Millar et al., 1991a; Gautier et al., 1991; Dunphy and Kumagai, 1991]. Members of this subgroup catalyze dephosphorylation of serine, threonine, and tyrosine residues of substrate proteins. Members of this subgroup are distantly related to the transmembrane and intracellular PTPs [Gautier et al., 1991].

The *cdc*25 gene was initially identified in the fission yeast, *Schizosacchoromyces pombe,* as a mitotic inducer that opposed the actions of *wee*1 and *mik*1, two redundant genes that suppress mitosis [Fantes, 1979; Russell and Nurse, 1986; Lundgren et al., 1991]. The Cdc25, Wee1, and Mik1 proteins regulate the activity of M-phase promoting factor (MPF), a protein that is the key regulator of the G_2 to M phase transition in all eukaryotes [Murray and Kirschner, 1989; Nurse, 1990; Murray, 1992]. The MPF protein consists of two nonidentical subunits, $p34^{cdc2}$ and cyclin B. Wee1 and Mik1 are members of the dual-specificity subgroup of protein kinases [Lindberg et al., 1992], and Wee1 (and probably also Mik1) phosphorylates the $p34^{cdc2}$ subunit of MPF [Lundgren et al., 1991; Parker et al., 1991]. The protein kinase, $p34^{cdc2}$, is inactive in the G_2 phase but becomes active in the M phase. The phosphorylation of $p34^{cdc2}$ occurs on tyrosine-15 in yeast [Gould and Nurse, 1989], and on threonine-14 and tyrosine-15 in animal cells [Krek and Nigg, 1991a; Norbury et al., 1991]. Both of these residues are in the ATP-binding site and phosphorylation of this site inactivates $p34^{cdc2}$ and maintains MPF in an inactive state in the G_2 phase. In yeast, phosphorylation of tyrosine-15 is sufficient to inactivate MPF [Gould and Nurse, 1989], whereas in animal cells, both phosphorylations are necessary for full inactivation [Krek and Nigg, 1991b; Norbury et al., 1991]. At the G_2/M boundary, Cdc25 dephosphorylates $p34^{cdc2}$; this results in activation of the histone H1 kinase activity of $p34^{cdc2}$ and of the M-phase-promoting activity of MPF. Several lines of evidence indicate that Cdc25 is the phosphatase responsible for activating MPF. The Cdc25 phosphatase has been shown to dephosphorylate and to activate yeast and *Xenopus laevis* $p34^{cdc2}$ *in vitro* [Millar et al., 1991b; Kumagai and Dunphy, 1991; Gautier et al., 1991]. Point mutations that result in substitutions of the critical cysteine residue in the HC motif abolish *in vitro* phosphatase activity and block the ability of Cdc25 to act as an M-phase inducer [Millar et al., 1991a; Dunphy and Kumagai, 1991; Gautier et al., 1991]. In yeast, mutations that inactivate Cdc25 result in G_2 arrest, which is coincident with hyperphosphorylation of Cdc2 on tyrosine-15 [Gould and Nurse, 1989]. Also in yeast, Cdc25 function is not needed in cells where tyrosine-15 of $p34^{cdc2}$ is replaced with phenylalanine.

The regulatory system comprised of Cdc25, Wee1, and Mik1 is involved in coordinating the completion of S phase with the beginning of M phase, and in delaying M phase until damaged DNA has been repaired [see Murray, 1992, for review and complete references]. In wild-type yeast cells, the presence of unreplicated or damaged DNA arrests cells in G_2 phase. Overexpression of Cdc25, null mutations in *wee*1 plus *mik*1, or mutations in *cdc*2 that uncouple it from this regulatory system also block the ability of unreplicated DNA to arrest cells in G_2 phase. In the case of damaged DNA, the G_2 phase arrest is completely lost in *wee*1⁻ cells and is partially lost in cells that overexpress Cdc25.

The Cdc25 phosphatase is found in a wide range of eukaryotic species from yeast to man [Millar and Russell, 1992]. All Cdc25s have a related catalytic domain of about 20 kD near the C-terminus but are more divergent in their N-terminal sequences.

In contrast to many protein phosphatases, Cdc25 shows an extreme preference for $p34^{cdc2}$ as a substrate, although it does have weak activity towards *p*-nitrophenyl phosphate and towards certain peptide and protein substrates [Millar et al., 1991a; Dunphy and Kumagai, 1991]. B-type cyclins stimulate the activity of Cdc25 by about five-fold, whereas cyclins A and D have no effect on Cdc25 activity [Galaktionov and Beach, 1991]. This effect of B-type cyclins may account in part for their preference for MPF.

In view of the importance of Cdc25 to cell cycle regulation there is considerable interest in mechanisms for regulating the expression or activity of this protein phosphatase. There are conflicting observations concerning the regulation of expression of Cdc25. In fission yeast, the levels of Cdc25 mRNA and protein product oscillate during the cell cycle, reaching a maximum at mitosis [Moreno et al., 1990; Ducommun et al., 1990]. The level of Cdc25 mRNA also oscillates in postblastoderm embryonic cell cycles of *Drosophila* [Edgar and O'Farrell, 1989]. By contrast, in frogs and humans [*Xenopus laevis* and HeLa cells) the level of the Cdc25 protein does not vary during the cell cycle, even though in HeLa cells the level of one form of Cdc25 mRNA (Cdc25-C) does vary during the cell cycle [Jessus and Beach, 1992; Sadhu et al., 1990; Millar et al., 1991b].

There is now strong evidence that the activity of Cdc25 is controlled via phosphorylation and dephosphorylation and that this control may be of critical importance for control of the cell cycle. The Cdc25 phosphatase has been found to be phosphorylated in *S. pombe, Xenopus laevis* eggs and early embryos, and HeLa cell extracts [Ducommun et al., 1990; Moreno et al., 1990; Kumagai and Dunphy, 1992; Hoffmann et al., 1993]. The N-terminal domain of the protein becomes hyperphosphorylated on serine and threonine residues during mitosis [Kumagai and Dunphy, 1992; Hoffmann et al., 1993]. This phosphorylation increases Cdc25 activity three- to six-fold when it is assayed using either $p34^{cdc2}$ or *p*-nitrophenyl phosphate as the substrate. The Cdc25 protein is also phosphorylated by the $p34^{cdc2}$ subunit of MPF; however, protein kinases associated with cyclin A do not phosphorylate Cdc25 [Hoffmann et al., 1993]. Human Cdc25-C protein has been shown to activate MPF in *Xenopus laevis* egg extracts and to induce oocyte maturation, but this activity requires stable thiophosphorylation of Cdc25-C [Hoffmann et al., 1993]. The phosphorylation and activation of Cdc25 by MPF provides a positive feedback loop which may greatly accelerate MPF activation at mitosis. This regulatory mechanism may also account for the MPF self-amplification observed in *Xenopus laevis* oocytes injected with small amounts of MPF.

Recently another PTP, termed Pyp3, that promotes MPF activation in *S. pombe* has been identified [Millar et al., 1992a]. Surprisingly, this PTP is more closely related to the intracellular class of PTPs than to the dual-specificity subclass of Cdc25. The *pyp*3 gene was identified as a high-copy suppressor of a *cdc*25-mutation. This PTP dephosphorylates and activates $p34^{cdc2}$ *in vitro*. When Cdc25 is present, Pyp3 is dispensable for cell cycle progression (only a slight delay in *pyp*3⁻ cells), but when Cdc25 function is disrupted, Pyp3 is essential for cell cycle progression. Thus in *S. pombe*, Cdc25 is the dominant $p34^{cdc2}$ phosphatase (for tyr-15) and Pyp3 only has a significant role when Cdc25 is absent.

2. Negative regulation of M-phase entry by PTPs. The Pyp1 (60 kD) and Pyp2 (85 kD) proteins are two additional intracellular type PTPs that have been identified in *S. pombe* [Ottilie et al., 1991; Ottilie et al., 1992; Millar et al., 1992b]. These two PTPs have overlapping functions; deletion of both genes results in the loss of cell viability, whereas the deletion of either gene by itself has no effect on cell viability [Ottilie et al., 1992; Millar et al., 1992b]. The functions of Pyp1 and Pyp2 do not overlap with that of Pyp3, a third intracellular PTP that promotes MPF activation (see above) [Millar et al., 1992a].

The Pyp1 protein (and to a lesser extent Pyp2) is a negative regulator of mitosis that promotes the inhibitory Wee1 pathway. Overexpression of Pyp1 or Pyp2 delays mitosis [Ottilie et al., 1992]. This delay is not observed in the absence of functional Wee1 or in cells in which Cdc2 is unresponsive to Wee1. Disruption of either Pyp1 or Pyp2 rescues the G_2 arrested phenotype of *cdc25*$^-$ cells [Millar et al., 1992b]. In addition, disruption of Pyp1 advances mitosis, supresses overexpression of Wee1, and causes a lethal mitotic catastrophe in Cdc25 overproducer cells. Deletion of Pyp1 has no effect in *wee1*$^-$ cells.

A negative effect of tyrosine dephosphorylation on the progression from G_2 to M phase has also been seen in the meiotic maturation of *Xenopus laevis* oocytes [Tonks et al., 1990b]. Microinjection of a mammalian intracellular PTP, human placental PTP-1B, delayed oocyte maturation induced by progesterone, insulin, and MPF.

3. Negative regulation of M-phase entry by the protein serine/threonine phosphatase PP-2A. The protein phosphatase inhibitor okadaic acid was found to stimulate the progression from G_2 to M in several animal species. In particular, okadaic acid induced $p34^{cdc2}$ kinase activation in interphase, shortened the lag period associated with activation of $p34^{cdc2}$ by cyclin B, and eliminated the requirement for a threshold level of cyclin for activation of $p34^{cdc2}$ in *Xenopus laevis* egg extracts [Felix et al., 1990; Solomon et al., 1990]. Okadaic acid was also found to stimulate MPF activation, germinal vesicle breakdown, and chromatin condensation in *Xenopus*, starfish, and mouse oocytes [Goris et al., 1989; Picard et al., 1989; Rime and Ozon, 1990; Gavin et al., 1991; Alexandre et al., 1991]. Okadaic acid also transiently activated MPF and induced a mitosis-like state in BHK21 cells [Yamashita et al., 1990]. Because okadaic acid is a very specific inhibitor of PP-1 and PP-2A (see above), this suggested that one or both protein phosphatases function(s) as an inhibitor of MPF activation.

Two lines of evidence point towards PP-2A as an inhibitor of MPF activation. First, an activity, termed INH, that inhibited MPF-induced activation of preMPF was found in Xenopus oocyte extracts [Cyert and Kirschner, 1988]. The INH activity also delayed activation of $p34^{cdc2}$ by cyclin B in Xenopus egg extracts [Solomon et al., 1990]. Purification of INH revealed that it consisted of the catalytic subunit of PP-2A plus two additional polypeptides of 52 kD and 55 kD [Lee et al., 1991]. Second, deletion of one of the two PP-2A isozymes in *S. pombe* resulted in a small cell size; this suggests that the PP-2A$^-$ cells were prematurely entering mitosis [Kinoshita et al., 1990].

There are two potential targets for PP-2A. One is a threonine phosphorylation site of $p34^{cdc2}$. Phosphorylation of this site (thr-167 in *S. pombe* and thr-161 in *Xenopus laevis*) is necessary for expression of kinase activity [Gould et al., 1991; Lorca et al., 1992; Krek and Nigg, 1991a]. Protein phosphatase-2A (INH) was found to inactivate *Xenopus* MPF *in vitro,* and the progress of inactivation correlated with the progress of dephosphorylation of the threonine-161 site [Lee et al., 1991]. A second potential target of PP-2A is Cdc25, because serine/threonine phosphorylation at mitosis greatly activates the latter enzyme (see above). Treatment of *Xenopus* egg extracts with okadaic acid produced premature phosphorylation and activation of Cdc25, a result which suggests that PP-1 or PP-2A is involved in Cdc25 dephosphorylation [Kumagai and Dunphy, 1992].

4. Role of PP-1. The protein serine/threonine phosphatase, PP-1, has also been implicated in regulating progression from G_2 to M phase, but its precise role is not well defined. In *S. pombe,* overexpression of PP-1 produces a phenotype in which MPF activation becomes dependent on Cdc25 activity, even though Wee1 is absent. A gene, termed *bsw*1, was identified that bypassed the *wee*1 suppression of a temperature-sensitive *cdc*25 mutation when Bsw1 was expressed from a plasmid [Booher and Beach, 1989]. Further characterization of the gene revealed that it corresponded to one of two PP-1 genes in *S. pombe*. The *bws*1 gene was also identified by another group using a different genetic screen (see below) and was termed *dis*2 [Ohkura et al., 1989].

Protein phosphatase-1 also antagonizes the ability of PP-2A (INH) to inhibit MPF activation in *Xenopus* oocyte extracts [Cyert and Kirschner, 1988]. Inhibitor-1, a specific PP-1 inhibitor (see above), markedly increased the ability of PP-2A to inhibit MPF activation but had no effect on MPF activation in the absence of PP-2A. The mechanism underlying this effect is unknown.

B. Protein Serine/Threonine Phosphatases Are Required for Mitotic Exit

1. Protein phosphatase 1. Genetic experiments in fission yeast, *Aspergillus*, and *Drosophila*, as well as cell biology experiments in mammalian cells, indicate that PP-1 is essential for eukaryotic cells to exit from mitosis. In *S. pombe* two PP-1 genes have been identified [Ohkura et al., 1989]. One gene, termed *dis*2, was identified because a cold-sensitive mutant exhibited a phenotype in which the cells entered mitosis normally but failed to exit from mitosis, due, at least in part, to a block in chromosome disjunction. The second PP-1 gene, *sds*21, was identified as a high-dosage suppressor of the cold-sensitive *dis*2 mutation. Deletion of both *sds*21 and *dis*2 was lethal, but deletion of either gene alone was not.

An additional gene, termed *sds*22, that acts as a high-dosage suppressor of the cold-sensitive *dis*2 mutation, has been identified [Ohkura and Yanagida, 1991]. This gene may be involved in PP-1 regulation or in substrate targeting. The protein encoded by the *sds*22 gene has an unusual structure consisting of 11 internal repeats of a 22-residue leucine-rich amino acid sequence. The protein is enriched in the nuclear fraction and is essential for the midmitotic metaphase–anaphase transition. This function of the gene is dispensable in the presence of a high dosage of PP-1 genes. Thus Sds22 seems to facilitate PP-1 action but does not substitute for PP-1 function.

The *bim*G gene encodes for PP-1 in *Aspergillus* [Doonan and Morris, 1989]. Mutation of *bim*G blocks cells in mitosis, and affects anaphase spindle elongation, spindle pole, and nuclear phosphorylation, and disrupts normal cell elongation. The *bim*G mutation can be suppressed by mammalian PP-1 [Doonan et al., 1991].

One of four *Drosophila* PP-1 genes, PP-1 87B, also seems to be involved in regulating the exit from mitosis [Axton et al., 1990; Dombradi et al., 1990]. Flies with mutations at this locus die at the larval-pupal boundary with little or no imaginal cell proliferation. Neuroblasts in these mutant flies are delayed in their progress through mitosis, are defective in spindle organization, exhibit abnormal sister chromatid separation, are hyperploid, and have excessive chromosome condensation.

In mammalian cells, microinjection of PP-1 antibodies was found to block cells in metaphase and to accelerate cytokinesis in anaphase cells [Fernandez et al., 1992]. The subcellular distribution of PP-1 was also found to vary in a cell cycle–dependent manner. In G_1 and S phases PP-1 was localized in the cytoplasm, whereas in G_2 and M phases it was found in the nucleus. Protein phosphatase-1 colocalized with individual chromosomes at mitosis.

2. Protein phosphatase 2A. There is evidence in *Drosophila* that PP-2A may also be involved in the exit from mitosis. The Drosophila gene for the 55-kD regulatory subunit of PP-2A corresponds to the locus *abnormal anaphase resolution* (*aar*) [Mayer-Jaekel et al., 1993]. Abnormalities associated with this mutation are overcondensed chromosomes that are frequently hyperploid, anaphases with lagging chromosomes, and anaphase figures that show bridged or stretched chromosomes.

C. SIT4 Regulates the Progression From G_1 Into S Phase

The SIT4 gene of the budding yeast *S. cerevisiae* encodes a 35.5-kD protein serine/threonine phosphatase that is 55% identical in amino acid sequence to the catalytic subunit of PP-2A and 40% identical to the catalytic subunit of PP-1 [Arndt et al., 1989]. The SIT4 gene was originally identified because of its involvement in regulating the transcription of the HIS4 gene. Subsequent studies indicated that when temperature-sensitive SIT4 mutants were incubated at the nonpermissive temperature they became arrested in late G_1 phase near the execution point for CDC28, the *S. cerevisiae* homologue of the *S. pombe* Cdc2, [Sutton et al., 1991]. Moreover SIT4 was found to be essential for the normal accumulation of two S-phase cyclin genes CLN1 and CLN2; a possible additional S-phase cyclin, HCS26; and SW14, a transcriptional regulator implicated in the control of CLN1 and CLN2 expression [Fernandez-Sarabia et al., 1992]. These results suggest that SIT4 is involved in regulating the G_1 to S phase transition in *S. cerevisiae.*

VI. PROTEIN PHOSPHATASES IN DIFFERENTIATION AND MORPHOGENESIS

A. Muscle Differentiation

Treatment of mouse myoblasts with okadaic acid reversibly altered the morphology of the cells and blocked differentiation [Park et al., 1992]. At the molecular level, okadaic acid completely suppressed the expression of myogenic determination genes (*myo*D and *myogenin*) but induced the expression of an inhibitor of muscle differentiation. The effect of okadaic acid on the expression of MyoD1 occurs at the level of transcription [Kim et al., 1992]. These data suggest that protein phosphatases play an important role during myogenic differentiation.

B. Yeast Morphogenesis

A variety of morphogenetic changes occur during the life cycle of the yeast, *S. cerevisiae;* these include polarization of cell surface growth which controls bud formation, positioning of the internal organelles in preparation for cell division, septum formation, and cell separation [Healy et al., 1991]. Cytoskeletal rearrangements are at the heart of these processes. Many mutants defective in these processes have been isolated. Among them is a recently characterized cold-sensitive mutant which exhibited defective cytokinesis and cell septation, as well as abnormal outgrowths, at the restrictive temperature. This has recently been demonstrated to be a novel CDC gene (CDC55) which encodes a protein that is homologous to the regulatory subunit A of PP-2A [Healy et al., 1991].

Two homologues of the PP-2A catalytic subunit, PPH21 and PPH22, and a PP-2A related gene, PPH3, have been cloned from budding yeast. Disruption of both PPH21 and PPH22 resulted in spores that gave rise to smaller colonies, whereas disruption of either gene alone had no effect; thus the two *S. cerevisiae* PP-2A genes have overlapping function. Disruption of PPH3 alone had little effect on cell growth, but disruption of this gene together with deletion of PPH21 and PPH22 completely prevented spore growth, so that PPH3 also has PP-2A-like function. Overexpression of PP-2A resulted in elongated cells or in a balloonlike phenotype, characterized by extremely large cells filled with large vacuoles [Ronne et al., 1991]. Results from these studies indicate PP-2A has a crucial role in yeast cell growth and bud morphogenesis.

VII. PROTEIN PHOSPHATASES IN DEVELOPMENT

A. Protein Serine/Threonine Phosphatases

1. *Drosophila* retinal degeneration C (*rdg*C) gene. The *rdg*C gene encodes a protein serine/threonine phosphatase, and genetic experiments have shown that it is necessary for the prevention of light-induced retinal degeneration in *Drosophila*. Molecular analysis of the gene indicated that *rdg*C codes for a novel protein serine/threonine phosphatase. It exhibits about 30% homology to PP-1, PP-2A, and PP-2B and contains a putative Ca^{2+}-binding domain, which suggests that this enzyme may be regulated by Ca^{2+} *in vivo* [Steele et al., 1992]. The *rdg*C gene product seems to be required in a pathway that is initiated by rhodopsin stimulation; however, it does not require the action of other known downstream elements of the established rhodopsin-initiated pathway (e.g., *norp*A that encodes a phospholipase C). This suggests that the RdgC phosphatase may be involved in an alternate rhodopsin-initiated pathway.

2. Protein phosphatase-2B. Protein phosphatase-2B also shows tissue-specific variations in its isozyme expression during differentiation and development. It is most abundantly expressed in brain tissues [Klee et al., 1979]. The expression of PP-2B mRNA is subject to regional and temporal control during brain development, which suggests that specific synaptic connections may influence PP-2B gene expression [Polli et al., 1991]. A recently cloned nonneural isozyme of the PP-2B gene has been shown to increase dramatically in its expression during testicular development in the mouse [Muramatsu et al., 1992].

B. Protein Tyrosine Phosphatases

1. *Drosophila corkscrew* gene (*csw*). The *csw* gene is required for the formation of terminal structures of the *Drosophila* embryo. The gene has recently been shown to encode an intracellular PTP with two amino-terminal SH2 domains [Perkins et al., 1992]. The Csw protein acts downstream of the receptor tyrosine kinase, Torso, and positively regulates the Torso signalling pathway in conjunction with the Raf (*pole hole*) protein serine/threonine kinase. The Csw protein transduces signals to downstream effectors encoded by the *huckebein* and *tailless* genes that are putative transcription factors. There is no biochemical evidence to show that Csw directly interacts with the Torso protein kinase. However, in view of Csws SH2 domains, it is expected that it interacts with a phosphotyrosyl protein.

Several SH2-containing phosphatases have been identified in mammals. These include the variously named PTP-1C/SH-PTP1/HCP/HSP [Shen et al., 1991; Feng et al., 1993] and PTP-1D/SH-PTP2/Syp PTP [Freeman et al., 1992; Feng et al., 1993; Vogel et al., 1993] The close resemblance of protein tyrosine phosphatase-1D to Csw and its extensive expression in early mouse embryos suggests a Csw-like role for PTP-1D in mammalian development. Protein tyrosine phosphatase-1D has been shown to associate with the EGFR and the PDGF receptor through its SH2 domains and is rapidly phosphorylated by these kinases [Feng et al., 1993]. This phosphorylation increases the PTP activity of PTP-1D [Vogel et al., 1993].

2. A *Dictyostelium discoideum* PTP is involved in developmental regulation. Protein tyrosine phosphatase-1 of *Dictyostelium discoideum* has about 38%–50% homology to mammalian PTPs and its expression is regulated in a spatiotemporal manner during development [Howard et al., 1992]. Vegetative cells express the PTP at low levels, and maximal expression is observed at the tight aggregate stage. Further, the expression is predominantly localized in anterior-like cells and stalk cells. Both disruption and overexpression of the PTP-1 gene caused abnormal development. These observations suggest an important role for PTP-1 in *Dictyostelium* development.

3. Transmembrane PTPs. Leukocyte common antigen-related protein (Lar) and several other transmembrane PTPs have extracellular motifs that resemble those of immunoglobulins and fibronectin type II and are reminiscent of neural cell adhesion molecules (NCAM). The NCAMs may be important for the cell–cell interactions occurring during developmental and postdevelopmental stages of the nervous system. It has been hypothesized that LAR and related phosphatases may behave as NCAMs that also can tranduce signals using their phosphatase activity [Charbonneau and Tonks, 1992]. Recent studies with *Drosophila* membrane PTPs (DLAR, DPTP99A, and DPTP10D) give credence to this hypothesis [Yang et al., 1991; Tian et al., 1991; Nairn and Shenolikar, 1992]. The results of analysis of mRNAs by *in situ* hybridization, and of proteins by immunocytochemistry, demonstrated distinct but overlapping patterns of expression in the developing nervous system of *Drosophila* embryos. Expression of DPTP99A mRNA precedes that of DPTP10D mRNA. Both proteins are mainly located in axons. The temporal coincidence of the transient expression of these proteins with axon outgrowth suggests a role for these PTPs in signal transduction pathways that are involved in axon outgrowth and guidance, and in the establishment of the embryonic central nervous system scaffold.

VIII. CONCLUSIONS

Three gene families of protein phosphatases remove phosphate from serine, threonine, and/or tyrosine residues of cell proteins. Over the past few years the identification and molecular characterization of protein phosphatases in eukaryotes has proceeded at a breathtaking pace. Progress has also been made in understanding mechanisms for regulating the activity and the subcellular localization of these enzymes. Knowledge of protein phosphatase function has been greatly extended by genetic approaches especially in yeast and *Drosophila,* and by the discovery of pharmacological inhibitors (e.g., okadiac acid) of protein serine/threonine phosphatases. These studies have shown that protein phosphatases play key roles in signal transduction pathways that stimulate and inhibit cell proliferation, and that control the entrance and exit from M phase in eukaryotic cells. Additionally there is growing evidence for their involvement in differentiation and development.

REFERENCES

Alexandre H, Van Cauwenberge A, Tsukitani Y, Mulnard J (1991): Pleiotropic effect of okadaic acid on maturing mouse oocytes. Development 112:971–980.

Arndt KT, Styles CA, Fink GR (1989): A suppressor of a *HIS*4 transcriptional defect encodes a protein with homology to the catalytic subunit of protein phosphatases. Cell 56:527–537.

Axton JM, Dombradi V, Cohen PTW, Glover DM (1990): One of the protein phosphatase 1 isozymes in *Drosophila* is essential for mitosis. Cell 63:33–46.

Beullens M, Van Eynde A, Stalmans W, Bollen M (1992): The isolation of novel inhibitory polypeptides of protein phosphatase1 from bovine thymus nuclei. J Biol Chem 267:16538–16544.

Bollen M, Stalmans W (1992): The structure, role, and regulation of type 1 protein phosphatases. Crit Rev Biochem Mol Biol 27:227–281.

Booher R, Beach D (1989): Involvement of a type 1 protein phosphatase encoded by bws1$^+$ in fission yeast mitotic control. Cell 57:1009–1016.

Brown-Shimer S, Johnson KA, Hill DE, Bruskin AM (1992): Effect of protein-tyrosine-phosphatase 1B on transformation by human neu oncogene. Cancer Res 52:478–482.

Buchkovich KL, Duffy LA, Harlow E (1989): The retinoblastoma protein is phosphorylated during specific phases of the cell cycle. Cell 58:1097–1105.

Cambier JC (1992): Signal transduction by T- and B- cell antigen receptors: converging structures and concepts. Curr Opin Immunol 4:257–264.

Carmichael GG, Schaffhausen BS, Dorsky DI, Oliver DB, Benjamin TL (1982): Carboxy terminus of polyoma middle-sized tumor antigen is required for attachment to membranes, associated protein kinase activities and cell transformation. Proc Natl Acad Sci USA 79:3579–3583.

Cayla X, Goris J, Hermann J, Hendrix P, Ozon R, Merlevede W (1990): Isolation and characterization of a tyrosyl phosphatase activator from rabbit skeletal muscle and Xenopus laevis oocytes. Biochemistry 29:658–667.

Chan CP, McNall SJ, Krebs EG, Fischer EH (1988): Stimulation of protein phosphatase activity by insulin and growth factors in 3T3 cells. Proc Natl Acad Sci USA 85:6257–6261.

Chan AC, Irving BA, Fraser JD, Weiss A (1991): The ζ chain is associated with a tyrosine kinase and upon T-cell antigen stimulation associates with ZAP-70, a 70 kD tyrosine phosphoprotein. Proc Natl Acad Sci USA 88:9166–9170.

Charbonneau H, Tonks NK (1992): 1002 protein phosphatases? Annu Rev Cell Biol 8:463–493.

Charbonneau H, Tonks NK, Walsh KA, Fischer EH (1988): The leukocyte common antigen (CD45): a putative receptor linked tyrosine phosphatase. Proc Natl Acad Sci USA 85:7182–7186.

Charles CH, Aber AS, Lau LF (1992): cDNA sequence of a growth factor inducible immediate early gene and characterization of its encoded protein. Oncogene 7:187–190.

Chen J, Martin BL, Brautigan DL (1992): Regulation of protein serine-threonine phosphatase type-2A by tyrosine phosphorylation. Science 257:1261–1264.

Cho H, Krishnaraj R, Itoh M, Kitas E, Bannwarth W, Saito H, Walsh CT (1993): Substrate specificities of catalytic fragments of protein tyrosine phosphatases (HPTPβ, LAR, CD45) toward phosphotyrosyl peptide substrates and thiophosphorylated peptides as inhibitors. Protein Sci 2:977–984.

Clevers H, Alarcon B, Wileman T, Terhorst C (1988): The T-cell receptor/CD3 complex: a dynamic protein ensemble. Annu Rev Immunol 6:629–662.

Clipstone NA, Crabtree GR (1992): Identification of calcineurin as a key signalling enzyme in T-lymphocyte activation. Nature 357:695–697.

Cohen P (1989): The structure and regulation of protein phosphatases. Annu Rev Biochem 58:453–508.

Cohen P (1991): Classification of protein serine/threonine phosphatases: identification and quantitation in cell extracts. Methods Enzymol 201:389–398.

Cohen P, Holmes CFB, Tsukitani Y (1990a): Okadaic acid: a new probe for the study of cellular regulation. Trends Biochem Sci 15:98–102.

Cohen PTW, Brewis ND, Hughes V, Mann DJ (1990b): Protein serine/threonine phosphatases; an expanding family. FEBS Lett 268:355–359.

Cooke MP, Abraham KM, Forbush KA, Perlmutter RM (1991): Regulation of T-cell receptor signalling by a src family protein tyrosine kinase (p59fyn). Cell 65:281–291.

Cool DE, Tonks NK, Charbonneau H, Walsh KA, Fischer EH, Krebs EG (1989): cDNA isolated from a human T-cell library encodes a member of the protein-tyrosine-phosphatase family. Proc Natl Acad Sci USA 86:5257–5261.

Cool DE, Tonks NK, Charbonneau H, Fischer EH, Krebs EG (1990): Expression of human T-cell protein tyrosine phosphatase in baby hamster kidney cells. Proc Natl Acad Sci USA 87:7280–7284.

Cool DE, Andreassen PR, Tonks NK, Krebs EG, Fischer EH, Margolis RL (1992): Cytokinetic failure and asynchronous nuclear division in BHK cells overexpressing a truncated protein-tyrosine-phosphatase. Proc Natl Acad Sci USA 89:5422–5426.

Cyert MS, Kirschner MW (1988): Regulation of MPF activity in vitro. Cell 53:185–195.

Cyert MS, Thorner J (1992): Regulatory subunit (CNB1 gene product) of yeast Ca^{2+}/calmodulin dependent phosphoprotein phosphatases is required for adaptation to pheromone. Mol Cell Biol 12:3460–3469.

DeCaprio JA, Furukawa Y, Achenbaum F, Griffin JD, Livingston DM (1992): The retinoblastoma susceptibility gene product becomes phosphorylated in multiple stages during cell cycle entry and progression. Proc Natl Acad Sci USA 89:1795–1798.

Dent P, Lavoinne A, Nakienly S, Claudwell FB, Watt P, Cohen P (1990): The molecular mechanism by which insulin stimulates glycogen synthesis in mammalian skeletal muscle. Nature 348:302–308.

Dombradi V, Axton JM, Barker HM, Cohen PTW (1990): Protein phosphatase 1 activity in *Drosophila* mutants with abnormalities in mitosis and chromosome condensation. FEBS Lett 275:39–43.

Doonan JH, Morris NR (1989): The *bim*G gene of *Aspergillus nidulans*, required for completion of anaphase, encodes a homolog of mammalian phosphoprotein phosphatase 1. Cell 57:987–996.

Doonan JH, MacKintosh C, Osmani S, Cohen P, Gai G, Lee EYC, Morris NR (1991): A cDNA encoding rabbit muscle protein phosphatase 1α complements the *Aspergillus* cell cycle mutation, *bim*G11. J Biol Chem 266:18889–18894.

Ducommun B, Draetta G, Young P, Beach D (1990): Fission yeast cdc25 is a cell-cycle regulated protein. Biochem Biophys Res Commun 167:301–309.

Dunphy, WG, Kumagai A (1991): The cdc25 protein contains an intrinsic phosphatase activity. Cell 67:189–196.

Edgar BA, O'Farrell PH (1989): Genetic control of cell division patterns in the Drosophila embryos. Cell 57:177–187.

Fantes, P (1979): Epistatic gene interactions in the control of division of fission yeast. Nature 279:428–430.

Felix MA, Cohen P, Karsenti E (1990): Cdc2 H1 kinase is negatively regulated by a type 2A phosphatase in the Xenopus early embryonic cell cycle: evidence from the effects of okadaic acid. EMBO J 9:675–683.

Feng G-S, Hui C-C, Pawson T (1993): SH2-containing phosphotyrosine phosphatase as target of protein-tyrosine kinases. Science 259:1607–1611.

Fernandez A, Brautigan DL, Lamb NJC (1992): Protein phosphatase type 1 in mammalian cell mitosis: chromosomal localization and involvement in mitotic exit. J Cell Biol 116:1421–1430.

Fernandez-Sarabia MJ, Sutton A, Zhong T, Arndt KT (1992): *SIT4* protein phosphatase is required for the normal accumulation of *SWI4, CLN1, CLN2* and *HCS26* RNAs during late G_1. Genes Dev 6:2417–2428.

Foor F, Parent SA, Morin N, Dahl AM, Ramadan N, Chrebet G, Bostain YA, Nielsen JB (1992): Calcineurin mediates inhibition by FK-506 and cyclosporin of recovery from α-factor arrest in yeast. Nature 360:682–684.

Foulkes JG, Cohen P (1980): The regulation of glycogen metabolism: purification and properties of protein phosphatase inhibitor-2 from rabbit skeletal muscle. Eur J Biochem 105:195–203.

Frangioni JV, Beahm PH, Shifrin V, Jost CA, Neel BG (1992): The nontransmembrane tyrosine phosphatase

PTP-1B localizes to the endoplasmic reticulum via its 35 amino acid C-terminal sequence. Cell 68:545–560.

Freeman RM Jr, Plutzky J, Neel BG (1992): Identification of a human src homology 2 containing protein-tyrosine-phosphatase: a putative homolog of Drosophila corkscrew. Proc Natl Acad Sci USA 89:11239–11243.

Fruman DA, Klee CB, Bierer BE, Burakoff SJ (1992): Calcineurin phosphatase activity in T-lymphocytes is inhibited by FK506 and cyclosporin A. Proc Natl Acad Sci USA 89:3656–3690.

Fuijiki H (1992): Is the inhibition of protein phosphatase 1 and 2A activities a general mechanism of tumour promotion in human cancer development? Mol Carcinogen 5:91–94.

Galaktionov K, Beach D (1991): Specific activation of cdc25 tyrosine phosphatases by B-type cyclins: evidence for multiple roles of mitotic cyclins. Cell 67:1181–1194.

Garcia-Morales P, Minami Y, Luong E, Klausner RD, Samelson LE (1990): Tyrosine phosphorylation in T cells is regulated by phosphatase activity: studies with phenylarsine oxide. Proc Natl Acad Sci USA 87:9255–9259.

Gautier J, Solomon MJ, Booher RN, Bazan JF, Kirschner, MW (1991): cdc25 is a specific tyrosine phosphatase that directly activates $p34^{cdc2}$. Cell 67:197–211.

Gavin AC, Tsukitani Y, Schorderet-Slatkine S (1991): Induction of M-phase entry of prophase-blocked mouse ooccytes through microinjection of okadaic acid, a specific phosphatase inhibitor. Exp Cell Res 192:75–81.

Gordon JA (1991): Use of vanadate as protein-phosphotyrosine phosphatase inhibitor. Methods Enzymol 201:477–482.

Goris J, Hermann J, Hendrix P, Ozon R, Merlevede W (1989): Okadaic acid, a specific protein phosphatase inhibitor induces maturation and MPF formation in Xenopus laevis oocytes. FEBS Lett 245:91–94.

Gould KL, Nurse P (1989): Tyrosine phosphorylation of the fission yeast $cdc2^+$ protein kinase regulates entry into mitosis. Nature 342:39–45.

Gould KL, Moreno S, Owen DJ, Sazer S, Nurse P (1991): Phosphorylation at thr 167 is required for Schizosaccharomyces pombe p34 cdc2 function. EMBO J 10:3297–3309.

Gruppuso PA, Mikumo R, Brautigan DL, Braun L (1991): Growth arrest induced by transforming growth factor β1 is accompanied by protein phosphatase activation in human keratinocytes. J Biol Chem 266:3444–3448.

Grussenmeyer T, Carbone-Wiley A, Scheidtmann KH, Walter G (1987): Interactions between polyoma medium T antigen and three cellular proteins of 88, 61, and 37 kilodaltons. J Virol 61:3902–3909.

Gu M, York JD, Warshawsky I, Majerus PW (1991): Identification, cloning, and expression of a cytosolic megakaryocyte protein-tyrosine-phosphatase with sequence homology to cytoskeletal protein 4.1. Proc Natl Acad Sci USA 88:5867–5871.

Guan K, Haun RS, Watson SJ, Geahlen RL, Dixon JE (1990): Cloning and expression of a protein-tyrosine-phosphatase. Proc Natl Acad Sci USA 87:1501–1505.

Hashimoto Y, Soderling TR (1989): Regulation of calcineurin by phosphorylation. J Biol Chem 264:16524–16529.

Healy AM, Zolnierowicz S, Stapleton AE, Goebl M, DePaoli-Roach AA, Pringle JR (1991): CDC55, a Saccharomyces cerevesiae gene involved in cellular morphogenesis: identification, characterization and homology to the B subunit of mammalian type 2A protein phosphatase. Mol Cell Biol 11:5767–5780.

Hemmings HC, Greengard P, Tung HYL, Cohen P (1984): DARPP-32, A dopamine regulated neuronal phosphoprotein is a potent inhibitor of protein phosphatase-1. Nature 310:503–505.

Hernandez-Sotomayor SM, Mumby M, Carpenter G (1991): Okadaic acid induced hyperphosphorylation of the epidermal growth factor receptor. J Biol Chem 266:21281–21286.

Hippen KL, Jakes S, Richards J, Jena BP, Beck BL, Tabatabai LB, Ingebritsen TS (1993): Acidic residues are involved in substrate recognition by two intracellular protein tyrosine phosphatases PTP-5 and rrbPTP-1. Biochemistry 32:12405–12412.

Hippen KL, Jakes S, Ingraham R, Ingebritsen TS (1994a): Comparison of the substrate specificities of three intracellular and two transmembrane protein tyrosine phosphatases. Biochemistry (submitted).

Hippen KL, Jakes S, Pot D, Zhang ZY, Ingebritsen TS (1994b): Acidic residues near phosphotyrosine influence substrate reactivity by intracellular and transmembrane protein tyrosine phosphatases. Biochemistry (submitted).

Hoffman I, Clarke PR, Marcote MJ, Karsenti E, Draetta G (1993): Phosphorylation and activation of human cdc25-C by cdc2-cyclin B and its involvement in the self amplification of MPF at mitosis. EMBO J 12:53–63.

Hovis RR, Donovan JA, Musci MA, Motto DG, Goldman FD, Ross SE, Koretzky GA (1993): Rescue of signalling by a chimeric protein containing the cytoplasmic domain of CD45. Science 260:544–546.

Howard PK, Sefton BM, Firtel RA (1992): Analysis of a spatially regulated phosphotyrosine phosphatase identifies tyrosine phosphorylation as a key regulatory pathway in Dictyostelium. Cell 71:637–647.

Huang FL, Glinsman WH (1976): Separation and characterization of two phosphorylase phosphatase inhibitors from rabbit skeletal muscle. Eur J Biochem 70:419–426.

Hubbard MJ, Cohen P (1993): On target with a new mechanism for the regulation of protein phosphorylation. Trends Biochem Sci 18:172–177.

Ingebritsen TS (1989): Phosphotyrosyl-protein phosphatases: identification and characterization of two heat-stable protein inhibitors. J Biol Chem 264:7754–7759.

Ingebritsen TS, Cohen P (1983): Protein phosphatases: properties and role in cellular regulation. Science 221:331–338.

Ishihara H, Martin L, Brautigan DL, Karaki H, Ozaki H, Kato Y, Fusetani N, Watabe S, Hashimoto K, Uemura D, Hartshorne DJ (1989): Calyculin A and okadaic acid: inhibitors of protein phosphatase activity. Biochem Biophys Res Commun 159:871-877.

Issinger OC, Martin T, Richter WW, Olson M, Fujiki H (1988): Hyperphosphorylation of N-60 a protein structurally and immunologically related to nucleolin after tumor promoter treatment. EMBO J 7:1621–1626.

Janeway CA Jr (1992): The T cell receptor as a multicomponent signalling machine: CD4/CD8 coreceptors and CD45 in T cell activation. Annu Rev Immunol 10:645–674.

Jessus C, Beach D (1992): Oscillation of MPF is accompanied by periodic association between cdc25 and cdc2-cyclin B. Cell 68:323–332.

Johansen JW, Ingebritsen TS (1986): Phosphorylation and inactivation of protein phosphatase 1 by $pp60^{v\text{-}src}$. Proc Natl Acad Sci USA 83:207–211.

Justement LB, Campbell KS, Chien NC, Cambier JC (1991): Regulation of B-cell antigen receptor signal transduction and phosphorylation by CD45. Science 252:1839–1842.

Kida Y, Esposito-Delpuente A, Bogardus C, Mott DM (1990): Insulin resistance is associated with reduced fasting and insulin stimulated glycogen synthase phosphatase activity in human skeletal muscle. J Clin Invest 85:476–481.

Kim SJ, Lafyatis R, Kim KY, Angel P, Fuijiki H, Karin M, Sporn MB, Roberts AB (1990): Regulation of collagenase gene expression by okadaic acid, an inhibitor of protein phosphatases. Cell Regul 1:269–278.

Kim SJ, Kim KY, Tapscott SJ, Winokur TS, Keunchi P, Fuijiki H, Weintraub H, Roberts AB (1992): Inhibition of protein phosphatases blocks myogenesis by first altering Myo D binding activity. J Biol Chem 267:15140–15145.

Kinkaid RL, Takayama H, Billingsley ML, Sitkovsky MV (1987): Differential expression of calmodulin binding proteins in B, T lymphocytes and thymocytes. Nature 330:176–178.

Kinoshita N, Ohkura H, Yanagida M (1990): Distinct, essential roles of type 1 and type 2A protein phosphatases in the control of the fission yeast cell division cycle. Cell 63:405–415.

Klausner RD, Samelson LE (1991): T cell antigen receptor activation pathways: the tyrosine kinase connection. Cell 64:875–878.

Klee CB, Crouch TH, Krinks MH (1979): Calcineurin: a calcium and calmodulin binding protein of the nervous system. Proc Natl Acad Sci USA 76:6270–6273.

Koretzky GA, Picus J, Thomas ML, Weiss A (1990): Tyrosine phosphatase CD45 is essential for coupling T-cell antigen receptor to the phosphotidylinositol pathway. Nature 346:66–68.

Koretzky GA, Picus J, Schultz T, Weiss A (1991): Tyrosine phosphatase CD45 is required for T-cell antigen receptor and CD2 mediated activation of a protein tyrosine kinase and interleukin-2 production. Proc Natl Acad Sci USA 88:2037–2041.

Krek W, Nigg EA (1991a): Differential phosphorylation of vertebrate $p34^{cdc2}$ kinase at the G_1/S and G_2/M transitions of the cell cycle: identification of major phosphorylation sites. EMBO J 10:305–316.

Krek W, Nigg EA (1991b): Mutations of $p34^{cdc2}$ phosphorylation sites induce premature mitotic events in HeLa cells: evidence for a double block to $p34^{cdc2}$ kinase activation in vertebrates. EMBO J 10:3331–3341.

Kumagai A, Dunphy WG (1991): The cdc25 protein controls tyrosine dephosphorylation of the cdc2 protein in a cell-free system. Cell 64:903–914.

Kumagai A, Dunphy WG (1992): Regulation of the cdc25 protein during the cell cycle in Xenopus extracts. Cell 70:139–151.

LaForgia S, Morse B, Levy J, Barnea G, Cannizzaro LA, Li F, Nowell PC, Boglosian-Cell L, Glick J, Weston A, Harris CC, Drabkin H, Patterson D, Croce CM, Schlessinger J, Huebner K (1991): Receptor protein-tyrosine phosphatase γ is a candidate tumor suppresser gene at human chromosome region 3p21. Proc Natl Acad Sci USA 88:5036–5040.

Lee TH, Solomon MJ, Mumby MC, Kirschner MW (1991): INH, a negative regulator of MPF, is a form of protein phosphatase 2A. Cell 64:415–423.

Li YM, Casida JE (1992): Cantharidin-binding protein: identification as protein phosphatase 2A. Proc Natl Acad Sci USA 89:11867–11870.

Liao K, Hoffman RD, Lane MD (1991): Phosphotyrosyl turnover in insulin signaling. J Biol Chem 266:6544–6553.

Lindberg RA, Quinn AM, Hunter T (1992): Dual-specificity protein kinases: Will any hydroxyl do? Trends Biochem Sci 17:114–118.

Liu J, Farmer JD, Lane WS, Friedman J, Weissman I, Schreiber SL (1991): Calcineurin is a common target of cyclophilin-cyclosporin A and FKBP-FK506 complexes. Cell 66:807–815.

Lokeshwar VB, Bourguignon LYW (1992): Tyrosine phosphatase activity of lymphoma CD45 (GP180) is regulated by a direct interaction with the cytoskeleton. J Biol Chem 267:21551–21557.

Lorca T, Labbe JC, Devault A, Fesquet D, Capony JP,

Cavadore JC, Le Bouffant F, Doree M (1992): Dephosphorylation of cdc2 on threonine 161 is required for cdc2 kinase inactivation and normal anaphase. EMBO J 11:2381–2390.

Ludlow JW, Glendening CL, Lingston DM, DeCaprio (1993): Specific enzymatic dephosphorylation of the retinoblastoma protein. Mol Cell Biol 13:367–372.

Lundgren K, Walworth N, Booher R, Dembski M, Kirschner M, Beach D (1991): mik1 and wee1 cooperate in the inhibitory tyrosine phosphorylation of cdc2. Cell 64:1111–1122.

MacKintosh C, Beattie K, Klumpp S, Cohen P, Codd GA (1990): Cyanobacterial microcystin-LR is a potent and specific inhibitor of protein phosphatases 1 and 2A from both mammals and higher plants. FEBS Lett 264:187–192.

Mahadevan LC, Willis AC, Baratt MJ(1991): Rapid histone H3 phosphorylation in response to growth factors, phorbol esters, okadaic acid, and protein synthesis inhibitors. Cell 65:775–783.

Mayer-Jaekel RE, Ohkura H, Gomes R, Sunkel CE, Baumgartner S, Hemmings BA, Glover DM (1993): The 55 kd regulatory subunit of Drosophila protein phosphatase 2A is required for anaphase. Cell 72:621–633.

Mihara K, Cao X-R, Yen A, Chandler S, Driscoll B, Murphree AL, T'ang A, Fung YKT (1989): Cell cycle dependent regulation of phosphorylation of the human retinoblastoma gene product. Science 246:1300–1303.

Millar JBA, Russell P (1992): The cdc25 M-phase inducer: an unconventional protein phosphatase. Cell 68:407–410.

Millar JBA, McGowan CH, Lenaers G, Jones R, Russel P (1991a): $p80^{cdc25}$ mitotic inducer is the tyrosine phosphatase that activates $p34^{cdc2}$ kinase in fission yeast. EMBO J 10:4301–4309.

Millar JBA, Blevitt J, Gerace L, Sadhu K, Featherstone C, Russell P (1991b): $P55^{cdc25}$ is a nuclear protein required for the initiation of mitosis in human cells. Proc Natl Acad Sci USA 88:10500–10504.

Millar JBA, Lenaers G, Russell P (1992a): Pyp3 PTPase acts as a mitotic inducer in fission yeast. EMBO J 11:4933–4941.

Millar JBA, Russell P, Dixon JE, Guan KL (1992b): Negative regulation of mitosis by two functionally overlapping PTPases in fission yeast. EMBO J 11:4943–4952.

Moreno S, Nurse P, Russell P (1990): Regulation of mitosis by cyclic accumulation of $p80^{cdc25}$ mitotic inducer in yeast. Nature 344:549–552.

Mumby MC, Walter G (1991): Protein phosphatases and DNA tumor viruses: transformation through the back door? Cell Regul 2:589–598.

Muramatsu T, Giri RP, Higuchi S, Kincaid RL (1992): Molecular cloning of a calmodulin dependent phosphatase from murine testis: identification of a developmentally expressed non-neural isozyme. Proc Natl Acad Sci USA 89:529–533.

Murray AW (1992): Creative blocks: cell cycle checkpoints and feedback controls. Nature 359:599–604.

Murray AW, Kirschner MW (1989): Dominoes and clocks: the union of two views of the cell cycle. Science 246:614–621.

Mustelin T, Altman A (1990): Dephosphorylation and activation of the T-cell tyrosine kinase p56lck by the leukocyte common antigen (CD45). Oncogene 5:101–105.

Nairn AC, Shenolikar S (1992): The role of protein phosphatases in synaptic transmission, plasticity and neuronal development. Curr Opin Neurobiol 2:296–301.

Nimmo GA, Cohen P (1978): The regulation of glycogen metabolism. Purification and characterization of protein phosphatase inhibitor-1 from rabbit skeletal muscle. Eur J Biochem 87:341–351.

Nishida E, Gotoh Y (1993): The MAP kinase cascade is essential for diverse signal transduction pathways. Trends Biochem Sci 18:128–131.

Norbury C, Blow J, Nurse P (1991): Regulatory phosphorylation of the $p34^{cdc2}$ protein kinase in vertebrates. EMBO J 10:3321–3329.

Nurse P, (1990): Universal control mechanism regulating onset of M-phase. Nature 344:503–508.

Ohkura H, Yanagida M (1991): S. pombe gene $sds22^+$ essential for a midmitotic transition encodes a leucine-rich repeat protein that positively modulates protein phosphatase-1. Cell 64:149–157.

Ohkura H, Kinoshita N, Miyatani S, Toda T, Yanagida M (1989): The fission yeast $dis2^+$ gene required for chromosome disjoining encodes one of two putative type protein phosphatases. Cell 57:997–1007.

O'Keefe SJ, Tamura J, Kincaid RL, Tocci MJ, O'Neil EA (1992): FK-506 and CsA sensitive activation of the interleukin-2 promoter by calcineurin. Nature 357:692–694.

Olivier AR, Ballou LM, Thomas G (1988): Differential regulation of S6 phosphorylation by insulin and epidermal growth factor in Swiss mouse 3T3 cell: Insulin activation of type I phosphatase. Proc Natl Acad Sci USA 85:4720–4724.

Ostergaard HL, Trowbridge IS (1991): Negative regulation of CD45 protein tyrosine phosphatase activity by ionomycin in T cells. Science 253:1423–1425.

Ottilie S, Chernoff J, Hannig G, Hoffman CS, Erikson RL (1991): A fission-yeast gene encoding a protein with features of protein-tyrosine-phosphatases. Proc Natl Acad Sci USA 88:3455–3459.

Ottilie S, Chernoff J, Hannig G, Hoffman CS, Erikson RL (1992): The fission yeast genes $pyp1^+$ and $pyp2^+$ encode protein tyrosine phosphatases that negatively regulate mitosis. Mol Cell Biol 12:5571–5580.

Ouimet CC, Miller P, Hemmings HC, Walaas SI, Greengard P (1984): DARPP-32, a dopamine- and adenosine 3′:5′-monophosphate-regulated phosphoprotein enriched in dopamine-innervated brain regions, III. Immunocytochemical localization. J Neurosci 4: 111–124.

Pan MG, Florio T, Stork PJS (1992): G-protein activation of a hormone-stimulated phosphatase in human tumor cells. Science 256:1215–1217.

Park K, Chung M, Kim S-J (1992): Inhibition of myogenesis by okadaic acid an inhibitor of protein phosphatases, 1 and 2A, correlates with the induction of AP1. J Biol Chem 267:10810–10815.

Parker LL, Atherton-Fessler S, Lee MS, Ogg S, Falk JL, Swenson KI, Piwnica-Worms H (1991): Cyclin promotes tyrosine phosphorylation of $p34^{cdc2}$ in a $wee1^{+}$ dependent manner. EMBO J 10:1255–1263.

Perkins LA, Larsen I, Perrimon N (1992): Corkscrew encodes a putative protein tyrosine phosphatase that functions to transduce the terminal signal from the receptor tyrosine kinase Torso. Cell 70:225–236.

Picard A, Capony JP, Brautigan DL, Doree M (1989): Involvement of protein phosphatase 1 and 2A in the control of M phase promoting factor (MPF) activity in starfish. J Cell Biol 109:3347–3354.

Pingel JT, Thomas ML (1989): Evidence that the leukocyte common antigen is required for antigen induced T-lymphocyte proliferation. Cell 58:1055–1065.

Polli JW, Billingsley ML, Kinkaid RL (1991): Expression of the calmodulin-dependent protein phosphatase, calcineurin, in rat brain: developmental patterns and the role of nigrostriatal innervation. Dev Brain Res 63:105–119.

Pot DA, Dixon JE (1992): A thousand and two protein tyrosine phosphatases. Biochim Biophy Acta 1136:35–43.

Pot DA, Woodford TA, Remboutsika E, Haun RS, Dixon JE (1991): Cloning, bacterial expression, purification, and characterization of the cytoplasmic domain of rat LAR, a receptor-like protein tyrosine phosphatase. J Biol Chem 266:19688–19696.

Reth M (1989): Antigen receptor tail clue. Nature 338:383–384.

Rime H, Ozon R (1990): Protein phosphatases are involved in the *in vivo* activation of histone H1 kinase in mouse oocyte. Dev Biol 141:115–122.

Rohan PJ, Davis P, Moskaluk CA, Kearns M, Krutzsch H, Siebenlist U, Kelly K (1993): PAC-1: a mitogen induced nuclear protein tyrosine phosphatase. Science 259:1763–1766.

Ronne H, Carlberg M, Hu G-Z, Nehlin JA (1991): Protein phosphatase 2A in *Saccharomyces cerevisiae:* effects on cell growth and bud morphogenesis. Mol Cell Biol 11:4876–4884.

Russell P, Nurse P (1986): $cdc25^{+}$ functions as an inducer in the mitotic control of fission yeast. Cell 45:145–153.

Ruzzene M, Donell-Deana A, Marin O, Perich JW, Rizza P, Borin G, Calderan A, Pinna LA (1993): Specificity of T-cell protein tyrosine phosphatase toward phosphorylated synthetic peptides. Eur J Biochem 211:289–295.

Sadhu K, Reed SI, Richardson H, Russell P (1990): Human homolog of fission yeast cdc25 mitotic inducer is predominantly expressed in G2. Proc Natl Acad Sci USA 87:5139–5143.

Samelson LE, Klausner RD (1992): Tyrosine kinases and tyrosine based activation motifs. J Biol Chem 267:24913–24916.

Samelson LE, Phillips AF, Luong E, Klausner RD (1990): Association of the fyn protein tyrosine kinase with the T-cell antigen receptor. Proc Natl Acad Sci USA 87:4358–4362.

Sasaki K, Shima H, Kitagawa Y, Irino S, Sugimura T, Nagao M (1990): Identification of members of protein phosphatase 1α gene in rat hepatocellular carcinomas. Jpn J Cancer Res 81:1272–1280.

Scheidtmann KH, Mumby MC, Rundell K, Walter G (1991): Dephosphorylation of SV40 virus large T-antigen and p53 protein by protein phosphatase 2A: inhibition by small T antigen. Mol Cell Biol 11:1996–2003.

Schneider HR, Meiskes G, Issinger OG (1989): Specific dephosphorylation by phosphatases 1 and 2A of a nuclear protein structurally and immunologically related to nucleolin. Eur J Biochem 180:449–455.

Schonthal A (1992): Okadaic acid—a valuable new tool for the study of signal transduction and cell cycle regulation? New Biol 4:16–21.

Schonthal A, Alberts AS, Frost JA, Feramisco JR (1991): Differential regulation of jun family gene expression by the tumor promoter okadaic acid. New Biol 3:862–868.

Schraven B, Kirchgessner H, Garber B, Samstag Y, Meuer S (1991): A functional complex is formed in human T lymphocytes between the protein tyrosine kinase p56lck and pp32, a possible common substrate. Eur J Immunol 21:2469–2477.

Schreiber SL (1992): Immunophillin-sensitive protein phosphatase action cell signalling pathways. Cell 70:365–368.

Sefton BM, Campbell MA (1991): The role of tyrosine phosphorylation in lymphocyte activation. Annu Rev Cell Biol 7:257–274.

Shaw A, Thomas ML (1991): Coordinate interaction of protein tyrosine kinases and protein tyrosine phosphatases in T-cell receptor mediated signalling. Curr Opin Cell Biol 3:862–868.

Shen SH, Bastein L, Posner BI, Chretien P (1991): A protein tyrosine phosphatase with sequence similarity to the SH2 domain of protein tyrosine kinases. Nature 352:736–739.

Shenolikar S, Nairn AC (1991): Protein phosphatases: recent progress. Adv Second Messenger Phosphoprotein Res 25:1–121.

Shiroo M, Goff L, Biffen M, Shivan E, Alexander D

(1992): CD45 tyrosine phosphatase activated p59fyn couples the T cell antigen receptor to pathways of diacylglycerol production, protein kinase C activation and calcium influx. EMBO J 11:4887–4897.

Sigal NH, Dumont FJ (1992): Cyclosporin A, FK-506 and rapamycin: pharmacologic probes of lymphocyte signal transduction. Annu Rev Immunol 10:519–560.

Solomon MJ, Glotzer M, Lee TH, Philippe M, Kirschner MW (1990): Cyclin activation of $p34^{cdc2}$. Cell 63:1013–1024.

Steele FR, Washburn T, Reiger R, O'Tusa (1992): Drosophila retinal degeneration C (rdgc) encodes a novel serine/threonine protein phosphatase. Cell 69:669–676.

Suganuma M, Fuijiki H, Suguri H, Yoshizawa S, Hirota M, Nakayasu M, Ojika M, Wakamatsu K, Yamada K, Sugimura T (1988): Okadaic acid: an additional non-phorbol-12-tetradecanoate-13-acetate type tumor promoter. Proc Natl Acad Sci USA 85:1768–1771.

Suganuma M, Fuijiki H, Okabe S, Nishiwaki S, Brautigan D, Ingebritsen TS, Rosner MR (1992): Structurally different members of okadaic acid class selectively inhibit protein serine/threonine but not tyrosine phosphatase activity. Toxicon 30:873–878.

Sutton A, Immanuel D, Arndt KT (1991): The *SIT4* protein phosphatase functions in late G_1 for progression into S phase. Mol Cell Biol 11:2133–2148.

Thevenin C, Kim SJ, Rieckmann R, Fuijiki H, Norcross MA, Sporn MB, Fauci AS, Kehrl JH (1990): Induction of nuclear factor kB and the human immunodeficiency virus long terminal repeat by okadaic acid, a specific inhibitor of phosphatases 1 and 2A. New Biol 2:793–800.

Thompson DB, Sommercorn J (1992): Use of multiple S1 nuclease protection assay to monitor changes in RNA levels for type 1 phosphatase and several proto-oncogenes in response to insulin. J Biol Chem 267:5921–5926.

Tian S-S, Tsoulfas P, Zinn K (1991): Three receptor linked protein-tyrosine phosphatases are selectively expressed on central nervous system axons in the Drosophila embryo. Cell 67:675–685.

Tonks NK, Diltz CD, Fischer EH (1991): Purification of protein-tyrosine phosphatases from human placenta. Methods Enzymol 201:427–442.

Tonks NK, Cicirelli MF, Diltz CD, Krebs EG, Fischer EH (1990): Effect of microinjection of a low-M_r human placenta protein tyrosine phosphatase on induction of meiotic cell division in Xenopus oocytes. Mol Cell Biol 10:458–463.

Toth B, Bollen M, Stalmans W (1988): Acute regulation of hepatic protein phosphatases by glucagon. J Biol Chem 263:14061–14066.

Trowbridge IS (1991): CD45. J Biol Chem 35:23517–23520.

Ulug ET, Cartwright AJ, Courteneidge SA (1992): Characterization of the interaction of polyoma virus middle T antigen with type 2A protein phosphatase. J Virol 66:1458–1467.

Vogel W, Lammers R, Huang J, Ullrich A (1993): Activation of a phosphotyrosine phosphatase by tyrosine phosphorylation. Science 259:1611–1614.

Volarevic S, Niklinska BB, Burns CM, June CH, Weissman AM, Ashwell JD (1993): Regulation of TCR signalling by CD45 lacking transmembrane and extracellular domains. Science 260:541–544.

Walter G, Ruediger R, Slaughter C, Mumby M (1990): Association of protein phosphatase 2A with polyoma virus medium tumor antigen. Proc Natl Acad Sci USA 87:2521–2525.

Weiss A (1993): T cell antigen receptor signal transduction: a tale of tails and cytoplasmic protein-tyrosine kinases. Cell 73:209–212.

Williams KR, Hemmings HC, LoPresti MB, Konigsberg WH, Greengard P (1986): DARPP-32, a dopamine- and cyclic AMP-regulated neuronal phosphoprotein: primary structure and homology with protein phosphatase inhibitor-1. J Biol Chem 261:1890–1903.

Woodford-Thomas TA, Rhodes JD, Dixon JE (1992): Expression of a protein-tyrosine phosphatase in normal and v-src-transformed mouse 3T3 fibroblasts. J Cell Biol 117:401–414.

Yamashita K, Yasuda H, Pines J, Yasumoto K, Nishitani H, Ohtsubo M, Hunter T, Sugimura T, Nishimoto T (1990): Okadaic acid, a potent inhibitor of type 1 and type 2A protein phosphatases, activates cdc2/H1 kinase and transiently induces a premature mitosis-like state in BHK21 cells. EMBO J 9:4331–4338.

Yang Q, Tonks NK (1991): Isolation of a cDNA clone encoding a human protein tyrosine phosphatase with homology to the cytoskeletal associated proteins band 4.1, ezrin and talin. Proc Natl Acad Sci USA 88:5949–5953.

Yang X, Seow KT, Bahri SM, Oon SH, Chia W (1991): Two Drosophila receptor-like tyrosine phosphatase genes are expressed in a subset of developing axons and pioneer neurons in the embryonic CNS. Cell 67:661–673.

Zafriri D, Argaman M, Canaani E, Kimchi A (1993): Induction of protein-tyrosine phosphatase activity of interleukin-6 in M1 myeloblastic cells and analysis of possible counteractions by the BCR-ABL oncogene. Proc Natl Acad Sci USA 90:477–481.

Zanke B, Suzuki H, Kishihara K, Mizzen L, Minden M, Pawson A, Mak TW (1992): Cloning and expression of an inducible lymphoid-specific protein tyrosine phosphatase (HePTPase). Eur J Immunol 22:235–239.

Zheng XM, Wang Y, Pallen CJ (1992): Cell transformation and activation of $pp60^{c\text{-}src}$ by overexpression of a protein tyrosine phosphatase. Nature 359:336–339.

ABOUT THE AUTHORS

THOMAS S. INGEBRITSEN is Associate Professor of Zoology and Genetics at Iowa State University in Ames, Iowa, where he teaches courses in endocrinology, signal transduction, and biotechnology. He received his B.S. degree in chemistry from Oregon State University in 1968. After working for several years in industry, he continued his training at the Indiana University School of Medicine. He received a Ph.D. in biochemistry in 1979 working with Professor David M. Gibson on the regulation of HMG-CoA reductase by reversible phosphorylation. HMG-CoA reductase is a key enzyme in the cholesterol biosynthetic pathway. Following this, Dr. Ingebritsen moved to the University of Dundee, Scotland, as a postdoctoral fellow in the laboratory of Professor Philip Cohen. During this period Dr. Ingebritsen developed methodology for identifying and classifying protein serine/threonine phosphatases. Since 1982, his work has focused on the enzymology, regulation, and physiological functions of protein tyrosine phosphatases. Dr. Ingebritsen was an Established Investigator of the Amerian Heart Association from 1984–1989 and currently is a member of the Editoral Board of the *Journal of Biological Chemistry.*

KELAGINAMANE T. HIRIYANNA is on the adjunct faculty and scientific staff of the Department of Zoology and Genetics at Iowa State University of Science and Technology, Ames, Iowa. After receiving his M.Sc. degree from the University of Mysore, Mysore, India, he pursued his doctoral research at the Indian Institute of Science in Bangalore, India, where he earned a Ph.D. in 1984 for his work on DNA replication in *Mycobacterium.* This was followed by postdoctoral research: on viral replicative enzymes at the Molecular Biology and Virus Laboratory, University of California, Berkeley, in association with Dr. Heinz Fraenkel-Conrat; and on DNA replication and cell cycle in *Xenopus* at the Department of Biology, Johns Hopkins University, Baltimore, and at Iowa State University, in association with Dr. Robert M. Benbow and Dr. Thomas Ingebritsen. At present he is pursuing research on the role of reversible protein-tyrosine phosphorylation in cell cycle and cellular signal transduction.

KELI HIPPEN is a postdoctoral fellow at the National Jewish Center for Immunology and Respiratory Medicine in Denver, Colorado. He is studying signal transduction mechanisms in B cells with Dr. John Cambier. After graduating from Iowa State University with a B.S. in biochemistry in 1989, he began a doctoral program in the laboratory of Dr. Ingebritsen. He received his Ph.D. degree in 1993 for work on substrate recognition by protein tyrosine phosphatases.

Growth Factors and Signal Transduction in Development: 165–177

Growth Factors and Signal Transduction in *Drosophila*

Eric C. Liebl and F. Michael Hoffmann

I. INTRODUCTION

The fruit fly, *Drosophila melanogaster*, chosen for its experimental advantages by T. H. Morgan and colleagues at the beginning of the century, was used by Morgan's group and others to establish basic principles of genetics. During the 1980s successful efforts were launched by many laboratories to use the experimental advantages of *Drosophila* to understand basic principles of developmental biology at a molecular level. These include the pioneering work of E. B. Lewis on the *Bithorax* complex, and the work of C. Nusslein-Volhard and E. Weishaus and their colleagues on *Drosophila* genes affecting the anterior-posterior and dorsal-ventral patterns of the embryonic cuticle. Some of the genes identified in these studies encode molecules similar to mammalian growth factors or to proteins that participate in signal transduction, such as receptors, cytoplasmic protein kinases, guanine nucleotide–binding proteins (G-proteins), or transcription factors responsive to intercellular signals. We believe that in the coming years *Drosophila* will prove useful in elucidating the molecular mechanisms by which growth factors and signal transduction cascades mediate intercellular communication during normal development. The high degree of evolutionary conservation of the genes encoding proteins involved in these processes makes it likely, in our opinion, that much of what is learned about the molecular interactions in *Drosophila* will be applicable in other species, both with regard to their roles in normal development and the perturbation of these regulatory mechanisms in cancer cells [Hoffmann et al., 1992].

Intercellular communication is important for the development of all metazoan organisms. The regulated expression and release of growth factors from one group of cells can impart developmental information to a responsive (growth factor receptor–bearing) cell. This mechanism is used to induce specific cell fate changes in an adjacent tissue layer, to establish a morphogenetic concentration or activity gradient of growth factor along which cells respond differently, or to coordinate patterns of gene expression among a group of cells: the so-called community effect [Gurdon, 1992]. In *Drosophila*, growth factors and signal transduction cascades are essential for many different developmental events in which one cell or a group of cells influences the developmental potential and cell fates of another cell or group of cells.

Specific examples of such intercellular communication are the induction of the R7 cell in the developing eye, the induction of dorsal-ventral polarity in the embryo, and the subdivision of cell fates in the developing embryo. Proper development of the 700 R7 retinal cells in the eye imaginal disk requires a signal from each of the 700 adjacent R8 retinal cells; the signal is received by a tyrosine kinase receptor, called sevenless, on the R7 cell [Rubin,

1991]. The signal, provided by a transmembrane protein on the R8 cell called bride of sevenless (Boss), acts locally, and requires juxtaposition of each R7 cell with an R8 cell. Proper embryonic development in *Drosophila* requires an inductive signal between the soma and the germline to establish dorsal-to-ventral polarity [Schupbach et al., 1991]. The signal is generated by proteolytic activation of a ligand in the fluid-filled perivitelline space that surrounds the plasma membrane of the syncytial blastoderm. This ligand is bound by the toll protein in the plasma membrane of the embryo and this initiates a signal transduction cascade which ultimately leads to translocation of the dorsal protein into the nuclei on the ventral side of the embryo. Further subdivision of cell fates along the dorsal-to-ventral embryonic axis requires a gradient of the growth factor decapentaplegic (Dpp) [Ferguson and Anderson, 1992a,b]. Later in embryogenesis, Dpp mediates an inductive interaction between the visceral mesoderm and the endoderm of the embryonic midgut [Panganiban et al., 1990b].

In this chapter we will briefly review recent studies pertaining to four different classes of growth factors or growth factor receptors in *Drosophila,* with emphasis on the importance of genetic strategies in elucidating these signal transduction pathways. The Drosophila proteins discovered thus far show remarkable similarity to growth factors, receptors, and other signal transduction molecules in mammalian species. The topics covered here include the protein tyrosine kinase receptors, maternal-effect dorsal-ventral pattern mediated through the receptor toll, the decapentaplegic (Dpp) member of the transforming growth factor beta (TGF-β) superfamily and the wnt family factor wingless. Two different genetic strategies possible in *Drosophila* have added to our knowledge about these signal transduction pathways: (1) isolation of mutations in different genes that produce the same phenotype or the opposite phenotype and (2) isolation of second-site mutations that modify a phenotype caused by mutations in a gene encoding one component of a signal transduction pathway.

II. RECEPTOR TYROSINE KINASES

Transmembrane tyrosine kinase receptors play pivotal roles in inter- and intracellular signalling pathways, by allowing for the action of molecules such as epidermal growth factor, nerve growth factor, and insulin. In *Drosophila,* the proteins encoded by the genes *torso, torpedo, Dtrk, fibroblast-growth-factor receptor,* and *sevenless* (*sev*) have been found to be transmembrane tyrosine kinase receptors. This has opened the door to the analysis of these receptors' signal transduction networks by genetic means [Shilo, 1992]. Of these genes, *sevenless* has proven to be the most amenable to genetic analysis.

The correct spatial and temporal activation of Sev leads to the development of a photoreceptor sensitive to UV light, the R7 cell [Tomlinson and Ready, 1987]. As mutations in the *sev* gene affect only eye development, leaving viability and fertility intact, and since the presence or absence of the R7 cell can be screened for in living, anesthetized flies [Franceschini and Kirschfeld, 1971], genetic strategies have been undertaken to dissect the signal transduction pathway(s) utilized by this transmembrane receptor tyrosine kinase. Early genetic screens isolated new mutations which resulted in the absence of R7 cells, and which were not allelic to *sev.* These screens identified the putative ligand of Sev, which is bride of sevenless (Boss) [Reinke and Zipursky, 1988]. Mutations in a gene encoding a nuclear protein, *seven in absentia* (*sina*), were also identified [Carthew and Rubin, 1990]. Sina, due to its nuclear localization, may be at or near the endpoint of the Sev signalling pathway.

Recent genetic screens have identified additional components of the signal transduction pathway affected by Sev by recovering second-site modifier mutations. These mutations genetically tag the genes encoding other proteins on the Sev-mediated signal transduction pathway. The screens have relied on Sev receptors that are mutated in such a way that their activity is either just below or just above a threshold required to produce R7 cells; thus the Sev-initiated signal transduction pathway is sensitized to additional genetic insults [Simon et al., 1991, 1993; Gaul et al., 1992; Olivier et al., 1993]. For example, a heterozygous loss-of-function mutation in a gene encoding a component of the signal transduction pathway will reduce the level of that protein by half. If the signal is already compromised by mutations in the receptor, reduction in the level of one protein in the pathway could reduce the transduction of the signal below a necessary threshold. The phenotypic consequence would be a failure of R7 to take on the proper cell fate. Such mutations that exacerbate a mutant phenotype are termed enhancer mutations. Similarly, if the Sev kinase activity is set just under the needed threshold, a heterozygous loss-of-function mutation in a component of the signal transduction pathway which normally negatively affects or attenuates signalling could elevate signalling above the needed threshold. R7 cells would result. Such mutations that alleviate the mutant phenotype are termed suppressor mutations.

Such genetic strategies have identified a number of components of the signal transduction network influenced by the Sev receptor. Epistasis tests, biochemical analyses, and comparisons to receptor tyrosine kinase signal transduction networks elucidated in other experimental systems have allowed for the ordering of these components into a tentative linear signal transduction pathway. This presumptive pathway, from its extracellular ligand signal to its current known endpoint, is outlined below.

A. *bride of sevenless*

As mentioned previously, the *boss* gene is thought to encode the ligand for the Sev receptor. Evidence that the Boss and Sev proteins physically interact comes from studies showing that cell lines expressing either Boss or Sev form heterotypic aggregates [Kraemer et al., 1991]. In the developing eye, the Boss protein is presented to the extracellular ligand-binding domain of the Sev receptor, and then both Boss and Sev are internalized by the developing R7 cell [Kraemer et al., 1991].

B. *sevenless*

The Sev protein contains a tyrosine kinase domain within its intracellular domain. The protein can phosphorylate both itself and exogenous substrates on tyrosine in *in vitro* kinase reactions. While a transgene encoding wild-type Sev can fully rescue the *sev* mutant phenotype, a transgene encoding Sev with a point mutation in its kinase domain which renders it an inactive kinase, cannot [Basler and Hafen, 1988]. Thus the tyrosine kinase activity of the Sev protein is essential to its ability to induce R7 cell development. Although it has not been directly demonstrated, by analogy to other transmembrane protein tyrosine kinase receptors such as the epidermal growth factor receptor (EGFR), it is believed that binding of Boss to Sev activates Sev tyrosine kinase activity. Again, by analogy to the EGFR, this then allows for both receptor autophosphorylation and phosphorylation of other substrates.

C. *corkscrew*

An X-linked enhancer of *sevenless* identified by Simon et al. [1991] has been shown to be allelic to *corkscrew* (*csw*) [Perkins et al., 1992; Tsuda et al., 1993]. The *csw* gene encodes a nonreceptor protein tyrosine phosphatase containing two src homology 2 (SH2) domains. While the precise role of Csw in the Sev signalling pathway has not been reported, Perkins et al. [1992] have shown that Csw functions downstream of the Torso receptor tyrosine kinase, perhaps by dephosphorylating autophosphorylated Torso. More discussion of Csw and other tyrosine phosphatases can be found in Ingebritsen et al. (Chapter 7, this volume).

D. *downstream of receptor kinase*

Mutations in the *downstream of receptor kinase* (*drk*) gene have been recovered as enhancers of *sevenless* [Simon et al., 1993; Olivier et al., 1993]. *Drosophila drk* is homologous to mammalian GRB-2 and *C. elegans sem*-5; its open reading frame is made up almost entirely of two src homology 3 (SH3) domains and one SH2 domain. SH2 domains are modular protein motifs which allow for the stable association of proteins to specific phosphotyrosine residues [Songyang et al., 1993]. SH3 domains allow for stable protein–protein interaction between themselves and proline-rich motifs in target proteins [Ren et al., 1993]. *In vitro*, autophosphorylation of the Sev kinase allows for its association with Drk. That this association occurs via the Drk SH2 domain was demonstrated by sequencing the *drk* alleles which were recovered as enhancers of *sevenless*. This analysis has shown these alleles' mutant lesions to be point mutations in the Drk SH2 domain. Analogous point mutations in bacterially synthesized Drk abolish its *in vitro* association with autophosphorylated Sev [Olivier et al., 1993]. Thus it seems likely that Drk binding to autophosphorylated Sev receptors is critical to the signalling cascade specifying R7 cell development.

E. *ras*

The *ras* genes have long been believed to be involved in signal transduction networks regulated by transmembrane receptor protein tyrosine kinases. Ras proteins are small (about 20-kD) G-proteins which are active when bound to GTP, and inactive when bound to GDP. Microinjection of neutralizing anti-Ras antibodies, or overexpression of dominant negative forms of Ras, block signalling from transmembrane receptor tyrosine kinases in a variety of mammalian cell culture systems. Mutations in the Drosophila *ras*1 gene have been recovered as enhancers of *sevenless* [Simon et al., 1991]. In addition, expression of constitutively activated *ras1* in the *sevenless* expression pattern (*sevRas1*^{v12}) mimics the phenotypic effect of rough eyes due to supernumery R7 cells seen as a result of expression of constitutively active sevenless. These effects are blocked, however, by the introduction of *sina* mutations [Fortini et al., 1992]. Together, these data strongly suggest that activation of Ras is a key event in signalling through Sev. More discussion of Ras signalling pathways can be found in the chapter by Hamilton (this volume).

F. *son of sevenless* and *gap1*, Regulators of *ras*

Mutations in two genes which are believed to regulate Ras activity have been isolated as mutations affecting R7 cell development. Alleles of the *son of sevenless* (*sos*) gene have been recovered as both enhancers and suppressors of *sevenless* [Rogge et al., 1991; Simon et al., 1991]. The *sos* gene is homologous to *S. cerevisiae* CDC25. CDC25 protein is an exchange factor which facilitates the activation of Ras by catalyzing the exchange of GDP for GTP [Crechet et al., 1990].

Alleles of the *gap1* gene have been isolated as suppressors of *sevenless* [Gaul et al., 1992]. *Drosophila* Gap1 is homologous to bovine RasGAP, which binds Ras and stimulates its GTPase activity, thus inactivating Ras by hydrolysing the bound GTP to GDP [McCormick, 1989].

Genetic evidence has shown that activation of Ras1 is central to signalling through Sev. Mutations in *sos* and *gap1* hint at how Sev may be affecting Ras1 activity. But how is Sev tyrosine kinase activity linked to Sos or Gap1? A critical piece of evidence has been provided by biochemical analysis of the Drk protein. As reviewed above, the SH2 domain of Drk can bind to autophosphorylated Sev. In addition, the SH3 domains of Drk can bind to the C-terminus of the Sos protein *in vitro* (Olivier et al., 1993]. Thus Drk may be an adaptor molecule, physically linking autophosphorylated Sev to Sos. It has not been shown whether Drk association with Sos stimulates Sos activity. It is tempting to speculate, however, that such binding is responsible for Sos (and subsequent Ras1) activation, either by simply recruiting Sos to the plasma membrane and allowing it access to the farnesylated Ras1, or perhaps by allowing Sev to tyrosine phosphorylate, and enzymatically activate Sos.

G. *raf*

Epistatic and biochemical evidence from mammalian tissue culture systems has implicated the Raf serine/threonine kinase as being a downstream effector of both receptor tyrosine kinases and Ras. The *Drosophila* homologue of Raf is Draf1. A weak allele of *Draf1* (*raf^{HM7}*) blocks R7 development in either wild-type Sev backgrounds or constitutively activated Sev backgrounds [Dickson et al., 1992]. In addition, expression of activated Draf1 in the *sevenless* expression pattern mimics the phenotypic effect of rough eyes due to supernumery R7 cells that are seen as a result of the expression of constitutively activated Sev. This effect depends on Sina [Dickson et al., 1992]. Thus Draf1 activity is needed for R7 cell induction. Mutations in *Draf1* were not discovered as enhancers of *sevenless* [Simon et al., 1991], most likely because loss-of-function *Draf1* mutations affect general cell proliferation. The weak *raf^{HM7}* allele suppresses the rough eye phenotype obtained by expressing activated Ras1 in the *sevenless* expression pattern (*sevRas1^{v12}*, above) [Dickson et al., 1992]. This observation supports the hypothesis that Draf1 acts downstream of Ras1 in the Sev signalling network.

H. *suppressor of raf*

The protein encoded by the *suppressor of raf* (*Dsor1*) locus has been hypothesized to act downstream of Draf1, because a presumptive gain-of-function allele of *Dsor1* (*Dsor1^{Su1}*) suppresses the lethality resulting from a weak *Draf1* mutation [Tsuda et al., 1993]. Dsor1^{Su1} suppresses the *sevenless* mutant phenotype, as well as those of other receptor tyrosine kinases (discussed next). These data suggest that Dsor1 acts in the sevenless signalling pathway downstream of Draf1. The *Dsor1* gene encodes a kinase homologous to MEK1 (MAP kinase or ERK; MAP kinase [mitogen-activated protein kinase] is identical to ERK [extracellular signal-regulated kinase]) [Tsuda et al., 1993; Crews et al., 1992]. MEK1 is an example of a kinase that phosphorylates other kinases, or a kinase kinase. MEK1 activates MAP kinase (also known as ERK) by phosphorylating it [Crews et al., 1992]. Biochemical data from mammalian tissue culture systems has shown that Raf1 activates MEK1 through direct phosphorylation [Howe et al., 1992], which makes Raf1 a kinase kinase kinase, and further supports the hypothesis that Dsor1 is directly downstream of Draf1.

I. Summary

A number of the components of the Sev signalling pathway have been found to affect signalling through other receptor tyrosine kinases in *Drosophila*. Mutations in *drk, sos,* and *ras1* affect signalling through ellipse [Simon et al., 1991], a gain-of-function allele of the *torpedo* locus. Mutations in *sos, ras1, csw, Draf1,* and *Dsor1* affect signalling through *torso* [Doyle and Bishop, 1993; Lu et al., 1993; Ambrosio et al., 1989; Perkins et al., 1992; Tsuda et al., 1993; Sprenger et al., 1993]. Indeed, the Ras-dependence of receptor tyrosine kinase signalling has been observed in *C. elegans* and mammalian cells as well (Tuck and Greenwald, Chapter 9, this volume; Hamilton, Chapter 6, this volume; Han and Sternberg, 1990; Szeberenyi et al., 1990]. Why is it then that the receptor tyrosine kinase in the Drosophila eye encoded by *ellipse* does not also lead to the development of R7 cells? Three of the loci identified as enhancers of *sevenless* had no effect on signalling through ellipse [Simon et al., 1991]. This suggests that these mutations may be in genes whose products give R7 cell–specificity to the receptor tyrosine kinase signal.

As mentioned above, the genetic analysis of signalling through the Sev receptor tyrosine kinase has been facilitated by the generation of mutant Sev proteins that are near the signalling threshold; these mutant proteins potentiate the system for mutational analysis. Similar strategies, involving other components of the network, can now be used to further dissect the signalling pathway. For example, expressing a slightly hyperactive allele of Draf 1 in the *sevenless* expression pattern results in roughened eyes. Recovering second site mutations that either enhance or suppress this phenotype may identify further components of the pathway downstream of raf kinase. Thus the genetic analysis of the signal transduction network affected by the sevenless receptor tyrosine kinase should continue to be fruitful.

III. DORSAL-VENTRAL PATTERN FORMATION

Genes that are essential during oogenesis or whose products are packaged in the egg during oogenesis for essential functions during early embryogenesis have been identified by isolating mutations that make females sterile. In general, these screens are limited to the recovery of mutations that do not alter development of the somatic tissues and do not affect viability of the zygote, because homozygous mutant females must be recovered in order to test for either their fertility or for maternal effects on the development of their progeny. Despite these limitations, screens to recover all of the mutations in maternal-effect genes that alter the dorsal-to-ventral polarity of the embryo have been remarkably successful. These genetic strategies have identified an elaborate signalling system between the soma and the germline.

Eleven maternal-effect mutations were found that caused all progeny of a mutant female to develop as dorsalized embryos. Other mutations recovered caused all progeny to develop as ventralized embryos [reviewed by Govind and Stewart, 1991]. Subsequent screens also recovered mutations in which the eggshell itself had lost normal dorsal-ventral polarity [Schupbach et al., 1991]. It was possible to order the functions of these genes in a genetic pathway by combining, in the same female, mutations that ventralized the embryo with mutations that dorsalized the embryo, and determining which phenotype "wins." For example, when ventralizing mutations in the gene *toll* and dorsalizing mutations in the gene *dorsal* were present in the same female, the embryos were dorsalized. It was concluded that the function of *dorsal* is "epistatic" to the function of *toll,* meaning that dorsal is downstream of toll in a regulatory pathway and that for mutations in *toll* to cause a ventralized phenotype requires the presence of the *dorsal* product.

A second technique that was valuable in sorting out the components of this pathway was the ability to transplant the germline precursor cells, the pole cells, into host embryos of different genotypes. In this way, it was determined that some of the maternal-effect mutations acted on the somatic cells of the ovary and others acted in the germ cells of the ovary [reviewed by Schupbach et al., 1991].

A third technique that has provided insights into this soma-to-germline communication was the transfer of the "extracellular" perivitelline fluid to demonstrate the presence of an activated signal in the fluid specifically along the ventral surface of the oocyte [Stein et al., 1991; Stein and Nusslein-Volhard, 1992].

The molecular cloning of the genes identified through their maternal-effect phenotypes on dorsal-to-ventral polarity is revealing the kinds of molecules that mediate the communication between soma and germline. It is this communication which imparts dorsal-to-ventral polarity to the embryo. The asymmetrical dorsal-ventral polarity is generated not by the asymmetrical localization of the signalling pathway components, but by the asymmetrical activation of the signalling pathway. This is accomplished by activating the toll transmembrane receptor only on the ventral side of the oocyte as outlined below.

A. *windbeutel, pipe,* and *nudel*

The products of these three genes are required in the somatic cells of the ovary, known as the follicle cells. One or more of them is thought to be responsible for a signal sent specifically from the follicle cells on the ventral surface of the oocyte to the perivitelline fluid; the signal activates the easter protease in the perivitelline fluid specifically on the ventral side of the oocyte (next section).

B. *easter, snake,* and *gastrulation defective*

These three genes all encode proteins with homology to serine proteases. It has been proposed that they act in a cascade to activate the easter protease specifically on the ventral side of the embryo. Activation of the easter protease acts to locally activate spatzle [next section; Hecht and Anderson, 1992]. A mutant form of easter lacking the amino-terminal domain causes ventralization of the embryo and therefore is active at all dorsal-ventral positions throughout the perivitelline fluid [Chasen et al., 1992]. The ventralization caused by this mutant form of easter does not require pipe, nudel, windbeutel, snake, or gastrulation defective, but it does require the product of the *spatzle* gene.

C. *spatzle*

The product of the *spatzle* gene is the primary candidate for the proligand that activates the toll receptor protein. This proligand requires processing to its active form, however; probably by the easter protease. Thus localized (ventral) easter activity results in locally (ventral) active spatzle ligand.

D. *toll*

The *toll* gene encodes a transmembrane protein with an extracellular domain that includes leucine-rich repeats [Schneider et al., 1991]. The toll protein is uniformly distributed on the surface of the early embryo but is activated only on the ventral surface, presumably by localized, activated spatzle. Mutant forms of toll that are constitutively activated have been recovered as dominant, female sterile mutations. These mutant females produce embryos with ventralized cuticles [Anderson et al., 1985].

E. *dorsal, cactus,* and *pelle*

The dorsal protein has homology to a family of mammalian transcription factors including nuclear transcription factor-kB (NF-kB) and the c-Rel protooncogene [Govind and Stewart, 1991]. Ventral activation of toll results in transport of the dorsal transcription factor into the nuclei of cells only on the ventral side of the oocyte. The role of dorsal is to repress the transcription of some genes, including *decapentaplegic* (*dpp*) (Section IV) and *zerknullt* (*zen*), in the ventral nuclei; and to activate the transcription of other genes, including *twist,* in the ventral nuclei. Specific DNA binding sites for the dorsal protein have been found in all three of these genes [Huang et al., 1993]. The mechanism by which dorsal can be an activator of some genes and a transcriptional repressor of other genes is a focus of research in several laboratories. The localized activation and repression of these genes gives rise to the dorsal-ventral polarity of the embryo.

The nuclear translocation of dorsal is mediated by the cactus and pelle proteins. Loss-of-function mutations in *cactus* cause a maternal-effect ventralization of the embryonic cuticle, the opposite phenotype as loss-of-function mutations in *dorsal* [Roth et al., 1991]. Molecular cloning of *cactus* has revealed a sequence encoding a protein with homology to I-kB, a mammalian protein that antagonizes the action of NF-kB by sequestering the transcription factor in the cytoplasm [Kidd, 1992]. Disruption of this protein complex by phosphorylation permits translocation of the released transcription factor into the nucleus, where it can bind to specific sequences and alter transcription. The *pelle* gene encodes a protein kinase [Shelton and Wasserman, 1993]. An attractive hypothesis which parallels the findings for NF-kB and I-kB in mammalian cells is that localized (ventral) activation of toll leads to localized activation of pelle; this, in turn, leads to localized disruption of the cactus–dorsal complex and allows localized (ventral) nuclear translocation of the dorsal transcription factor.

IV. DECAPENTAPLEGIC

One of the zygotic genes that is a direct target for regulation by the dorsal protein is *decapentaplegic* (*dpp*) [Huang et al., 1993]. Molecular cloning of the gene and biochemical characterization of the gene product has demonstrated similarity to members of the TGF-β family, particularly to the bone morphogenetic proteins (BMPs) 2 and 4 [St. Johnston et al., 1990; Padgett et al., 1987; Panganiban et al., 1990a]. Functional evidence of the similarity between Dpp and BMP has come from recent studies in which BMP expression in *Drosophila* rescues *dpp* mutant phenotypes, and *dpp* expression in rats causes ectopic bone growth [Padgett et al., 1993; Sampath et al., 1993]. The mRNA encoded by *dpp* is expressed initially along the dorsal surface of the embryo. Evidence that a morphogenetic gradient of the Dpp protein may be formed from this dorsal source of protein has come from studies in which Dpp mRNA is injected into embryos to cause a concentration-dependent specification of dorsal-to-lateral cell fates [Ferguson and Anderson, 1992a,b].

The activity of the Dpp protein along the dorsal-to-ventral axis is modulated by at least two other gene products, by thus far unknown mechanisms. The tolloid protein is believed to enhance the activity of Dpp along the embryo's dorsal surface, and the protein encoded by the *short gastrulation* gene is believed to inactivate the Dpp protein along the ventral surface of the embryo [Ferguson and Anderson, 1992b].

Although the mechanism by which Dpp specifies dorsal-to-ventral cell fates in the ectoderm is not understood at a molecular level, Dpp expression is required to maintain expression of the homeodomain protein Zen in the most dorsal cells of the embryo, which become a squamous epithelial layer called the amnioserosa [Ferguson and Anderson, 1992a]. Identification of other target genes and of the components of the signal transduction pathway that Dpp activates are current goals of research in several laboratories.

An extremely interesting problem is how cells respond differently to different concentrations or activities of a morphogenic gradient. One possibility is that the cells express multiple receptors with different affinities for the ligand, and that at different positions along a gradient, combinations of receptors would be activated that would cause distinct changes in cell fates. Identification of receptors for Dpp will be needed to test this hypothesis. Receptors for members of the TGF-β family are transmembrane serine kinases [reviewed by Lin and Lodish, 1993]. The identification of serine kinase receptors in *Drosophila* should recover receptors for Dpp and other members of the TGF-β family of growth factors in *Drosophila*. The TGFβ family is discussed in more detail in the chapter by Akhurst (this volume).

The Dpp protein is not only used to specify dorsal-ventral cell fate in the developing embryo; it is required to specify cell fate in the imaginal disks (tissues that give rise to adult cuticle structures), as well as in the embryonic gut. The *dpp* gene was identified initially through the phenotypes of mutations which caused the absence of structures from the appendages of the adult cuticle, with increasingly more severe mutant alleles causing the absence of more structures along the distal-to-proximal axis of the appendages [Spencer et al., 1982]. As in the embryo, where *dpp* expression is localized in specific groups of cells [St. Johnston and Gelbart, 1987], *dpp* expression in the imaginal disks is localized to a narrow band of cells adjacent to the anterior-posterior compartment boundary, a cell lineage restriction boundary in the imaginal disks [Masucci et al., 1990; Blackman et al., 1991; Raftery et al., 1991].

The *dpp* gene is also expressed during embryonic development of the gut tube. The interactions between homeotic genes and *dpp* in the *Drosophila* midgut has made this tissue of particular interest to those studying target genes for the homeodomain proteins and target genes for Dpp action [Immergluck et al., 1990; Panganiban et al., 1990b; Reuter et al., 1990; Hursh et al., 1993; Masucci and Hoffmann, 1993]. Expression of *dpp* in the parasegment seven region of the visceral mesoderm is increased by the homeodomain protein ultrabithorax (Ubx) and is repressed by the homeodomain protein abdominal A. The Dpp protein acts to maintain expression of *ubx* in the viseral mesoderm and to enhance expression of the homeotic gene *labial* in the cells of the underlying endoderm. This latter observation, supported by visualization of Dpp produced in the visceral mesoderm that eventually surrounds the adjacent endodermal cells [Panganiban et al., 1990b], is an example of a growth factor produced in one germ layer altering gene expression in an adjacent germ layer. The extracellular distribution and action of Dpp is consistent with very limited movement or diffusion of this protein in the extracellular space and might be accomplished by the association of Dpp with other extracellular or cell surface proteins.

V. THE WINGLESS GENE FAMILY

As in the case of *dpp*, the *wingless* gene product in *Drosophila* is involved in multiple developmental processes. The first mutation in *wingless* was recovered because it caused the loss of wing blade structures from adult flies. Recent studies have confirmed the importance of *wingless* expression during imaginal disk development [Struhl and Basler, 1993]. The *wingless* gene has been most thoroughly studied for its role in maintaining the segment polarity pattern in the embryo. The *wingless* gene product is a member of the wnt family of factors [reviewed by Nusse and Varmus, 1992]. Two other members of this family have been reported in *Drosophila* and many more related genes are present in vertebrate species [Russell et al., 1992; McMahon, 1992]. The wnt gene products are secreted proteins but have been difficult to solubilize and characterize in biologically active forms.

In the *Drosophila* embryo, *wingless* is expressed in a segmented pattern in the epithelial cells and its expression is required to maintain expression of the homeodomain protein engrailed in the adjacent row of cells. Although early studies focused on a potential role for wingless as a secreted morphogen, recent studies indicate that the wingless protein is probably not an instructive signal or a morphogen, but is more likely to be involved in maintaining the parasegment boundaries [Sampedro et al., 1993]. Recent studies in which *wingless* was expressed ectopically in imaginal disks indicate that the wingless protein may be able to specify ventral cell fates during patterning of the imaginal disks [Struhl and Basler, 1993].

Receptors for wingless or other wnt proteins have not been identified to date. Some clues to molecules involved in the transduction of a wingless signal have come from the molecular identification of other genes in which mutations also affect segment polarity pattern formation. Gene products potentially on the signal transduction pathway include armadillo, disheveled, porcupine, and zeste-white-3. The best-characterized target gene is *engrailed,* which encodes a homeodomain protein. Two other genes that are involved in the segment polarity functions of wingless are *patched* and *hedgehog,* which encode two membrane proteins that determine which cells express *wingless* [Ingham and Hidalgo, 1993].

A. *hedgehog* and *patched*

The *hedgehog* gene encodes a putative secreted protein that is thought to be expressed by the same cells that express *engrailed* [Lee et al., 1992]. Its function appears to be to signal to adjacent cells so that they maintain expression of *wingless.* However, recent studies indicate that the *hedgehog* mutant phenotype cannot be rescued by ubiquitous *wingless* expression; this observation leads to the suggestion that hedgehog may function downstream of wingless [Sampedro et al., 1993]. The *patched* gene encodes a transmembrane protein that represses *wingless* expression in all cells except those acted on by hedgehog.

B. *zeste-white-3 (shaggy)*

The *zeste-white-3* gene (also known as *shaggy*) encodes a serine/threonine kinase homologous to the mammalian glycogen synthase kinase-3 (GSK-3) which has been implicated in the regulation of several transcription factors including c-Jun [Hughes et al., 1993]. Embryos in which both the maternal and zygotic contribution of *zeste-white-3* are eliminated by mutation lack the ventral hairs or denticles on each segment. This cuticular phenotype is associated with an increased width in the bands of cells in each segment of the embryo expressing the homeodomain protein engrailed. Tests of the epistatic relationships of mutant alleles of *zeste-white-3* and *wingless* led to the conclusion that wingless activity normally inactivates zeste-white-3, so that zeste-white-3 is prevented from inhibiting engrailed from autoactivating its own expression [Siegfried et al., 1992]. This genetic mechanism is consistent with the recent biochemical characterization of GSK-3. The mammalian GSK-3 and *Drosophila* zeste-white-3 proteins are phosphorylated on tyrosine, and the phosphorylated forms of the proteins are active serine/threonine kinases [Hughes et al., 1993]. The active, tyrosine phosphorylated form of GSK-3 inactivates the c-Jun transcription factor by maintaining it in a phosphorylated state. Activation of either engrailed or c-Jun might therefore require removal of phosphotyrosine from zeste-white-3 or GSK-3, a transformation which would inactivate their serine kinase activities and allow the engrailed or c-Jun to be dephosphorylated and thereby activated. Perhaps wingless activates a phosphotyrosine phosphatase to accomplish this. Maintenance of *engrailed* expression by wingless from adjacent cells has been recreated in a culture system which may allow further biochemical analysis of the signal transduction pathway involved [Cumberledge and Krasnow, 1993].

C. *armadillo, disheveled,* and *porcupine*

Mutations in these three genes produce phenotypes identical to *wingless* mutations, suggesting that they may all be on the wingless signalling pathway [Peifer and Bejsovec, 1992]. The *armadillo* gene encodes a protein with homology to β-catenin, a protein which links cadherins to the cytoskeleton, and therefore may be involved in cell adhesion. How cell adhesion is related to the wingless pathway is still an open question. The molecular characterization of *disheveled* and *porcupine* is currently underway.

VI. SUMMARY

These examples of signal transduction pathways in *Drosophila* cover a wide range of mechanisms involving receptor tyrosine kinases, localized activation of serine proteases leading to localized nuclear translocation of a transcription factor, and activity gradients of secreted growth factors. Future analysis of these pathways in *Drosophila* will have several beneficial outcomes. First, we will learn more about the ways in which growth factors and signal transduction pathways are used in normal development. Second, we will identify new molecules essential for specific signal transduction pathways by further application of genetic screens for specific phenotypes, by genetic screens for second-site modifier mutations, and by the discovery of new proteins through their protein–protein interactions with known components. Further elucidation of the signal transduction processes in *Drosophila* will provide conceptual insights for those interested in similar pathways in mammalian systems and will provide molecular reagents with which to identify the mammalian homologues and to test for the conservation of the molecular interactions. Identification of the critical protein–protein interactions in signal transduction pathways will be important for future strategies to selectively disrupt signal transduction pathways in order to control cell behavior, for example, to control the unregulated proliferation of cancer cells [Brugge, 1993].

REFERENCES

Ambrosio L, Mahowald AP, Perrimon N (1989): Requirement of the Drosophila raf homolog for torso function. Nature 342:288–291.

Anderson KV, Juergens G, Nuesslein-Volhard C (1985): Establishment of dorsal-ventral polarity in the Drosophila embryo: genetic studies on the role of the Toll gene product. Cell 42:779–789.

Basler K, Hafen E (1988): Control of photoreceptor cell fate by the sevenless protein requires a functional tyrosine kinase domain. Cell 54:299–311.

Blackman RK, Sanicola M, Raftery LA, Gillevet T, Gelbart WM (1991): An extensive 3′ cis-regulatory region directs the imaginal disk expression of *decapentaplegic*, a member of the TGF-β family in Drosophila. Development 111:657–665.

Brugge JS (1993): New intracellular targets for therapeutic drug design. Science 260:918–919.

Carthew RW, Rubin GM (1990): seven in absentia, a gene required for the specification of R7 cell fate in the Drosophila eye. Cell 63:561–577.

Chasen R, Jin Y, Anderson KV (1992): Activation of the easter zymogen is regulated by five other genes to define dorsal-ventral polarity in the Drosophila embryo. Development 115:607–616.

Crechet JB, Poullet P, Mistou MY, Paraeggiani A, Camonis J, Boy-Marcotte E, Damak F, Jacquet M (1990): Enhancement of the GDP-GTP exchange of RAS proteins by the carboxy-terminal domain of SCD25. Science 248:866–868.

Crews CM, Alessandrini A, Erikson RL (1992): The primary structure of MEK, a protein kinase that phosphorylates the ERK gene product. Science 258:478–480.

Cumberledge S, Krasnow MA (1993): Intercellular signalling in Drosophila segment formation reconstructed in vitro. Nature 363:549–552.

Dickson B, Sprenger F, Morrison D, Hafen E (1992): Raf functions downstream of ras1 in the sevenless signal transduction pathway. Nature 360:600–603.

Doyle HJ, Bishop JM (1993): Torso, a receptor tyrosine kinase required for embryonic pattern formation, shares substrates with the Sevenless and EGF-R pathways in Drosophila. Genes Dev 7:633–646.

Ferguson EL, Anderson KV (1992a): Decapentaplegic acts as a morphogen to organize dorsal-ventral pattern formation in the Drosophila embryo. Cell 71:451–461.

Ferguson EL, Anderson KV (1992b): Localized enhancement and repression of the activity of the TGF-β family member, decapentaplegic, is necessary for dorsal-ventral pattern formation in the Drosophila embryo. Development 114:583–597.

Fortini ME, Simon MA, Rubin GM (1992): Signalling by the sevenless protein tyrosine kinase is mimicked by ras1 activation. Nature 355:559–561.

Franceshini N, Kirschfeld K (1971): Pseudopupil phenomena in the compound eye of Drosophila. Kybernetik 9:159–182.

Gaul U, Mardon G, Rubin GM (1992): A putative ras GTPase activating protein acts as a negative regulator of signaling by the sevenless receptor tyrosine kinase. Cell 68:1007–1019.

Govind S, Steward R (1991): Dorsoventral pattern formation in Drosophila. Trends Genet 7:119–125.

Gurdon JB (1992): The generation of diversity and pattern in animal development. Cell 68:185–199.

Han M, Sternberg PW (1990): let-60, a gene that specifies cell fate during vulval induction, encodes a ras protein. Cell 63:921–931.

Hecht PM, Anderson KV (1992): Extracellular proteases and embryonic pattern formation. Trends Cell Biol 2:197–202.

Hoffmann, FM, Sternberg PW, Herskowitz I (1992): Learning about cancer genes through invertebrate genetics. Current Opinions in Genetics and Development 2:45–52.

Howe LR, Leevers SJ, Gomez N, Nakielny S, Cohen P, Marshall CJ (1992): Activation of the MAP kinase pathway by the protein kinase raf. Cell 71:335–342.

Huang J-D, Schwyter DH, Shirokawa JM, Courey AJ (1993): The interplay between multiple enhancer and silencer elements defines the pattern of *decapentaplegic* expression. Genes Dev 7:694–704.

Hughes K, Nikolakaki E, Plyte SE, Totty NF, Woodgett JR (1993): Modulation of glycogen synthase kinase-3 family by tyrosine phosphorylation. EMBO J 12:803–808.

Hursh DA, Padgett RW, Gelbart WM (1993): Cross regulation of decapentaplegic and Ultrabithorax transcription in the embryonic visceral mesoderm of Drosophila. Development (in press).

Immergluck K, Lawrence PA, Bienz M (1990): Induction across germ layers in Drosophila mediated by a genetic cascade. Cell 62:261–268.

Ingham PW, Hidalgo A (1993): Regulation of wingless transcription in the Drosophila embryo. Development 117:283–291.

Kidd S (1992): Characterization of the Drosophila cactus locus and analysis of interactions between cactus and dorsal proteins. Cell 71:623–635.

Kraemer H, Cagan RL, Zipursky SL (1991): Interaction of bride of sevenless membrane-bound ligand and the sevenless tyrosine-kinase receptor. Nature 352:207–212.

Lee JJ, Von Kessler DP, Parks S, and Beachy PA (1992): Secretion and localized transcription suggest a role in positional signaling for products of the segmentation gene hedgehog. Cell 71:33–50.

Lin HY, Lodish HF (1993): Receptors for the TGF-β superfamily: multiple polypeptides and serine/threonine kinases. Trends Cell Biol. 3:14–19.

Lu X, Chou T-B, Williams NG, Roberts T, Perrimon N (1993). Control of cell fate determination by p21ras/Ras1, an essential component of *torso* signaling in Drosophila. Genes Dev 7:621–632.

Masucci JD, Hoffmann FM (1993): Identification of two regions from the Drosophila decapentaplegic gene required for embryonic midgut development and larval viability. Dev Biol (in press).

Masucci JD, Miltenberger RJ, Hoffmann FM (1990): Pattern-specific expression of the Drosophila decapentaplegic gene in imaginal disks is regulated by 3′ cis-regulatory elements. Genes Dev 4:2011–2023.

McCormick F (1989): ras GTPase activating protein: signal transmitter and signal terminator. Cell 56:5–8.

McMahon AP (1992): The Wnt family of developmental regulators. Trends Genet 8:236–242.

Nusse R, Varmus HE (1992): *Wnt* genes. Cell 69:1073–1087.

Olivier JP, Raabe T, Henkemeyer M, Dickson B, Mbamalu G, Margolis B, Schlessinger J, Hafen E,

Pawson T (1993): A Drosophila SH2-SH3 adaptor protein implicated in coupling the sevenless tyrosine kinase to an activator of ras guanine nucleotide exchange, Sos. Cell 73:179–191.

Padgett RW, St. Johnston RD, Gelbart WM (1987): A transcript from a Drosophila pattern gene predicts a protein homologous to the transforming growth factor-β family. Nature 325:81–84.

Padgett RW, Wozney JM, Gelbart WM (1993): Human BMP sequences can confer normal dorsal-ventral patterning in the Drosophila embryo. Proc Natl Acad Sci USA 90:2905–2909.

Panganiban GEF, Rashka KE, Neitzel MD, Hoffmann FM (1990a): Biochemical characterization of the Drosophila dpp protein, a member of the transforming growth factor-β family. Mol Cell Biol 10:2669–2677.

Panganiban GEF, Reuter R, Scott MP, Hoffmann FM (1990b): A Drosophila growth factor homolog, decapentaplegic, regulates homeotic gene expression within and across germ layers during midgut morphogenesis. Development 110:1041–1050.

Peifer M, Bejsovec A (1992): Knowing your neighbors: cell interactions determine intrasegmental patterning in Drosophila. Trends Genet 8:243–249.

Perkins LA, Larsen I, Perrimon N (1992): corkscrew encodes a putative protein tyrosine phosphatase that functions to transduce the terminal signal from the receptor tyrosine kinase torso. Cell 70:225–236.

Raftery LA, Sanicola M, Blackman RK, Gelbart WM (1991): The relationship of *decapentaplegic* and *engrailed* expression in Drosophila imaginal disks: Do these genes mark the anterior-posterior compartment boundary? Development 113:27–33.

Reinke R, Zipursky SL (1988): Cell-cell interaction in the Drosophila retina: the bride of sevenless gene is required in photoreceptor cell R8 for R7 development. Cell 55:321–330.

Ren R, Mayer BJ, Cicchetti P, Baltimore D (1993): Identification of a ten-amino acid proline-rich SH3 binding site. Science 259:1157–1161.

Reuter R, Panganiban GEF, Hoffman FM, Scott MP (1990): Homeotic genes regulate the spatial expression of putative growth factors in the visceral mesoderm of Drosophila embryos. Development 110:1031–1040.

Rogge RD, Karlovich CA, Barnerjee U (1991): Genetic dissection of a neurodevelopmental pathway: Son of sevenless functions downstream of the sevenless and EGF receptor tyrosine kinases. Cell 64:39–48.

Roth S, Hiromi Y, Godt D, Nusslein-Volhard C (1991): cactus, a maternal gene required for proper formation of the dorsoventral morphogen gradient in Drosophila embryos. Development 112:371-388.

Rubin GM (1991): Signal transduction and the fate of the R7 photoreceptor in Drosophila. Trends Genet 7:372–377.

Russell J, Gennissen A, Nusse R (1992): Isolation and expression of two novel Wnt/wingless gene homologues in Drosophila. Development 115:475–485.

Sampath TK, Rashka KE, Doctor JS, Tucker RF and Hoffmann FM (1993): Drosophila transforming growth factor β superfamily proteins induce endochondral bone formation in mammals. Proc Natl Acad Sci USA 90:6004–6008.

Sampedro J, Johnston P, Lawrence PA (1993): A role for wingless in the segmental gradient of Drosophila? Development 117:677–687.

Schneider DS, Hudson KL, Lin TY, Anderson KV (1991): Dominant and recessive mutations define functional domains of Toll, a transmembrane protein required for dorsal-ventral polarity in the Drosophila embryo. Genes Dev 5:797–807.

Schupbach T, Clifford RJ, Manseau LJ, Price JV (1991): Dorsoventral signaling processes in Drosophila oogenesis. In Cell-cell Interactions in Early Development. New York: Wiley-Liss, pp. 163–174.

Shelton CA, Wasserman SA (1993): Pelle encodes a protein kinase required to establish dorsoventral polarity in the Drosophila embryo. Cell 72:515–525.

Shilo B-Z (1992): Roles of receptor tyrosine kinases in Drosophila development. FASEB 6:2915-2922.

Siegfried E, Chou T-B, Perrimon N (1992): wingless signaling acts through zeste-white 3, the Drosophila homolog of glycogen synthase kinase-3, to regulate engrailed and establish cell fate. Cell 71:1167–1179.

Simon MA, Botwell DDL, Dodson GS, Laverty TR, Rubin GM (1991): Ras1 and a putative guanine nucleotide exchange factor perform crucial steps in signaling by the sevenless protein tyrosine kinase. Cell 67:701–716.

Simon MA, Dodson GS, Rubin GM (1993): An SH3-SH2-SH3 protein is required for p21Ras1 activation and binds to sevenless and Sos proteins *in vitro*. Cell 73:169–177.

Songyang Z, Shoelson SE, Chaudhuri M, Gish G, Pawson T, Haser WG, King F, Roberts T, Ratnofsky S, Lechleider RJ, Neel BG, Birge RB, Fajardo JE, Chou MM, Hanafusa H, Schaffhausen B, Cantley LC (1993): SH2 domains recognize specific phosphopeptide sequences. Cell 72:767–778.

Spencer FA, Hoffmann FM, Gelbart WM (1982): *Decapentaplegic:* A gene complex affecting morphogenesis in *Drosophila melanogaster*. Cell 28:451–461.

Sprenger F, Trosclair MM, Morrison DK (1993): Biochemical analysis of Torso and D-raf during Drosophila embryogenesis: implications for terminal signal transduction. Mol Cell Biol 13:1163–1172.

St. Johnston RD, Gelbert WM (1987): *Decapentaplegic* transcripts are localized along the dorsal-ventral axis of the *Drosophila* embryo. EMBO J 6:2785–2791.

St. Johnston RD, Hoffmann FM, Blackman RK, Segal D, Grimaila R, Padgett RW, Irick HA, Gelbart WM (1990): The molecular organization of the decapen-

taplegic gene in Drosophila melanogaster. Genes Dev 4:1114–1127.
Stein D, Nusslein-Volhard C (1992): Multiple extracellular activities in Drosophila egg perivitelline fluid are required for establishment of embryonic dorsal-ventral polarity. Cell 68:429–440.
Stein D, Roth S, Vogelsang E, Nusslein-Volhard C (1991): The polarity of the dorsoventral axis in the Drosophila embryo is defined by an extracellular signal. Cell 65:725–735.
Struhl G, Basler K (1993): Organizing activity of wingless protein in Drosophila. Cell 72:527–540.
Szeberenyi J, Cai H, Cooper GM (1990): Effect of a dominant inhibitory Ha-ras mutation on neuronal differentiation of PC12 cells. Mol Cell Bio 10:5324–5332.
Tomlinson A, Ready DF (1987): Cell fate in the Drosophila ommatidium. Dev Biol 123:264–275.
Tsuda L, Inoue YH, Yoo M-A, Mizuno M, Hata M, Lim Y-M, Adachi-Yamada T, Ryo H, Masamune Y, Nishoda Y (1993): A protein kinase similar to MAP kinase activator acts downstream of the raf kinase in Drosophila. Cell 72:407–414.

ABOUT THE AUTHORS

ERIC C. LIEBL is a Damon Runyon–Walter Winchell Postdoctoral Fellow in the laboratory of F. Michael Hoffmann at the McArdle Laboratory for Cancer Research, University of Wisconsin-Madison. He is supported by a Damon Runyon–Walter Winchell Cancer Research Fund Fellowship, DRG-1126. He received baccalaureate degrees in both molecular biology and German from the University of Wisconsin-Madison in 1985. He earned his Ph.D. in molecular biology, awarded 1991, at the University of California-Berkeley in the laboratory of Professor G. Steven Martin, studying neoplastic transformation by tyrosine kinase oncogenes. As a postdoctoral fellow in Dr. Hoffmann's laboratory, he is using *Drosophila* genetics to investigate signal transduction pathways utilized by protooncogene tyrosine kinases.

F. MICHAEL HOFFMANN is Associate Professor of Oncology and Genetics at the McArdle Laboratory for Cancer Research and the University of Wisconsin-Madison. He has taught in the undergraduate BioCore curriculum and in graduate level courses on carcinogenesis and tumor cell biology and kinases in cell growth and development. He received his B.S. in chemistry in 1973 from Rensselaer Polytechnic Institute in Troy, New York; in 1979 he was awarded a Ph.D. in biochemistry from Cornell University in Ithaca, New York, for work on the dopamine-sensitive adenylate cyclase from bovine brain. This work was carried out under the direction of Professors Efraim Racker and June Fessenden-Raden. He joined the laboratory of Professor William M. Gelbart at Harvard University to learn Drosophila developmental genetics. During this postdoctoral training, Dr. Hoffman studied the genetic and molecular organization of the Drosophila *decapentaplegic* gene and began a fruitful collaboration with Dr. Ben-Zion Shilo to identify homologues of vertebrate protooncogenes in *Drosophila*. He joined the faculty at the McArdle Laboratory in 1984 and has continued to study the roles of protooncogene and growth factor homologues in Drosophila development. In 1991, Dr. Hoffmann was awarded a Faculty Research Award by the American Cancer Society.

Growth Factors and Signal Transduction in Development: 179–197

Cell Interactions and Signal Transduction in *C. elegans* Development

Simon Tuck and Iva Greenwald

I. INTRODUCTION

During the development of any multicellular organism, the behavior of any given cell can be influenced in two ways: by its ancestry, i.e., by the particular pattern of determinants it inherits (lineal programming); or by its environment, i.e., the signals it receives from other cells. In *C. elegans,* the relative importance of these two factors for the development of any given cell can be examined with an unusually high degree of precision. There are a number of reasons for this, but perhaps the most important is that the cell lineage, the particular pattern of cell divisions and differentiations that occur in development, is known, and is largely the same from animal to animal [Sulston and Horvitz, 1977; Kimble and Hirsh, 1979; Sulston et al., 1983]. Alterations in the lineage, therefore, can be understood in terms of altered developmental decisions of individual cells.

Experiments designed to investigate the importance or otherwise of cell interactions for the development of a given cell are conceptually straightforward: A change is made in the environment of that cell, by altering the number or configuration of its neighbors, and then its development is followed to determine whether or not it occurs normally. If it does not then we can infer that the cell can be influenced by communication with others. Such experiments have shown that many cells in *C. elegans* require interactions to develop properly. Thus the invariance of the lineage, at least in part, reflects the reproducibility of the cell communication events that control it.

The techniques used to change the environment of a cell can be either genetic or physical. One particularly powerful procedure, which has been used extensively, is to kill selected cells with a fine beam of laser light. In the early embryo, it is also possible to manipulate cells with micropipets; transplantation experiments, however, are not presently feasible. Experiments using these techniques have shown that cell interactions can influence a range of developmental phenomena in *C. elegans,* including cell fate determination, cell migration, the outgrowth of neuronal processes, and morphogenesis.

For several of the cell communication events identified, one or more of the gene products involved in cell signalling have been characterized. In most cases this has been achieved by isolating mutations that alter the fates (or migrations, etc.) of the cells requiring interactions for their proper development. The roles played in the communication process by the products of the genes defined by these mutations have then been analyzed by a combination of classical and molecular genetic techniques.

The molecular genetic analysis of genes involved in the development of the vulva, the somatic gonad, and the germline has shown that genes which control cell fate determination in *C. elegans* are often similar to genes which have been identified in vertebrate systems on the basis of their ability to induce neoplasia *in vivo* or transformation *in vitro*. For example, Lin-3, which is a ligand involved in the development of the *C. elegans* vulva contains a motif resembling epidermal growth factor (EGF) and is thought to bind to a receptor of the epidermal growth factor receptor (EGFR) subfamily [Aroian et al., 1990; Hill and Sternberg, 1992]. It seems clear, however, from the results of experiments in many organisms, that ligands that are able to induce proliferation in one cell type often induce differentiation (or at least differential gene expression in the absence of proliferation) in another. In many cases, therefore, the term "growth factor" is probably misleading, in that the response a particular ligand can induce is determined, not by the nature of the ligand itself, but by the developmental state of the cell on which it acts. Similar arguments apply to components of the signal transduction pathways activated by these ligands. For this reason, in this review, we have not restricted ourselves to describing cell communication events in which one cell (or cells) induces the growth of other cells, but have also documented cases where cell interactions cause the cell on the receiving end to differentiate or divide less than it would in the absence of the signal.

We have focused our attention primarily on how cell communication can determine cell fate. We have also described one case in which cell communication influences cell migration (of the sex myoblasts) because some of the molecules that regulate the migration of these cells are used elsewhere in the worm to determine cell fate. The influence of cell interactions on the development of the *C. elegans* nervous system has been reviewed recently elsewhere [Wadsworth and Hedgecock, 1992]. For reasons of space, we have described in detail only those cell interactions for which at least one of the components of the signal transduction pathway has been molecularly cloned.

II. VULVAL PRECURSOR CELL (VPC) FATE SPECIFICATION

A. General

Of the many instances in *C. elegans* where pattern formation is known to depend on cell interactions the development of the vulva is presently one of the best understood. The genetic analysis of vulval development has led to the identification of several components of a signal transduction pathway involving, among other things, a ligand which contains an EGF-like motif, a tyrosine kinase type receptor, and a *ras* gene. It is likely that the molecular characterization of other components of this pathway defined by classical genetic analysis will greatly aid our understanding of cell signalling in all multicellular eukaryotes.

In the wild-type *C. elegans* hermaphrodite the vulva is composed of 22 cells descended from the 3 ventral hypodermal cells: P5.p, P6.p, and P7.p (Fig. 1). These cells undergo three rounds of division over a period of approximately 5 hours during the third larval (L3) stage [Sulston, 1976; Sulston and Horvitz, 1977]. During normal development, P3.p, P4.p, and P8.p, the three immediate neighbours of P5.p–P7.p do not contribute to the vulva; instead, they divide once and their daughters fuse with the large syncytial cell, hyp7, which constitutes much of the hypodermis of the animal.

Each of the lineages generated by the six cells P3.p–P8.p constitutes a cell fate. Thus, for example, the pattern of cell divisions which P6.p gives rise to and the nature of the cells that are born at the terminal divisions constitute the 1° cell fate. Similarly, the lineage generated by P5.p during normal development constitutes the 2° fate, and that generated by P3.p, P4.p, and P8.p, the 3° fate. The lineage of P7.p is simply a mirror image of that of P5.p, and so P7.p is also said to adopt the 2° fate. (The 3° fate is also called the hypodermal fate, and the 1° and 2° fates are collectively termed vulval fates). During the development of wild-type hermaphrodites the fate adopted

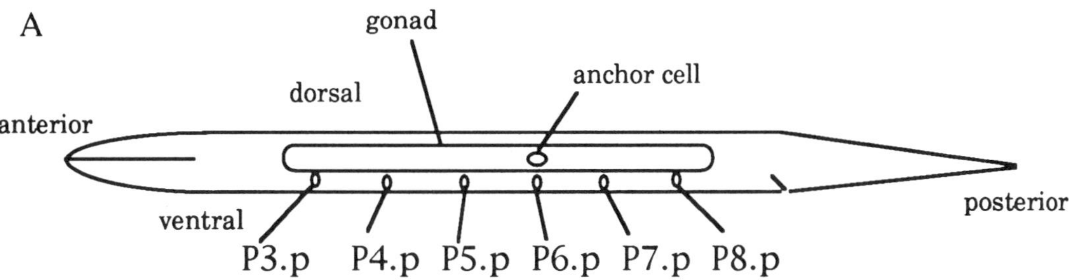

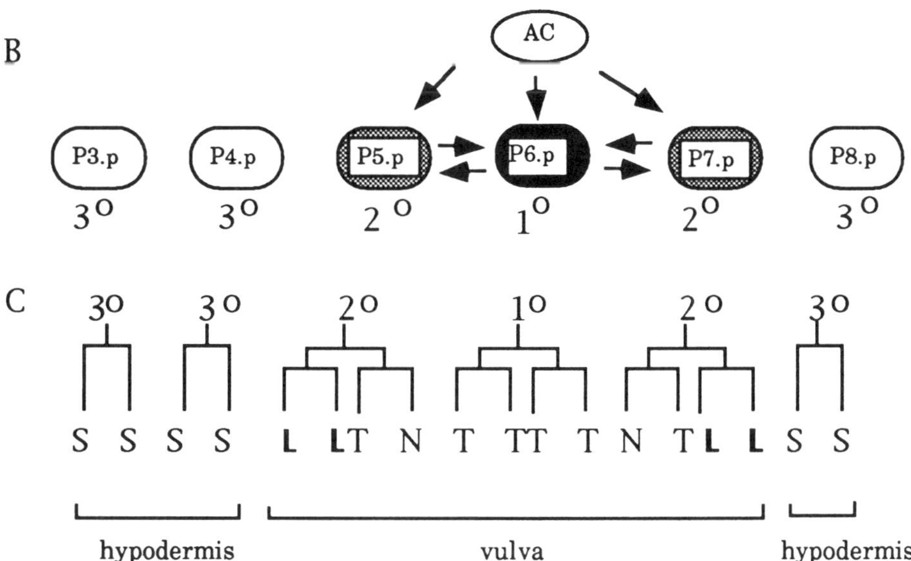

Fig. 1. **(A)** *Six multipotent hypodermal cells, P3.p, P4.p, P5,p, P6.p, P7.p, P8.p (the VPCs) are located just ventral to the gonad.* (B) *A gonadal cell, the anchor cell (AC), induces P5.p, P6.p, and P7.p to adopt vulval cell fates [Kimble, 1981]. VPCs adopting the primary fate signal to their neighbors to adopt the secondary fate [Sternberg, 1988].* (C) *Each fate is to generate a lineage, a particular pattern of cell divisions which produces a characteristic set of progeny cell types. T and **L** designate the orientation in which nuclei produced by the first two rounds of division divide: T indicates a transverse (left-right) division (with respect to the symmetry of the animal), **L** a longitudinal division. N indicates that the cell does not divide. If the cells produced by a longitudinal (anterior-posterior) division adhere to the ventral cuticle this is denoted by bold face type* (i.e., ***L***). *S indicates that the cell fuses with the large hypodermal cell, hyp7, which surrounds the VPCs. Diagram has been adapted from Sternberg and Horvitz, 1989.*

by each of these six cells is always the same. Moving from anterior to posterior, therefore, the pattern is always: 3°, 3°, 2°, 1°, 2°, 3°. Laser ablation experiments have shown that these six cells require cell interactions for their proper fate determination [Sulston and White, 1980; Kimble, 1981; Sternberg and Horvitz, 1989]. These experiments have also shown that initially, each of these six cells has the potential to adopt any of the three fates: primary, secondary, or tertiary (1°, 2°, or 3°). The cells P3.p–P8.p are therefore referred to collectively as the vulval precursor cells (VPCs).

The fate adopted by each VPC appears to be determined by (at least) three different signals [Sternberg and Horvitz, 1986; 1989; Sternberg, 1988; Herman and Hedgecock, 1990]. First, a signal from the anchor cell (AC) normally induces the cells closest to it to adopt a vulval fate (1° or 2°) [Kimble, 1981]. Secondly, cells that have been induced to adopt a vulval cell fate signal to one another to adopt the secondary fate [Sternberg, 1988]. (Not all of them adopt this fate.) Thirdly, there is evidence that the VPCs normally receive an inhibitory signal (i.e, one that inhibits them from adopting vulval cell fates) from hyp7, the large syncytial cell which surrounds them [Herman and Hedgecock, 1990]. The temporal order and/or relative strengths of these signals may be important for specifying different fates, but this aspect is not yet clear. Ablating cells after any of the VPCs have divided does not affect the behavior of the cells that remain. This suggests that once a VPC becomes committed to a particular fate the subsequent expression of that fate is independent of extrinsic factors [Sternberg and Horvitz, 1986].

A large number of mutations that affect VPC fate specification have been isolated from a variety of different genetic screens [Horvitz and Sulston, 1980; Sulston and Horvitz, 1981; Trent et al., 1983; Ferguson and Horvitz, 1985; Han et al., 1990; Clark et al., 1992,1993]. These mutations fall into two classes: those that cause a multivulva phenotype and those that cause a vulvaless phenotype. Multivulva mutations cause one or more of the VPCs that normally adopt the hypodermal (3°) fate to adopt instead a vulval fate (1° or 2°). Hermaphrodites carrying multivulva mutations develop multiple ectopic pseudovulvae along their ventral sides. Conversely, vulvaless mutations cause cells that normally adopt a vulval fate to instead adopt the hypodermal one. Animals carrying these mutations do not develop a functional vulva and are consequently unable to lay eggs. (Hermaphrodites carrying vulvaless mutations become filled with eggs, produced by self-fertilization, which eventually hatch *in utero*. The larvae devour their parent from the inside to form a "bag of worms"). Epistasis tests have been used to construct a genetic pathway for vulval cell fate specification which is shown in Figure 2 [Ferguson et al., 1987; Clark et al., 1992b]. Molecular analysis of several of these genes is entirely consistent with their involvement in signal transduction.

B. Induction by the Anchor Cell

The vulva-inducing signal is encoded by the *lin-3* gene. The predicted sequence of the Lin-3 protein is similar to that of EGF and other ligands of the EGFR [Hill and Sternberg, 1992]. Overexpression of Lin-3 from a transgene causes a dominant multivulva phenotype, whereas mutations that reduce *lin-3* activity cause a vulvaless phenotype. Regulatory sequences from the *lin-3* gene can direct the expression of a *lac*Z reporter gene in the AC. The receptor for Lin-3 appears to be encoded by the *let-23* gene [Aroian et al., 1990]. The *let-23* gene encodes a putative receptor-type tyrosine kinase of the EGFR subfamily. Mutations that reduce *let-23* gene activity abolish the ability of the VPCs to respond to the vulval-inducing signal and therefore cause a vulvaless phenotype [Ferguson and Horvitz, 1985; Aroian and Sternberg, 1991].

The *sem-5* gene, which is also characterized by alleles that cause a vulvaless phenotype, encodes a protein that consists almost entirely of one SH2 (src homology region 2) and two SH3 (src homology region 3) domains [Clark et al., 1992a]. SH2 domains can interact with phosphotyrosine-containing proteins, and it seems likely that Sem-5 binds to phosphotyrosine residues on Let-23. A human homologue of *sem-5*, GRB2 (growth factor receptor binding protein 2), has recently been isolated on the basis of its ability to bind to human EGFR [Lowenstein et al., 1992], and it has been shown that Sem-5 can also complex with hEGFR *in vitro* [Stern et al., 1993]. Furthermore, GRB2, under the control of *sem-5* gene regulatory sequences, can rescue *sem-5* mutant defects in *C. elegans* [Stern et al., 1993]. Six *sem-5* mutant alleles have been sequenced

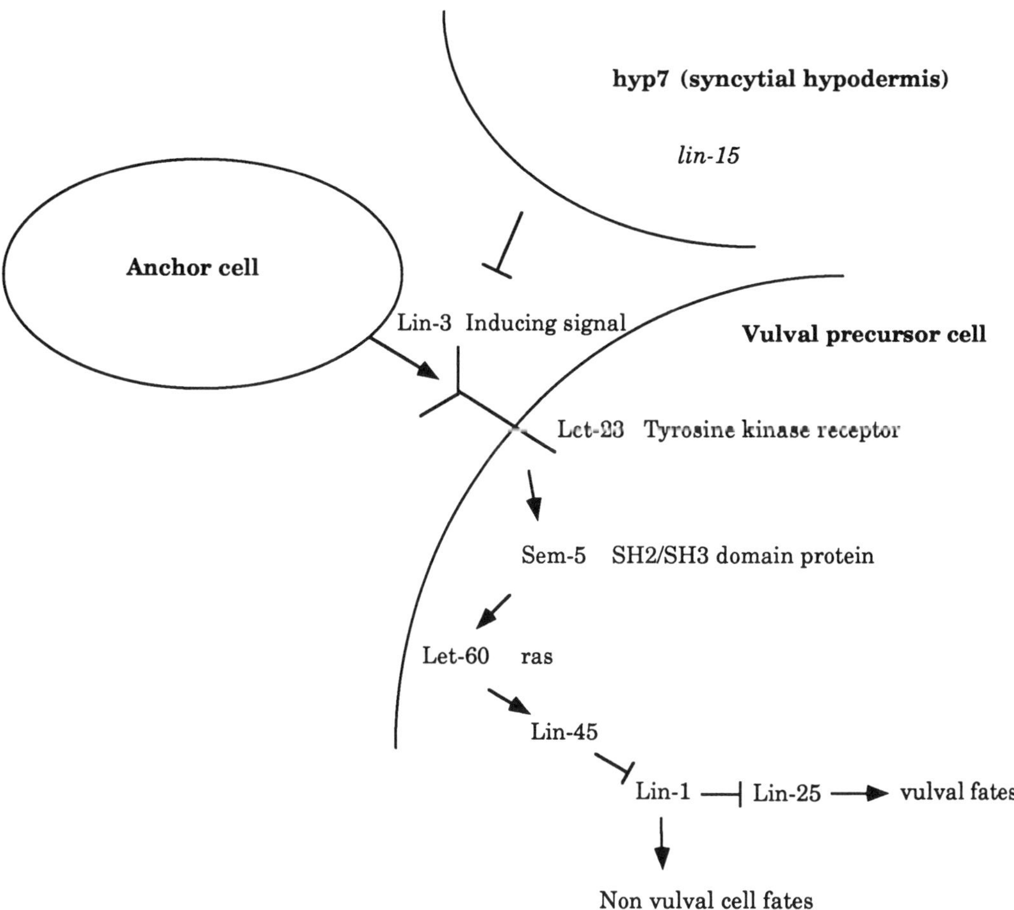

Fig. 2. *Proposed vulval induction pathway. The order of genes in this pathway is based on genetic epistasis data. The cellular focus for genes other than lin-3 [Hill and Sternberg, 1992] and lin-15 [Herman and Hedgecock, 1990] is not known but is postulated to be as shown, based on the nature of the gene products and nature of the cell interactions [e.g., Aroian et al., 1990]. Several genes known to be involved in vulval induction are not shown. Lin-31 functions after let-60 [Miller et al., 1993], but it is not known in which order lin-31 and lin-1 function. The let-341 gene functions before let-60 and probably transduces information from the activated* Let-23 *receptor to* let-60 *[Clark et al., 1992b]. Arrows do not necessarily indicate a direct biochemical interaction.*

[Clark et al., 1993]. One allele results from a change in the codon for a highly conserved proline residue in the amino-terminal SH3 domain, and another alters a conserved glycine in the carboxy-terminal SH3 domain. Thus mutations in either SH3 domain can decrease Sem-5 activity implying that each of these domains is essential for normal function, i.e., they are not redundant.

In addition to binding EGFR, GRB2 can also form a complex with a pair of proteins encoded by the *shc* gene [Rozakis-Adcock et al., 1992]. In cultured rat fibroblasts these proteins are phosphorylated on tyrosine *in vivo* in response to treatment with EGF, or transformation by *v-src*. Perhaps the function of GRB2 and Sem-5, which are not themselves phosphorylated on tyrosine, is to couple receptor-type tyrosine kinases to their downstream targets. It will be interesting to see whether any of the genes which interact genetically with *sem-5,* such as *clr-1* and *egl-15* (see sex myoblast migration), encode *C. elegans* homologues of Shc.

The *let-60* gene encodes a Ras protein very similar in sequence to mammalian N-Ras (over the first 164 amino acids, 84% are identical) [Han and Sternberg, 1990]. Genetic analysis indicates that *let-60* acts as a switch in the inductive signalling pathway which initiates vulva formation [Beitel et al., 1990; Han et al., 1990]. Hermaphrodites carrying dominant, hypermorphic (elevated activity) alleles of *let-60* display a multivulva phenotype, whereas those carrying recessive, hypomorphic (reduced activity) or dominant, antimorphic (dominant negative) alleles are vulvaless. Genetic epistasis experiments indicate that *let-60* functions downstream of *lin-3* and *let-23* in vulval fate specification; hypermorphic mutations in *let-60* are epistatic to Vul mutations in *lin-3* and *let-23* [Han et al., 1990].

All five independently isolated hypermorphic mutations result from the same sequence change in codon 13 in which G-A transition causes a substitution of glutamic acid for glycine [Beitel et al., 1990]. Glycine 13 is one of several amino acids that can cause oncogenic activation of mammalian Ras [Barbacid, 1987]. It is not yet clear why the other mutations which activate mammalian Ras have not also been observed in Let-60. One possibility is that this reflects the specificity of the mutagen (ethyl methanesulfonate); alternatively, it is possible that other mutations, such as those in codon 12, activate *let-60* too much and cause lethality. It is interesting in this regard that simply overexpressing wild-type Let-60 gives rise to the multivulva phenotype and that higher levels still cause lethality [Han and Sternberg, 1990]. This suggests that the cells in *C. elegans* that respond to activated Ras are much more sensitive to levels of the protein than are mammalian cells to mammalian Ras.

The sequence changes associated with eight hypomorphic and nine antimorphic *let-60* mutations have been determined [Beitel et al., 1990; Han and Sternberg, 1991]. All but one of the recessive mutations are in regions of the protein that are not necessary for the transforming activity of activated v-Ha-Ras, a finding which suggests that they might be required for the proper activation of *let-60* by upstream regulators. The other recessive mutation alters part of the "effector" region, which is thought to be required for activation of downstream targets by Ras proteins. Most of these mutations alter regions of the protein that are highly evolutionarily conserved but, rather surprisingly perhaps, two of the strongest alleles, *s1155* and *s59,* result from changes in amino acids which are not conserved. The dominant-negative mutations affect five different amino acids, all of which are conserved in evolution [Han and Sternberg, 1991]. Studies on mammalian Ras have shown that four of these amino acids are in regions required for binding of guanine nucleotides [Der et al., 1986; Sigal et al., 1986]. The fifth is predicted to lie close to the binding site and so may also be involved in nucleotide binding. The effects of these dominant-negative alleles are all completely suppressed by a hypermorphic allele of *let-60* suggesting that the proteins encoded by the dominant-negative alleles act by sequestering an upstream activator of *let-60* rather than a downstream target. Given the nature of the dominant-negative mutations, a good candidate for such an upstream activator is guanine nucleotide exchange factor (GNEF). Dominant-negative alleles of yeast Ras function by sequestering the GNEF encoded by the CDC25 gene of *S. cerevisiae* [Powers et al., 1989; Jones et al., 1991]. The *C. elegans* gene *let 341* may encode a GNEF. Hypomorphic mutations in *let-341* confer a vulvaless phenotype identical to that conferred by hypomorphic alleles of *lin-3, let-23, sem-5,* and *let-60* [Clark et al., 1993]. Furthermore, hypermorphic, gain-of-function (gf) mutations in *let-60* suppress the vulvaless phenotype of *let-341* alleles [Clark et al., 1993].

Two of the *let-60* dominant-negative mutations, which result in changes at codons 16 and

119 respectively, give rise to proteins which retain some wild-type activity; it is therefore possible to separate the dominant-negative effect from the ability to be activated.

The genes *lin-45, lin-1,* and *lin-25* may all function after *let-60* in the genetic pathway for VPC fate specification. Hypomorphic mutations in *lin-45* confer a vulvaless phenotype which is epistatic to the multivulva phenotype conferred by *let-60(gf)*. Thus *lin-45* acts after, or at the same time as, *let-60*. Mutations which reduce or eliminate *lin-1* activity suppress the lethality associated with hypomorphic alleles of *let 60* and *lin 45,* and confer a multivulva phenotype in a wild-type background. The *lin-1* gene is therefore thought to act downstream of both these genes [Han et al., 1990; Clark et al., 1992b]. Null mutations in *lin-25* suppress the multivulva phenotype of *lin-1* and *let-60(gf),* an observation which suggests that *lin-25* functions after or at the same time as *lin-1*.

The *lin-31* gene also appears to function after *let-60*. Animals homozygous for *lin-31* null mutations frequently show aspects of both the multivulva and vulvaless phenotypes [Ferguson et al., 1987; Miller et al., 1993]. Specifically, P3.p, P4.p, and P8.p, which normally adopt the hypodermal fate, often adopt vulval fates; whereas P5.p, P6.p, and P7.p, which normally adopt vulval fates, can adopt the hypodermal fate. Both types of fate transformation can sometimes be seen within the same animal. The gene has recently been found to encode a transcription factor of the HNF3/Forkhead family [Miller et al., 1993]. One way to explain the *lin-31* mutant phenotype, therefore, would be to suppose that transcription complexes exist in the VPCs that can simultaneously activate the vulval specific genes and repress hypodermal ones, and that others exist which repress vulval specific genes and activate hypodermal ones [Miller et al., 1993]. If Lin-31 were to be part of one or both of such complexes, disruption of the gene could upset the balance of positive and negative regulators and lead to deregulation of cell fate specification. There would be inefficient repression of vulval specific genes in P3.p, P4.p, and P8.p and, at the same time, inefficient activation of vulval genes in P5.p, P6.p and P7.p. This model requires that *lin-31* acts in concert with other factors in transcriptional regulation and that these can function to some extent in the absence of *lin-31*. This is consistent with the observation that normal 1°, 2°, and 3° lineages are observed in *lin-31* mutant animals.

At the present time it is not known what roles are played by *lin-13, lin-2, lin-7,* and *lin-10*. Null mutations in *lin-2, lin-7,* and *lin-10* confer a vulvaless phenotype which, in each case, is suppressed by hypermorphic mutations in *let-60*. The *lin-10* gene has been cloned [Kim and Horvitz, 1990] but it is not similar to any other genes in the data bases. Lin-10 mRNA appears to be expressed throughout development and is not restricted to the VPCs.

C. The Lateral Signalling Pathway

A lateral signal from P6.p causes P5.p and P7.p to adopt the 2° cell fate [Sternberg, 1988]. This lateral signalling acts over short distances and is thought to be mediated by the product of the *lin-12* gene. Mutations in *lin-12* cause reciprocal cell fate transformations in the VPCs [Greenwald et al., 1983]. In animals carrying dominant alleles of *lin-12,* all six VPCs adopt the 2° fate, whereas in animals homozygous for mutations which eliminate *lin-12* activity, no 2° cells are specified, and P5.p, P6.p, and P7.p all adopt the 1° fate. By analogy with its postulated role as a receptor in the AC/VU decision (Section IV, this chapter), the *lin-12* gene product may be the receptor for the lateral signal. It is curious that mutations that activate *lin-12* can cause a multivulva phenotype even in the absence of an AC [Greenwald et al., 1983; Greenwald and Seydoux, 1990]. This implies that, for the 2° cell fate, *lin-12* activation bypasses the need for the *let-60* Ras-mediated inductive signalling pathway.

Three genes that function downstream of *lin-12* have been identified: *lin-11, lin-17,* and *lin-18*. Mutations in these genes cause abnormal 2° lineages [Ferguson et al., 1987]. These genes do not function in VPC determination, however, but are required for the execution of the 2° fate. Double mutants between loss-of-function mutations in these genes and mutations that activate *lin-12* display ectopic expression of abnormal 2° fates [Sternberg and Horvitz, 1989].

D. Evidence for a Third Signalling Pathway in VPC Fate Specification

Genetic mosaic analysis of *lin-15* suggests that besides the AC signalling pathway which originates in the AC, and the lateral signalling pathway which mediates interactions between adjacent VPCs, a third pathway may exist between the VPCs and hyp7, the large syncytial cell which surrounds them [Herman and Hedgecock, 1990]. Loss-of-function (lf) mutations in *lin-15* confer a highly penetrant multivulva phenotype, which is independent of the AC. Hermaphrodites homozygous for null alleles of *lin-15* are multivulva even if the entire gonad has been ablated at the first larval (L1) stage [Ferguson et al., 1987]. Genetic mosaic analysis of *lin-15* indicates that it functions non-autonomously. In mosaic animals, a given VPC may be genotypically wild-type for *lin-15,* yet adopt a mutant fate; and genotypically mutant, yet adopt a wild-type fate.

Herman and Hedgecock [1990] have examined the fates adopted by the VPCs in a number of mosaic animals and concluded that the focus of *lin-15* gene activity is hyp7. They proposed a model for cell signalling in the developing vulva based upon their results. In this model, the fate adopted by a VPC in the absence of all extrinsic signals would be vulval. The hyp7 cell normally inhibits the VPCs from adopting a vulval fate so that in the absence of the AC all the VPCs adopt the hypodermal fate. The function of the AC, therefore, is to repress the inhibitory signal from hyp7.

E. Synthetic Multivulva Genes

A number of *lin-15* mutations exist which do not confer a multivulva phenotype in a wild-type genetic background [Ferguson and Horvitz, 1989]. These "silent" *lin-15* mutations often show strong genetic interactions with mutations in a number of genes including those in *lin-8, lin-9, lin-35, lin-36, lin-37,* and *lin-38*. Hermaphrodites homozygous for null alleles of any of these other genes, but otherwise genotypically wild-type, also do not display vulval lineage defects; those homozygous for mutations in two of them, however, are often multivulval. The genes defined by these mutations have thus been termed synthetic multivulva (*syn muv*) genes. By constructing every possible double-mutant combination it has been shown that these *syn muv* mutations fall into two different classes, A and B. Hermaphrodites homozygous for both a class A and a class B mutation are multivulval whereas those homozygous for two different class B ones (or two different class A mutations) show no vulval lineage defects. All mutations in *lin-8* and *lin-38* fall into class A and all those in *lin-9, lin-35, lin-36,* and *lin-37* fall into class B. The precise roles of the *syn muv* genes are not yet known. It seems very likely, however, that they function in the same signalling event affected by *lin-15* mutations. To date, *lin-15* is the only gene characterized which can mutate to give both class A and class B alleles; it is also the only gene which can give rise to alleles which confer a vulval lineage defect on their own.

F. Many of the Genes Required in the Inductive Signalling Pathway Are Also Required at Other Stages of Development

Animals homozygous for null mutations in several of the genes described above—*lin-3, let-23, sem-5, let-341, lin-45,* and *let-60*—die as larvae. This implies that these genes (and perhaps other *lin* genes for which the null phenotypes have not yet been determined) mediate at least one signalling event during early development which is required for viability. Animals carrying a null allele of *let-60,* for example, arrest at the early L1 stage [Han and Sternberg, 1991], a phenomenon which suggests that a signalling event occurs during embryogenesis or at the very beginning of the L1 stage. It is not yet known which cells are involved.

A number of postembryonic developmental defects, besides those in vulval development, are observed in animals carrying certain hypomorphic alleles of these genes [Aroian and Sternberg, 1991; Clark et al., 1993] for example, the fate of P12 is often incorrectly specified. Laser ablation experiments have shown that this cell requires cell interactions for its proper fate determination [Sulston and White, 1980], and it seems likely that much, if not all, of the vulval induction pathway might function in an intercellular signalling system that specifies the fate of this cell. Other defects observed in animals carrying mutations in these genes include grossly abnormal male tail development, and hermaphrodite sterility. It is not known yet what causes the sterility but it may be due to defects in oogenesis. Oocytes are produced initially but they later degenerate; they cannot be fertilized by sperm from wild-type males [Aroian and Sternberg, 1991].

The *sem-5* gene is required not only for viability of larvae and for vulval development but also for the proper migration of a pair of cells, the sex myoblasts, which generate the muscles that open the vulva during egg-laying (see sex-myoblast migration, below). Interestingly, hypomorphic mutations in several other genes required for proper vulval development, including *lin-3, let-23,* and *let-60,* do not affect sex-myoblast migration, so that *sem-5* may function in more than one type of signal transduction pathway.

III. SEX-MYOBLAST MIGRATION

The ability to lay eggs in *C. elegans* requires a number of different structures: the gonad, the uterus, the vulva, the "sex muscles" which open the vulva and uterus, and the neurons that innervate these muscles. Ablation experiments have shown that a cascade of interactions are required for these different components to be properly assembled [Thomas et al., 1990; Li and Chalfie, 1990]. The gonad induces the formation of the vulva [Kimble, 1981]; regulates the migration of the sex myoblasts, which are the progenitors of the sex muscles [Thomas et al., 1990; Li and Chalfie, 1990]; and is required for vulval morphogenesis [Kimble, 1981; Seydoux et al., 1993]. In turn, the vulva induces branching of the neuronal processes which innervate these muscles [Li and Chalfie, 1990].

The two sex myoblasts (SMs) give rise to the 16 vulval and uterine muscles required for opening of the vulva during egg-laying. The SMs are born in the L1-stage in bilaterally symmetric positions at the posterior end of the hermaphrodite. During the L2 stage they migrate anteriorly a distance of approximately 65 microns (20% of the worm length) toward the developing gonad and then move to positions precisely flanking the center of the gonad, close to the ventral side of the worm. During the L3 stage each SM produces 4 vulval and 4 uterine muscles [Sulston and Horvitz, 1977].

The migrations of the SMs occur in two distinct phases. An initial, gonad-independent migration allows them to reach the general vicinity of the gonad. In the second phase, a cell (or cells) in the somatic gonad attracts the SMs to their precise final positions [Thomas et al., 1990]. In a mutant in which the gonad can be displaced as far as the dorsal side of the animal, the SMs still migrate towards it and take up their final positions flanking it [Thomas et al., 1990]. In the absence of the gonad, however, the SMs undergo their initial migration but end up in a variety of different positions centered loosely about the position where the gonad would have been.

The sex muscles are required for proper egg-laying [Trent et al., 1983]. Hermaphrodites in which the SMs have been ablated become filled with fertilized eggs which eventually hatch *in utero* [Trent et al., 1983]. Mutations in three genes, *sem-5*, *egl-15*, and *egl-17*, disrupt the migration of the SMs and hence confer an egg-laying defective phenotype [Stern and Horvitz, 1992]. In hermaphrodites carrying mutations in these genes the second-phase migrations of the SMs do not occur. In consequence, the sex muscles are born in the wrong positions and are not able to make their proper attachments. The observation that the initial migration of the SMs occurs correctly in these mutants indicates that it is not the ability to migrate *per se* which is defective, but the regulation of migration. Mutations in *egl-15, egl-17,* and *sem-5* could conceivably disrupt the production of the signal from the gonad which regulates the migration of the SMs, the reception of that signal, or its interpretation (transduction). Ablation experiments point to a defect in signal transduction. Specifically, these experiments show that not only do the second-phase migrations of the SMs not occur in animals carrying mutations in *egl-15* and *egl-17,* but that the somatic gonad actively repels the SMs in these animals [Stern and Horvitz, 1992]. Thus the gonad still sends a signal and the SMs are still capable of receiving that signal; they respond, however, by migrating in the wrong direction. (One caveat is that it may not be the same signal in the mutants as in the wild-type organisms.)

A cell which migrates towards (or away from) the source of a signal presumably does so because it is able to sense the relative concentration of that signal on either side of itself. The information must be relayed in some way to influence the polymerization and/or depolymerization of cytoskeletal components used in migration. Perhaps *sem-5, egl-15,* and *egl-17* mutations disrupt the transduction of information from the activated signal receptor to the cytoskeleton. The identity of *sem-5* is consistent with this conjecture. As described above (see VPC specification), *sem-5* encodes a protein consisting almost entirely of two SH3 domains and one SH2 domain [Clark et al., 1992]. Sem-5 can bind to the activated human EGFR *in vitro* and therefore presumably binds to tyrosine kinase type receptors in *C. elegans* [Stern et al., submitted]. The binding of Sem-5 to hEGFR is mediated by the SH2 domain. Mutations in either SH3 domain which disrupt function *in vivo* do not affect binding *in vitro* indicating that the SH3 domains are required for some other function of the Sem-5 protein. Work in other systems suggests that SH3 domains mediate association with the cytoskeleton and the cell membrane [Rodaway et al., 1989; Koch et al., 1991].

Mutations in *sem-5* and *egl-15* show interesting genetic interactions with mutations in the gene *clr-1*. Null alleles of *egl-15* confer an L1 larval-arrest phenotype (indicating that *egl-15* is required for some aspect of development besides SM migration) [Clark et al., 1993]. Screens for suppressors of this phenotype produced alleles of a previously characterized gene, *clr-1* [Hedgecock et al., 1990; Clark et al., 1993]. In a wild-type genetic background, these suppressor alleles confer either a larval-lethal or a Clear (Clr) phenotype (Clr reflects the clarity with which cell boundaries can be observed in these mutants). Screens for suppressors of a lethal *clr-1* allele in turn produced alleles of *egl-15* and *sem-5* [Clark et al., 1993]. Homozygous viable *clr-1* mutations do not themselves confer SM migration defects; those genetic interactions with mutations which do, however, strongly suggest that this gene is also involved in regulating the migration of these cells. It will be very interesting to determine the nature of the products encoded by all these genes and the developmental foci of their action.

IV. THE AC/VU DECISION IN GONADOGENESIS

A simple example of intercellular signalling involves two cells in the hermaphrodite gonad, Z1.ppp and Z4.aaa. These cells are descendants of the somatic progenitor cells, Z1 and Z4, which are present in the first larval stage (L1) gonad primordium. Z1 and Z4 divide during early larval development, and the fates of most of their descendants, Z1.ppp and Z4.aaa, have a choice of cell fates: One of the two cells always becomes the anchor cell (the cell required for vulval induction) and the other becomes a ventral uterine precursor cell (VU). Each has an equal chance of becoming an AC, and if all gonadal cells except for Z1.ppp or Z4.aaa are ablated, then one of the two survivors becomes an AC and the other becomes a VU. This observation implies that interactions between Z1.ppp and Z4.aaa specify the VU fate, and that the AC/VU decision may be considered to involve a "signal" from the presumptive AC that is received by the presumptive VU [Seydoux and Greenwald, 1989].

Genetic analysis has demonstrated that activity of the *lin-12* gene controls the AC/VU decision. Both Z1.ppp and Z4.aaa become ACs in mutants that lack Lin-12 activity, and both become VUs in mutants that have elevated Lin-12 activity [Greenwald et al., 1983]. In addition, Lin-12 activity is needed or Lin-12 is synthesized at the time that Z1.ppp and Z4.aaa become determined [Greenwald et al., 1983; Seydoux and Greenwald, 1989].

Genetic mosaic analysis has revealed that Lin-12 acts at the receiving end in the AC/VU signalling event. Furthermore, mosaic analysis has also revealed that Lin-12 has an intimate role in the decision-making process, because in genetic mosaics in which Z1.ppp and Z4.aaa were of different *lin-12* genotypes, the cell fate choice of the *lin-12*(+) cell was biased, so that it always became a VU. This bias in cell fate choice implies that Z1.ppp and Z4.aaa assess their relative levels of Lin-12 activity as part of the decision-making process, before either cell commits to the default fate, and that a feedback mechanism exists that ensures that only one of the two cells will decide to express the default fate.

The predicted molecular structure of Lin-12 also suggests a direct involvement in the signalling process [Greenwald, 1985; Yochem et al., 1988]. The Lin-12 protein has been postulated to function as a receptor for intercellular signals in part because it is a transmembrane protein with a large extracellular domain and a large intracellular domain [Yochem et al., 1988]. A model for the AC/VU decision is shown in Figure 3.

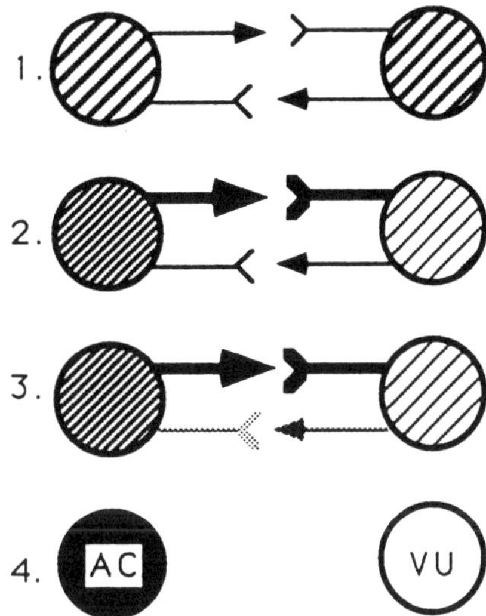

Fig. 3. *A model for the AC/VU decision of Z1.ppp and Z4.aaa [adapted from Seydoux and Greenwald, 1989]. Step 1: Z1.ppp and Z4.aaa are equivalent. This equivalence implies that they both have equivalent signalling (arrow) and receiving (inverted arrow) potential or level of activity. The receptor is proposed to be the product of the* lin-12 *gene. Step 2: A stochastic event causes one cell to produce more signal (bold arrow), which stimulates Lin-12 activity (bold inverted arrow) in the other cell. Step 3: When Lin-12 activity is stimulated in the receiving cell, the cells signalling potential or level is reduced (dotted arrow); the receiving potential of the signalling cell might also be reduced (dotted arrow). Step 4: The cells are determined to become an AC and a VU.*

Lin-12 is a member of the "Lin-12/Notch" family of proteins defined by the C. elegans *lin-12* gene and the Drosophila *Notch* gene [reviewed in Greenwald and Rubin, 1992]. All members of the Lin-12/Notch family share a similar arrangement of epidermal growth factor–like motifs and "Lin-12/Notch repeat" motifs in their extracellular domains, and Cdc 10/SW16 motifs [Breeden and Nasmyth, 1987] in their intracellular domains. Another member of this gene family, the *C. elegans* gene, *glp-1* [Austin and Kimble, 1989; Yochem and Greenwald, 1989], appears to function as a receptor in other cell fate decisions (see below). There are members of the *lin-12/Notch* gene family in vertebrates. Of particular interest are the mouse *int-3* [Gallahan and Callahan, 1987; Robbins et al., 1992] and human *TAN-1* [Ellisen et al., 1991] genes. These genes were defined by oncogenic mutations, implying that aberrant cell signalling involving Lin-12/Notch family members can play a role in oncogenesis.

The identification of potential ligands and targets for Lin-12 and Glp-1 has been facilitated by powerful genetic selections for extragenic suppressor mutations. Both elevation and reduction of Lin-12 activity causes an egg-laying defective phenotype [Greenwald et al., 1983; Sundaram and Greenwald, 1993]. Reversion of these egg-laying defects has led to the identification of a number of genes whose products might be involved in Lin-12-mediated cell fate specification [Ferguson and Horvitz, 1985; Sundaram and Greenwald, 1993; F. Tax, J. Thomas, and H. R. Horvitz, personal communication]. Some of the suppressor mutations isolated for their ability to suppress *lin-12* mutant defects also suppress alleles of *glp-1* [Sundaram and Greenwald, 1993; A. M. Howell and J. Priess, personal communication], which suggests that at least some ligands or targets of these two structurally similar proteins may be shared. Some suppressor mutations, however, interact genetically with either *lin-12* or *glp-1,* but not both [Maine and Kimble, 1989, 1993; F. Tax and J. Thomas, personal communication; A. M. Howell and J. Priess, personal communication]. Finally, using phenotypic criteria, Lambie and Kimble [1991] have identified two genes, *lag-1* and *lag-2,* which also appear to function in both *lin-12*- and *glp-1*-mediated cell signalling pathways.

V. INTERACTIONS BETWEEN SOMA AND GERMLINE

The somatic gonad has two reflexed arms and each has a distal-proximal axis with respect to the vulva. At the distal end of each arm, there is a distal tip cell (DTC) [Kimble and Hirsh, 1979]. Within each arm, there is a germ cell syncytium, with a distal-proximal axis of germ cell maturation: Germ nuclei at the distal end divide mitotically, germ nuclei located more proximally undergo meiosis, and most proximally, germ cells undergo differentiation into gametes [Hirsh et al., 1976]. If the DTCs are ablated with a laser microbeam, distal germ cells enter meiosis [Kimble and White, 1981]. This result implies that the DTCs produce an intercellular signal that maintains distal germline nuclei in mitosis (or inhibits them from undergoing meiosis).

The Glp-1 protein is likely to function as the receptor for the signal from the DTCs to the germline. Mutations in the *glp-1* (abnormal germ line proliferation) gene cause a phenotype reminiscent of DTC ablation: mitotic proliferation is minimal, and instead germ cells enter meiosis prematurely [Austin and Kimble, 1987; Priess et al., 1987]. Temperature-shift experiments demonstrated a continuous requirement for *glp-1* synthesis or function for normal germline development [Austin and Kimble, 1987]. In addition, genetic mosaic analysis demonstrated that Glp-1 functions in the germline, i.e. it functions in the receiving end of the DTC-germline communication [Austin and Kimble, 1987]. Finally, the protein encoded by the *glp-1* gene is a transmembrane protein with large extracellular and intracellular domains, and the hallmarks of the Lin-12/Notch family [Austin and Kimble, 1989; Yochem and Greenwald, 1989; see AC/VU decision in gonadogenesis].

The prevention of inappropriate cell–cell interactions also seems to be important for formal germline development [Seydoux et al., 1990]. Somatic gonadal cells seem to prevent the AC from serving as a source of signal for Glp-1 mediated germline proliferation: If all somatic gonadal cells except for the AC are ablated, germline mitosis is seen in the proximal region of the gonad. This abnormal interaction requires Glp-1 activity, and therefore appears to involve the same mechanism as the DTC-promoted proliferation described above. The fact that this abnormal interaction can also occur in an intact gonad in *lin-12* null mutants has led to the suggestion that the ligand produced by the AC that normally binds to Lin-12 to promote the VU fate can also bind to Glp-1 to promote germline proliferation; AC-to-VU signal binding to Glp-1 is normally prevented by its binding to Lin-12 on somatic gonadal cells [Seydoux et al., 1990].

VI. CELL INTERACTIONS DURING EARLY EMBRYOGENESIS

A. General

Embryonic development in *C. elegans* begins with a series of asymmetric divisions which give rise to six founder cells: AB, MS, C, D, E, and P_4 (Fig. 4). some of these founder cells generate progeny of a single type: D produces only body wall muscle, E is the intestinal precursor, and P4 is the progenitor of the germline [Sulston et al., 1983]. The other founder cells, however, generate many different cell types and most tissues are polyclonal in origin.

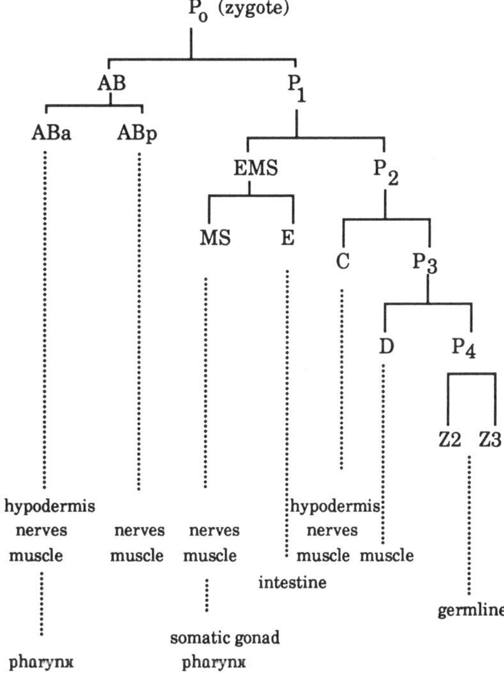

Fig. 4. *Generation of the embryonic founder cells and the cell types derived from them. The somatic founder cells, AB, MS, E, C, and D. AB, EMS, C, and D, are derived from stem cell–like cleavages of the P cells: EMS then cleaves into MS and E. P_4 is the progenitor of the germline. The intestine and the germ-line are derived clonally, from E and P_4, respectively; all other tissues are derived from more than one founder cell [Kimble and Hirsh, 1979; Sulston et al., 1983].*

Embryonic development in *C. elegans* was long thought to be the paradigm for autonomously determined (mosaic) development. Certain cells born relatively late in embryonic development were known to require cell interactions for their proper fate specification [Sulston et al., 1983], but it was thought that the fates of most of the others, in particular the early blastomeres, were specified by the inheritance of developmental determinants. In recent years, however, it has become clear that cell interactions are important for the specification of a large number of cells in the embryo, including the majority of the early blastomeres. It seems very likely that, as more cell markers become available for the different embryonic lineages, many other cells will also be shown to require cell interactions for their proper fate specification.

B. The Blastomeres ABa and ABp Are Initially Equivalent

The first indication that cell interactions are important during early embryogenesis was found by Priess and Thomson [1987]. Their experiments showed that the early blastomeres ABa and ABp are initially equivalent and strongly suggested that ABa requires a signal from EMS to adopt its proper fate. During wild-type development ABa and ABp adopt very different fates [Sulston et al., 1983]. Priess and Thomson [1987] showed that these two cells are initially developmentally equivalent by reversing their positions relative to the other two cells present in the 4-cell-stage embryo. Lineage analysis of the manipulated embryos showed that ABa in its new position adopts the fate of ABp and vice versa. This implies that when these two cells are born they are not committed to a particular fate by the inheritance of cytoplasmic determinants. The fact that they become different, therefore, must be because they (or their descendants) experience different interactions with other cells.

C. Descendants of EMS Are Required for the Descendants of ABa to Generate Pharyngeal Cells

The pharynx in *C. elegans* is a muscular organ used for feeding. It consists of several cell types including muscles, neurons, and gland cells [Albertson and Thomson, 1976]. All pharyngeal cells are descended from just five cells present in the 28-cell-stage embryo (ABalpa, ABaraa, ABararap, MSa, MSpa); each of these five-cells produces predominantly, but not solely, pharyngeal cells [Sulston et al., 1983]. All the AB-derived pharyngeal precursor cells are the descendants of ABa, whereas all the P_1-derived precursors are from EMS. Pharyngeal muscles are arranged in a series of eight sets of cells. The sets at the anterior tip are all derived from ABa, whereas the posterior sets are derived from EMS. The sets in the middle have cells from both AB, and EMS. The other components of the pharynx, such as the marginal cells and neurons, also show this anterior-posterior distribution. Micromanipulation experiments have revealed that cell interactions are required for ABa descendants to generate pharyngeal cells. If the posterior blastomere P_1 is removed from the 2-cell-stage embryo, then no pharyngeal cells are produced by the remaining AB blastomere (pharyngeal cells can be identified in several different ways, including staining with tissue specific antibodies and by their distinctive morphology) [Priess and Thomson, 1987]. This failure to differentiate is not simply the result of the partial embryos failing to develop, because cell division continues for several hours after P_1 ablation, and because several differentiated cell types such as hypodermal cells and nonpharyngeal neurons are produced. Conversely, if AB is extruded from the 2-cell-stage embryo, some pharyngeal cells are produced by the isolated P_1 blastomere. In these cases, as one would predict, only the pharyngeal cells which are derived from P_1 are formed. Taken together, these experiments show that one (or more) P_1 descendant is required for ABa descendants to generate pharyngeal cells. At the present time it is not known precisely with which P_1-derived cells the ABa descendants must interact, or exactly which ABa-derived cells require the interaction. However, experiments with 3-cell partial embryos containing either ABa, ABp, and EMS, or ABa, ABp, and P_2 show that the EMS daughter of P_1, and not P_2, is required for the descendants of ABa to produce pharyngeal cells. One possible explanation for these results is that, at the 4-cell stage, EMS normally sends a signal to ABa to induce it to adopt its proper fate. In the absence of EMS, ABa adopts a different fate—one that does not lead to the generation of pharyngeal cells. Evidence in support of this hypothesis has come from the analysis of the lineage generated by an isolated AB blastomere [Schnabel, 1991]. When P_1 is irradiated, the fates of cells throughout the ABa lineage are incorrectly specified, not just those of the pharyngeal precursor cells. In particular, many cell deaths fail to occur and a number of cells derived from ABar which normally differentiate as hypodermis fail to do so. The fact that so many ABa descendants do not generate their normal lineage suggests that in the absence of P_1 ABa itself suffered some cell fate transformation. Alternatively, one must argue that several different interactions are required between the descendants of P_1 and the descendants of ABa for the latter to adopt their correct fates.

Priess et al. [1987] performed a genetic screen for embryonic lethal mutations which cause aberrant pharyngeal development. Four recessive mutations were isolated, all of which were subsequently found to be alleles of *glp-1*. As we have described above, *glp-1* was first identified as a gene required for a cell interaction which occurs in postembryonic development. The phenotype of the embryonic lethal mutations of *glp-1* strongly suggests that the *glp-1* gene product also mediates the cell interaction (or interactions) which is required for ABa descendants to generate pharyngeal cells. In particular, *glp-1* mutant embryos produce P_1- but not ABa-derived pharyngeal cells. The *glp-1* mutant embryos, like the P_1-derived partial embryos described above, develop a partial pharynx which contains most of the cells from the posterior region but which lacks cells from the anterior region. The *glp-1* gene is not simply required for the proper differentiation of pharyngeal cells because the cells in the posterior region of the pharynx in *glp-1* mutant embryos are normal. It seems very likely that instead, the lack of anterior pharyngeal cells is caused by one or more cell fate transformations in the ABa lineage which cause either ABa itself or two or more of its descendants to be transformed to a cell type which does not generate pharyngeal cells. The *glp-1* mutant embryos also display a hypodermal defect that is consistent with a requirement for *glp-1* in an interaction between EMS and ABa. Both *glp-1* mutant embryos and 3-cell partial embryos formed from the ablation of EMS generate ectopic hypodermal cells [Priess et al., 1987; Schnabel, 1991]. In the case of the laser-ablated embryos, lineage analysis has revealed that, in the absence of EMS, ABal and ABaraa generate ectopic hypodermal cells. It will be interesting to determine whether or not this is also true of *glp-1* mutant embryos.

D. Multiple Cell Interactions Are Required for Cell Fate Specification in the Early Embryo

Experiments using the differentiation of gut as a marker for the EMS cell fate suggest that P_2 sends a signal to EMS to induce it to adopt its proper fate [Schierenberg, 1987; Priess and Thomson, 1987; Goldstein, 1992]. Furthermore, experiments by Bowerman et al. [1992] using two cells, the valve cell pair, as markers for the ABp lineage suggest that P_2 also sends a signal to ABp (or its daughters ABpl and ABpr) to induce it to adopt its proper fate. Thus it seems that, even at this early stage, cell interactions are crucial for correct cell fate specification. The cell interactions which may occur in the 4-cell-state embryo are shown in Figure 5.

It is clear that cell interactions are also required at later stages of embryonic development. Firstly, after ABp (or its daughters ABpl and ABpr) have been induced to give rise to valve cells, a second interaction must occur between two cells within the ABp lineage for the proper fate specification of the left homologue of the valve cell precursor [Bowerman et al., 1992]. Secondly, symmetry reversal experiments imply that an extensive series of interactions occurs at the 6-cell stage or later [Wood, 1991; Wood and Kershaw, 1991].

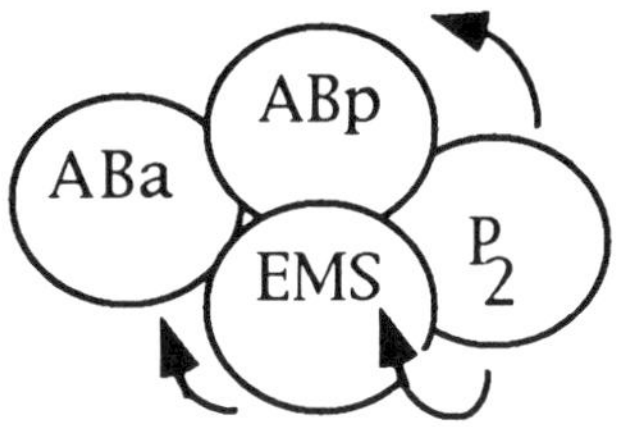

Fig. 5. *Possible cell interactions occurring in the 4-cell-stage embryo. P_2 may signal to ABp and EMS to cause them to adopt their appropriate fates [Goldstein, 1992; Bowerman et al., 1992]: EMS may then signal to ABa [Priess and Thomson, 1987]. It is also possible that signalling between P_2 and the ABp lineage occurs at the 6-cell stage i.e., between P_2 and the two ABp daughters, ABpl and ABpr [Bowerman et al., 1992]. Likewise, it is possible that the signal between the EMS lineage and the ABa lineage occurs at a later stage. The signal sent by P_2 to EMS may be the same as that sent to ABp (or its daughters ABpl and ABpr). EMS and ABp are not developmentally equivalent and therefore could respond to the same signal in different ways. The 4-cell-stage embryo is planar: P_2 does not contact ABa.*

VII. CELL INTERACTIONS IN SOMATIC SEX DETERMINATION

There are two sexes in *C. elegans*, hermaphrodites and males. Hermaphrodites can be considered as somatic females with a germ line that can make sperm as well as oocytes. (In this discussion we will be concerned only with somatic sex determination.) The ratio of the number of X chromosomes to the number of sets of autosomes (X:A ratio) specifies the sex of an individual [Madl and Herman, 1979]. A wild-type individual having two sets of autosomes and two X chromosomes develops as a somatic female (X:A = 1), and with one X chromosome develops as a male (X:A = 0.5). Sex determination mutations exist, however, that cause individuals with an X:A ratio of one to develop as somatic males (loss of function alleles of *tra* or *sdc* genes), and individuals with X:A ratio of 0.5 to develop as somatic females (loss of function alleles of *her, fem,* or *xol* genes). The genes defined by these mutations have been ordered into a regulatory hierarchy by examining the phenotypes of double-mutant combinations [see Hodgkin, 1990, and Villeneuve and Meyer, 1990a for review]. According to this hierarchy, an X:A ratio of one would lead to the activation of *tra* and *sdc* genes, resulting in female somatic development, and an X:A ratio of 0.5 would lead to the activation of *her, fem,* and *xol* genes, resulting in male somatic development.

It was generally assumed that somatic sex determination would be a cell-autonomous process: each cell would read its X:A ratio, and adopt either a female or male mode of sex determination gene expression. Indeed, genetic analysis of *tra-1* indicated that *tra-1* functions cell autonomously [Hunter and Wood, 1990]. Genetic mosaic analysis of the *sdc-1* [Villeneuve and Meyer, 1990b] and the *her-1* [Hunter and Wood, 1992] genes, however, revealed that somatic sex determination is not cell-autonomous.

DNA sequence analysis of the sex determination genes has suggested how the cell–cell interactions specifying the somatic sex may be mediated. The *her-1* gene encodes a cysteine-rich protein that has a signal sequence, and hence resembles a ligand [Perry et al., 1993]. The *tra-2* gene, which is immediately downstream of *her-1* in the genetic hierarchy, encodes a protein that is predicted to be transmembranous [Kuwabara et al., 1992]. The Her-1 protein has therefore been proposed to be the ligand for Tra-2. The model has a further interesting twist: Binding of Her-1 is proposed to reduce the activity of Tra-2. This has been inferred from the phenotypes conferred by *her-1* and *tra-2* mutations. Null mutations in *her-1* cause animals with an X:A = 0.5 to develop as somatic females, while *tra-2* mutations confer the opposite phenotype, i.e., they cause animals with an X:A = 1 to develop as somatic males. Genes that function downstream of *tra-2* according to the genetic hierarchy, the *fem* genes, could be involved in transducing a signal resulting from the reduction of Tra-2 activity to cause the terminal gene in the pathway, *tra-1*, to be turned off. The *tra-1* gene encodes a putative DNA-binding protein [Zarkower and Hodgkin, 1992]. Thus the Tra-1 protein could promote female somatic development by directly binding to target genes.

Kuwabara and Kimble [1992] and Perry et al. [1993] have proposed provocative reasons for why cell interactions are involved in somatic sex determination. Because a female soma consists of only 959 cells produced from a defined lineage [Sulston and Horvitz, 1977], and a male soma consists of only 1031 cells produced from a defined lineage [Sulston et al., 1983], any mistakes, particularly during early development, could lead to intersexuality. Cell interactions provide a way to coordinate the sexual specification of all cells, so that none choose an inappropriate fate: for example, by misreading the X:A ratio. In addition, the sexual fate of many parasitic nematode species is known to depend on environmental cues rather than being strictly dependent on the X:A ratio [Triantaphyllou, 1973]. The involvement of cell interactions in the specification of sexual fate may be necessary to prevent intersexuality under such conditions, and may be a relic of an ancestor that had an environmentally-influenced mechanism for sex determination.

VIII. CONCLUSIONS

Development in *C. elegans* requires a large number of cell interactions. Classical and molecular genetic techniques have been used to identify many components of several different signal transduction pathways and allowed the order in which the components function within each pathway to be established. Similar analyses in the future will lead to the identification of many other molecules which function in cell signalling and reveal the ways in which different signalling pathways may intersect.

REFERENCES

Albertson D, Thomson JN (1976): The pharynx of *Caenorhabditis elegans*. Philos Trans R Soc Lond [Biol] 275:299–325.

Aroian RV, Sternberg PW (1991): Multiple functions of *let-23*, a *Caenorhabditis elegans* receptor tyrosine kinase gene required for vulval induction. Genetics 128:251–267.

Aroian RV, Koga M, Mendel JE, Ohshima Y, Sternberg PW (1990): The *let-23* gene necessary for *Caenorhabditis elegans* vulval induction encodes a tyrosine kinase of the EGF receptor subfamily. Nature 348:693–699.

Austin J, Kimble J (1987): *glp-1* is required in the germ line for regulation of the decision between mitosis and meiosis in C. elegans. Cell 51:589–599.

Austin J, Kimble J (1989): Transcript analysis of *glp-1* and *lin-12*, homologous genes required for cell interactions during development of C. elegans. Cell 58:565–571.

Barbacid M (1987): *ras* genes. Annu Rev Biochem 56:779–827.

Beitel GJ, Clark SG, Horvitz HR (1990): *Caenorhabditis elegans* gene *let-60* acts as a switch in the pathway of vulval induction. Nature 348:503–509.

Bowerman B, Tax FE, Thomas JH, Priess JR (1992): Cell interactions involved in the development of bilaterally symmetrical intestinal valve cells during embryogenesis in *Caenorhabditis elegans*. Development 116:1113–1122.

Breeden L, Nasmyth K (1987): Similarity between cell-cycle genes of budding yeast and fission yeast and the *Notch* gene product of Drosophila. Nature 329:651–654.

Brenner S (1974): The genetics of *Caenorhabditis elegans*. Genetics 77:71–94.

Clark SG, Stern MJ, Horvitz HR (1992): *C. elegans* cell-signalling gene *sem-5* encodes a protein with SH2 and SH3 domains. Nature 356:340–356.

Clark SG, Stern MJ, Horvitz HR (1993): Genes involved in two *C. Elegans* cell signaling pathways. Cold Spring Harbor Symp Quant Biol 57:363–373.

Der CJ, Pan BT, Cooper GM (1986): ras^H mutants deficient in GTP binding. Mol Cell Biol 6:3291–3294.

Ellisen LW, Bird J, West DC, Soreng AL, Reynolds TC, Smith SD, Sklar J (1991): *TAN-1*, the human homolog of the Drosophila *Notch* gene is broken by chromosomal translocations in T lymphoblastic neoplasms. Cell 66:649–661.

Ferguson EL, Horvitz HR (1985): Identification and characterization of 22 genes that affect the vulval cell lineages of the nematode *Caenorhabditis elegans*. Genetics 110:17–72.

Ferguson EL, Horvitz HR (1989): The multivulva phenotype of certain *Caenorhabditis elegans* mutants results from defects in two functionally redundant pathways. Genetics 123:109–121.

Ferguson EL, Sternberg PW, Horvitz HR (1987): A genetic pathway for the specification of the vulval cell lineages of *Caenorhabditis elegans*. Nature 326:259–267.

Gallahan D, Callahan R (1987): Mammary tumorigenesis in feral mice: identification of a new *int* locus in mouse mammary tumor virus (Czech II)-induced mammary tumors. J Virol 61:66–74.

Goldstein B (1992): Induction of gut in *Caenorhabditis elegans* embryos. Nature 357:255–257.

Greenwald I (1985): lin-12, a nematode homeotic gene, is homologous to a set of mammalian proteins that includes epidermal growth factor. Cell 43:583–590.

Greenwald, I, Broach JR (1990): Cell fates in C. elegans: In medias *ras*. Cell 63:1113–1116.

Greenwald I, Rubin GM (1992): Making a difference: the role of cell-cell interactions in establishing separate identities for equivalent cells. Cell 68:271–281.

Greenwald, I, Seydoux G (1990): Analysis of gain-of-function mutations of the *lin-12* gene of *Caenorhabditis elegans*. Nature 346:197–199.

Greenwald IS, Sternberg PW, Horvitz HR (1983): The *lin-12* locus specifies cell fates in Caenorhabditis elegans. Cell 34:435–444.

Han M, Sternberg PW (1990): *let-60*, a gene that specifies cell fate during *C. elegans* vulval induction encodes a *ras* protein. Cell 63:921–931.

Han M, Sternberg PW (1991): Analysis of dominant-negative mutations of the *Caenorhabditis elegans let-60 ras* gene. Genes Dev 5:2188–2198.

Han M, Aroian R, Sternberg PW (1990): The *let-60* locus controls the switch between vulval and nonvulval cell fates in *Caenorhabditis elegans*. Genetics 126:899–913.

Hedgecock EM, Culloti JG, Hall DH (1990): The *unc-5*, *unc-6* and *unc-40* genes guide circumferential migrations of pioneer axons and mesodermal cells on the epidermis in *C. elegans*. Neuron 2:61–85.

Herman RK, Hedgecock EM (1990): Limitation of the size of the vulval primordium of *Caenorhabditis elegans* by *lin-15* expression in the surrounding hypodermis. Nature 348:169–171.

Hill RJ, Sternberg PW (1992): The gene *lin-3* encodes an inductive signal for vulval development in *C. elegans*. Nature 358:470–476.

Hirsh D, Oppenheim D, Klass M (1976): Development of the reproductive system of C. elegans. Dev Biol 49:200–219.

Hodgkin J (1990): Sex determination in Drosophila and Caenorhabditis compared. Nature 344:721–728.

Horvitz HR, Sulston JE (1980): Isolation and genetic characterisation of cell-lineage mutants of the nematode *Caenorhabditis elegans*. Genetics 96:435–454.
Horvitz HR (1988): Genetics of cell lineage. In Wood WB (ed): The Nematode *Caenorhabditis elegans*. Cold Spring Harbor Monograph Series. Cold Spring Harbor, NY: Cold Spring Harbor Laboratory, pp. 157–213.
Hunter CP, Wood WB (1990): The *tra-1* gene determines sexual phenotype cell-autonomously in *C. elegans*. Cell 63:1193–1204.
Hunter CP, Wood WB (1992): Evidence from mosaic analysis of the masculinizing gene *her-1* for cell interactions in *C elegans* sex determination. Nature 355:551–555.
Jones S, Vignais ML, Broach JR (1991): The *CDC25* protein of *Saccharomyces cerevisiae* promotes exchange of guanine nucleotides bound to Ras. Mol Cell Biol 11:2641–2646.
Kim SK, Horvitz, HR (1990): The *Caenorhabditis elegans* gene *lin-10* is broadly expressed while required specifically for the determination of vulval cell fates. Genes Dev 4:357–371.
Kimble J (1981): Alterations in cell lineage following laser ablation of cells in the somatic gonad of *Caenorhabditis elegans*. Dev Biol 87:286–300.
Kimble J, Hirsh D (1979): The postembryonic cell lineages of the hermaphrodite and male gonads in *Caenorhabditis elegans*. Dev Biol 70:396–417.
Kimble J, White JG (1981): On the control of germ cell development in *Caenorhabditis elegans*. Dev Biol 81:208–219.
Koch CA, Anderson D, Moran MF, Ellis C, Pawson T (1991): SH2 and SH3 domains: elements that control interactions of cytoplasmic signaling proteins. Science 252:668–674.
Kuwabara PE, Okkema PG, Kimble J (1992): Tra-2 encodes a membrane-protein and may mediate cell communication in the Caenorhabditis elegans sex determination pathway. Molecular Biology of the Cell 3:461–473.
Lambie EJ, Kimble J (1991): Two homologous regulatory genes, *lin-12* and *glp-1*, have overlapping functions. Development 112:231–240.
Li C, Chalfie M (1990): Organogenesis in *C. elegans:* positioning of neurons and muscles in the egg-laying system. Neuron 4:681–695.
Lowenstein EJ, Daly RJ, Batzer AG, Wi W, Margolis B, Lammers R, Ullrich A, Skolnik EY, Bar-Sagi D, Schlessinger J (1992): The SH2 and SH3 domain-containing protein GRB2 links receptor tyrosine kinases to ras signalling. Cell 70:431–442.
Madl J, Herman, R (1979): Polyploids and sex determination in *C. elegans*. Genetics 93:393–402.
Maine EM, Kimble J (1989): Identification of genes that interact with *glp-1*, a gene required for inductive cell interactions in *Caenorhabditis elegans*. Development 105:133–143.
Miller LM, Gallegos ME, Morisseau BA, Kim SK (1993): *lin-31*, a *Caenorhabditis elegans* HNF-3/*fork head* transcription factor homolog, specifies three alternative fates in vulval development. Genes Dev 7:933–947.
Perry MD, Li W, Trent C, Robertson B, Fire A, Hageman JM, Wood WB (1993): Molecular characterization of the *her-1* gene suggests a direct role in cell signaling during Caenorhabditis elegans sex determination Genes Dev 7:216–228.
Powers S, O'Neill K, Wigler M (1989): Dominant yeast and mammalian ras mutants that interfere with the CDC25-dependent activation of wild-type RAS in *Saccharomyces cerevisiae*. Mol Cell Biol 9:390–395.
Priess JR, Thomson JN (1987): Cellular interactions in early *C. elegans* embryos. Cell 48:241 250.
Priess JR, Schnabel H, Schnabel R (1987): The *glp-1* locus and cellular interactions in early *C. elegans* embryos. Cell 51:601–611.
Robbins J, Blondel BJ, Gallahan D, Callahan R (1992): Mouse mammary tumor gene int-3: a member of the notch gene family transforms mammary epithelial cells. J Virol 66:2594–2599.
Rodaway ARF, Sternberg MJE, Bentley DL (1989): Similarity in membrane proteins. Nature 342:624.
Rozakis-Adcock M, McGlade J, Mbamalu G, Pelicci G, Daly R, Li W, Batzer A, Thomas S, Brugge J, Pelicci PG, Schlessinger J, Pawson T (1992): Association of the Shc and GRB2/Sem5 SH2-containing proteins is implicated in the activation of the Ras pathway by tyrosine kinases. Nature 360:689–692.
Schierenberg E (1987): Reversal of cell polarity and early cell-cell interactions in the embryo of *Caenorhabditis elegans*. Dev Biol 122:452–463.
Schnabel R (1991): Cellular interactions involved in the determination of the early *C elegans* embryo. Mech Dev 34:85–100.
Seydoux G, Greenwald I (1989): Cell autonomy of *lin-12* function in a cell fate decision in *C. elegans*. Cell 57:1237–1245.
Seydoux G, Schedl T, Greenwald I (1990): Cell-cell interactions prevent a potential inductive interaction between soma and germ line in C. elegans. Cell 61:939–951.
Seydoux G, Savage C, Greenwald I (1993): Isolation and characterization of mutations causing abnormal eversion of the vulva in *Caenorhabditis elegans*. Dev Biol 157:423–436.
Sigal IS, Gibbs JB, Dalonzo JS, Temeles GL, Wolanski BS, Socher SH, Scolnick EM (1986): Mutant ras-encoded proteins with altered nucleotide binding exert dominant negative effects. Proc Natl Acad Sci USA 83:952–956.
Stern MJ, Horvitz HR (1992): A normally attractive cell interaction is repulsive in two *C. elegans* mesodermal cell migration mutants. Development 113:797–803.

Stern MJ, Marengere LEM, Daly RJ, Lowenstein EJ, Kokel M, Batzer AG, Olivier P, Pawson T, Schlessinger J (1993): The human GRB2 and Drosophila Drk genes can functionally replace the Caenorhabditis elegans cell signaling gene sem-5. Mol Biol Cell 4:1175–1188.

Sternberg PW, (1988): Lateral inhibition during vulval induction in *Caenorhabditis elegans*. Nature 335:551–554.

Sternberg PW, Horvitz HR (1986): Pattern formation during vulval development in *C. elegans*. Cell 44:761–772.

Sternberg PW, Horvitz HR (1989): The combined action of two intercellular signalling pathways specifies three cell fates during vulval induction in C. elegans. Cell 58:679–693.

Sulston JE (1976): Postembryonic development in the ventral cord of *Caenorhabditis elegans*. Philos Trans R Soc Lond [Biol] 275:287–298.

Sulston JE, Horvitz HR (1977): Post-embryonic cell lineages of the nematode, *Caenorhabditis elegans*. Dev Biol 56:110–156.

Sulston JE, Horvitz HR (1981): Abnormal cell lineage mutants of the nematode *Caenorhabditis elegans*. Dev Biol 82:41–55.

Sulston JE, White JG (1980): Regulation and cell autonomy during a postembryonic development of *Caenorhabditis elegans*. Dev Biol 78:577–597.

Sulston JE, Schierenberg E, White JG, Thomson JN (1983): The embryonic cell lineage of the nematode *Caenorhabditis elegans*. Dev Biol 100:64–119.

Sundaram M, Greenwald I (1993): Suppressors of Lin-12 hypomorph define genes that interact with both Lin-12 and glp-1 in *Caenorhabditis elegans*. Genetics 135:765–783.

Thomas JH, Stern MJ, Horvitz HR (1990): Cell interactions coordinate the development of the *C. elegans* egg-laying system. Cell 62:1041–1052.

Triantaphyllou AC (1973): Environmental sex determination of nematodes in relation to pest management. Annu Rev Phytopathol 11:441–462.

Trent C, Tsung N, Horvitz HR (1983): Egg-laying defective mutants of the nematode *Caenorhabditis elegans*. Genetics 104:619–647.

Villeneuve AM, Meyer BJ (1990a): The regulatory heirarchy controlling sex determination and dosage compensation in C. elegans. Adv Genet 27:117–188.

Villeneuve AM, Meyer BJ (1990b): The role of *sdc-1* in the sex determination and dosage compensation decisions in *Caenorhabditis elegans*. Genetics 124:91–114.

Wadsworth WG, Hedgecock EM (1992): Guidance of neuroblast migrations and axonal projections in Caenorhabditis elegans. Curr Opin Neurobiol 2:36–41.

Wood WB (1991): Evidence from reversal of handedness in *C. elegans* embryos for early cell interactions determining cell fates. Nature 349:536–538.

Wood WB, Kershaw D (1991): Handed asymmetry, handedness reversal and mechanisms of cell fate determination in nematode embryos. In Bock and Marsh (Ed): "Biological Asymmetry and Handedness." CIBA Found. Symp. Vol 162. Chichester: John Wiley & Sons, pp. 143–159.

Yochem J, Greenwald I (1989): *glp-1* and *lin-12*, genes implicated in distinct cell-cell interactions in *C. elegans* encode similar transmembrane proteins. Cell 58:553–563.

Yochem J, Weston K, Greenwald I (1988): The *Caenorhabditis elegans lin-12* gene encodes a transmembrane protein with overall similarity to *Drosophila Notch*. Nature 335: 547–550.

Zarkower D, Hodgkin J (1992): Molecular analysis of the *C. elegans* sex determination gene *tra-1*-a gene encoding 2 zinc finger proteins. Cell 70:237–249.

ABOUT THE AUTHORS

SIMON TUCK is a postdoctoral research scientist in Dr. Iva Greenwald's laboratory at the College of Physicians and Surgeons of Columbia University. After receiving a B.A. from Oxford University, he pursued doctoral research at the Imperial Cancer Research Fund, London in the laboratory of Dr. Lionel Crawford. He earned a Ph.D. in 1989 from University College, London University for his work on the human tumor suppressor gene, p53. He first joined Dr. Greenwald's laboratory as an Imperial Cancer Research Fund Travelling Fellow in the Department of Molecular Biology, Princeton University. Since that time he has been studying vulval development in *C. elegans;* in particular, the role played by the *lin-25* gene.

IVA GREENWALD is Associate Professor of Biochemistry and Molecular Biophysics at the College of Physicians and Surgeons of Columbia University. After receiving a B.S. from Cornell, she obtained her Ph.D. at M.I.T., where she conducted genetic studies of muscle structure and of the *lin-12* gene in the laboratory of H.R. Horvitz. She then conducted postdoctoral research at the MRC Laboratory of Molecular Biology in Cambridge, England, where she initiated the molecular characterization of the *lin-12* gene. Dr. Greenwald was on the faculty at Princeton University before moving to Columbia in 1993. Her research has been concerned with genetic and molecular studies of *C. elegans* development.

Growth Factors and Signal Transduction in Development: 199–227
Published 1994 by Wiley-Liss, Inc.

Embryonic Induction and Axis Formation in *Xenopus laevis:* Growth Factor Action and Early Response Genes

Igor B. Dawid

I. INTRODUCTION

Amphibians have been historically, and *Xenopus laevis* continues to be, an exceptionally useful system for the study of early vertebrate embryogenesis. The basic question at hand is how the single-celled zygote is transformed into the highly structured organism with its many cell types and tissues, all arranged in a functional pattern characteristic of the species [see Gurdon, 1992]. This question thus subsumes issues of cell differentiation and pattern formation, the latter being a major focus of research in developmental biology, both past and current. It has long been clear that the process of cell differentiation is closely linked to differential gene activity. The growing body of evidence documenting the view that pattern formation is likewise associated with differential gene activity has been at the core of recent progress in developmental biology. Genes responsible for pattern formation, and for specifying not just tissue types but regional identity, have been isolated and characterized, and their study has proved of great help not only in understanding the development of the particular animal under investigation but also in shedding light on the evolutionary relationships between groups, and on the ways in which regulatory pathways in development themselves have evolved.

Processes of pattern formation in the early embryogenesis of all animals utilize both regional information deposited in the egg and interactions between the cells and cell groups that form during cleavage divisions [for broad discussion of the subject see Wilson, 1925; Davidson, 1986]. It is characteristic of vertebrate embryos that cell interactions play a particularly large role in their development, and thus it is understandable that questions of cell communication have been a focus of studies in vertebrate embryology. The amphibian embryo has been especially favorable for the analysis of interactive mechanisms, a direction that has been central to embryology, at least since the discovery by Spemann and Mangold [1924] of the induction of a secondary axis in newt embryos by transplantation of the dorsal blastopore lip. The inductive, regulative, and self-differentiating properties of the dorsal lip earned it the name "organizer." Understanding organizer function has been a goal of many studies since 1924, with significant advances coming in the past few years. Regions comparable to the organizer function during gastrulation of vertebrates other than amphibians, and similar molecules and molecular mechanisms are probably involved (e.g., Kintner and Dodd, 1991; Blum et al., 1992; Zhou et al., 1993].

The organizer mediates the establishment of the anteroposterior axis during gastrulation, as expressed in the differentiation of axial mesoderm (notochord, somites) and the specification and patterning of the neural plate. How ever, these gastrulation events are not the first wave of cell interactions in amphibian embryogenesis. Studies primarily initiated by Nieuwkoop [reviewed by Nieuwkoop, 1973, 1977] and extended by many workers, led to the conclusion that the generation and action of the organizer are preceded by inductive events that specify the mesoderm and establish the dorsoventral axis of the embryo. As a consequence of this work, mesoderm induction and neural induction have been recognized as distinct, though closely connected, events in embryogenesis. This distinction has been heuristically very useful, allowing the elucidation, in part, of the molecular basis of mesoderm induction and the role of growth factors in it.

While earlier work on amphibian embryology was carried out primarily with newts, more recent studies have relied on *Xenopus laevis* as the primary research object [see Kay and Peng, 1991]. Work on this animal has contributed substantially to our understanding of vertebrate embryogenesis [reviewed by Smith, 1989; Dawid et al., 1990; Jessel and Melton, 1992; Dawid, 1991, 1992; Kimmelman et al., 1992; Sive, 1993], as well as leading to important progress in molecular biology [for review see Dawid and Sargent, 1988].

II. MESODERM INDUCTION AND THE ESTABLISHMENT OF DORSOVENTRAL POLARITY

A. The Dorsoventral Axis Is Established During Pregastrula Development

1. Initial dorsalization: the cortical rotation. The unfertilized *Xenopus* egg is radially symmetrical about its animal-vegetal axis. While this axis approximately corresponds to the future anteroposterior axis, the actual elaboration of the latter axis occurs during the complex events of gastrulation (see below). The first step of the pathway determining dorsoventral polarity, a rotation of the egg cortex relative to the deep cytoplasm, occurs between fertilization and first cleavage (Fig. 1A). The direction of this rotation is usually determined by the point of sperm entry, but it can be directed anywhere by experimental intervention. The direction of this rotation accurately predicts the future dorsoventral axis [Vincent et al., 1986; Vincent and Gerhart, 1987; Gerhart et al., 1989]. Complex cytoplasmic rearrangements take place at this time [Danilchik and Denegre, 1991].

Cortical rotation involves microtubules; interventions like UV irradiation and certain drugs that interfere with microtubule action lead to axis-deficient, ventralized embryos which, in their most extreme form, develop only ventral mesoderm, like blood, but no dorsal structures like notochord, skeletal muscle, or nervous system [Grant and Wacaster, 1972; Malacinski et al., 1975; Scharf and Gerhart 1983].

2. The fate map of the early embryo depends on differences between animal and vegetal, and between dorsal and ventral blastomeres. Animal versus vegetal differences are generated during oogenesis [Danilchik and Gerhart, 1987], generating the first axis of the egg. After fertilization and cortical rotation, the overall organization of the future embryo is specified, but this specification is not rigidly determined, and many fates can be changed by subsequent manipulations. Yet in the undisturbed embryo, the cells (blastomeres) arising from the different regions of the egg (Fig. 1B) reproducibly give rise to certain tissues and regions of the embryo, a process which allows the generation of a fate map that is shown in Figure 2 and schematically indicated in Figure 1C,D [Dale and Slack, 1987a; Moody, 1987a,b].

The fate map depends on the fact that future dorsal and ventral regions of the early embryo, and the cells that come to occupy these regions, have distinct properties. The 8-cell embryo may have three cell types. By removing blastomeres from embryos and determining the outcomes, Kageura and Yamana [1984] could show that an essentially normal embryo arises as long as at least one animal, one dorsal vegetal, and one ventral vegetal cell remains. Attention has focused on dorsal vegetal cells

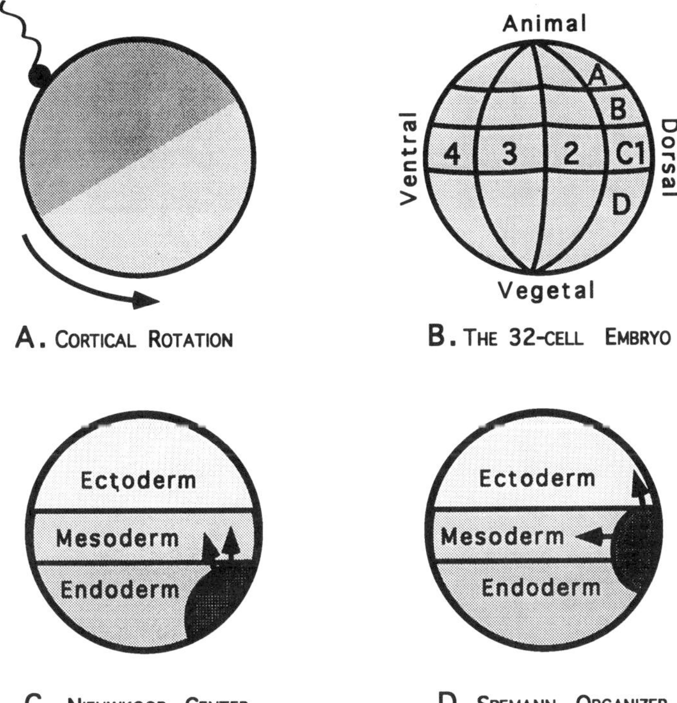

Fig. 1. *All drawings have the same orientation which is given in B.* **A:** *Cortical rotation relative to the deep cytoplasm determines the future dorsal-ventral axis [Vincent et al., 1986; Vincent and Gerhart, 1987].* **B:** *Nomenclature for cells (blastomeres) of the 32-cell embryo as used in this paper.* **C:** *Schematic fate map (see Fig. 2) of the early embryo, showing the axis-inducing Nieuwkoop center in the dorsovegetal region.* **D:** *At the beginning of gastrulation the Spemann organizer sends signals that dorsalize the mesoderm and participate in the formation of the neural plate, as indicated by the two arrows. See text for further discussion.*

because they are able to influence other cells most conspicuously in transplantation experiments. Two distinct, though probably related, biological properties have been ascribed to dorsal blastomeres (see Fig. 1B). The first is self-differentiation potency, i.e., the ability to form dorsal structures when placed in ectopic sites [Takasaki and Konishi, 1989; Gallagher et al., 1991]. This is true for all dorsal blastomeres (A1 through D1 in Fig. 1B), albeit to a differing extent. The second property of all vegetal blastomeres is the ability to induce animal hemisphere cells to form mesoderm, as shown in explant recombination experiments [Nieuwkoop, 1969a,b, 1973, 1977; Sudarwati and Nieuwkoop, 1971; Boterenbrood and Nieuwkoop, 1973]. Furthermore, dorsovegetal blastomeres (e.g., D1 in Fig. 1B) can induce an entire dorsal axis without themselves contributing to it. This has been shown in various ways, for example in the experiments of Gimlich and Gerhart [1984], where D1 cells (or their daughters at the 64-cell stage) were transplanted into the corresponding ventral position of host embryos, replacing D4 cells. Double-axis embryos resulted, in which the dorsal structures (notochord, somites, neural tube) of the duplicated embryo were derived from the host, while the progeny of the grafted D1 cells gave rise to endodermal products, as predicted from the fate map (Figs. 1,2). Thus, early dorsovegetal blastomeres can generate a dorsal axis by induction rather than by self-differentiation. The ability to induce a second-

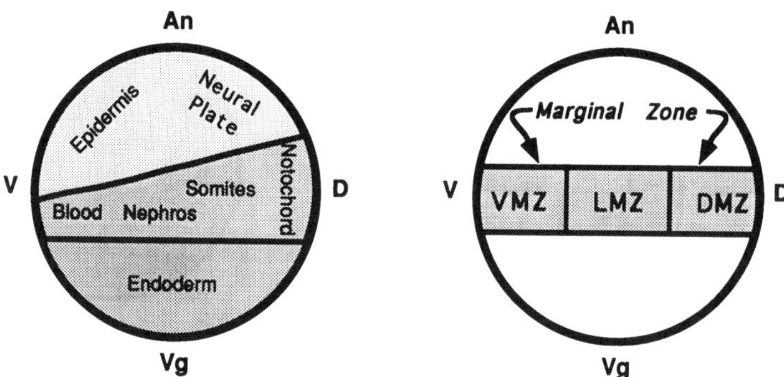

Fig. 2. *Left, fate map of the early* Xenopus *embryo based on Moody [1987a,b] and Dale and Slack [1987a]. The right drawing indicates schematically the division of the mesoderm-forming marginal zone into ventral (VMZ), lateral (LMZ), and dorsal (DMZ) regions.*

ary dorsal axis resides in all dorsal cells, with peak activity in C1 or D1 [Gimlich and Gerhart, 1984; Kageura and Yamana, 1986; Dale and Slack, 1987b; Kageura, 1990]. This early inducing region is often called the Nieuwkoop center (Fig. 1C).

As development progresses, dorsal inducing ability moves towards the animal pole. This region, called the dorsal marginal zone (DMZ, Fig. 2), can both contribute to dorsal structures and induce dorsal fates in other cells, as shown in blastula (Gimlich, 1986) and gastrula stages [Smith and Slack, 1983; Dale and Slack, 1987b; Lettice and Slack, 1993]. As gastrulation begins, the DMZ is equivalent to the Spemann organizer (Fig. 1D). Beyond semantics, the distinctions between Nieuwkoop center and Spemann organizer aid in thinking about possible mechanisms. Cells that induce but do not contribute to a structure must do so by sending an extracellular signal, whereas self-differentiation may involve both intracellular determinants and cell–cell interactions.

III. GASTRULATION: ANTEROPOSTERIOR PATTERNING AND THE FORMATION OF THE NEURAL PLATE

A. Extensive Cell Movements Take Place During Gastrulation

Dorsoventral axis specification becomes overt with the formation of the blastopore lip and the involution of dorsal mesoderm (Fig. 3), which together are known as the Spemann organizer. The advancing mesoderm undergoes complex cell movements in which the notochord and other parts of the axial system are laid down. Cell intercalation movements leading to elongation of dorsal mesoderm have been studied in detail and named convergent extension [reviewed in Keller, 1991]. In addition to forces generated within the mesoderm, interactions between migrating cells and the extracellular matrix of the blastocoel roof play a part in gastrulation [Nakatsuji and Johnson, 1983; Nakatsuji, 1984]. In the newt, *Pleurodeles waltl,* and in certain other amphibians, but not in *Xenopus,* gastrulation movements can be completely inhibited by interfering with fibronectin–integrin interactions [Boucaut et al., 1984, 1985; Darribère et al., 1988; Johnson et al., 1993]. In *Xenopus,* such interactions play an important but less decisive role in gastrulation movements [Smith et al., 1990a; Winklbauer, 1990; Howard et al., 1992].

B. Neural Induction

The elaboration of anteroposterior polarity during gastrulation is intimately linked with the specification and differentiation of the nervous system [reviewed by Slack and Tannahill, 1992]. The classical view holds that dorsal ectoderm is induced by the blastopore lip and advancing dorsal mesoderm (together named the organizer) to form the neural plate [reviewed by Sharpe, 1990; Yamada, 1990; Kintner, 1992]. In recent years a vigorous discussion has ensued over the relative importance of planar versus transverse (or vertical) induction (see Fig. 3). Planar signals are believed to emanate from the dorsal lip and travel through the ectoderm along a meridian towards the animal pole, while transverse signals travel from the dorsal mesoderm along a radius to the overlying ectoderm. Evidence for transverse induction is implicit in much of the older literature (see reviews above) and is supported in some recent papers [Sharpe and Gurdon, 1990; Hemmati-Brivanlou et al., 1990], while evidence for planar induction or a combination of both has been emphasized in several other studies [Dixon and Kintner, 1989; Ruiz i Altaba, 1992; Doniach et al., 1992]. Whatever the direction of the inductive stimulus, it appears that polarization of the neural plate along the anteroposterior axis proceeds gradually [Sive et al., 1989; Saha and Grainger, 1992].

1. Neural induction: signalling mechanisms. The nature of neural inducing factors is not known, but there is some evidence for the participation of secreted molecules, based on transfilter experiments [Saxén, 1961; Toivonen et al., 1975], and on the neuralizing ability of partially purified fractions derived from various sources [Janeczek et al., 1992, and refs. therein]. Some information is available on the signal transmission pathways in this system. Otte et al. [1988, 1989, 1991; Otte and Moon, 1992] have implicated the cyclic AMP and protein kinase C (PKC) pathways [see also Davids et al., 1987] in neural induction, and presented evidence that the expression of different forms of PKC is responsible for the differential competence of dorsal versus ventral ectoderm to respond to neuralizing stimuli.

2. Neural differentiation without (apparent) induction. While many studies have demonstrated the influence of the organizer region in the formation of the neural plate, several studies have suggested that ectoderm cells can differentiate certain neural properties of their own. Culture of dissociated ectodermal cells from blastula or gastrula stages of *Xenopus* or *Pleurodeles* in simple media led to the differentiation of neuronal morphology and expression of specific antigens [Grunz and Tacke, 1989; Sato and Sargent, 1989; Godsave and Slack, 1989, 1991; Saint-Jeannet et al., 1990]. However, the repertoire of specific neuronal types differentiated in such cultures was limited [J.-P. Saint-Jeannet, S. Huang, F. Pituello, F. Foulquier, A.-M. Duprat, personal communication].

In a different approach, Hemmati-Brivanlou and Melton (1992) showed that at least one neural marker (NCAM; neural cell adhesion molecule) was expressed in embryos in which mesoderm induction was suppressed (discussed further below). This result, like the single-cell culture experiments referred to above, suggests that basic neural differentiation may be the "default state" of ectodermal cells. It may be speculated that this neural ground state is suppressed by interactions within the ectoderm leading to epidermis, while organizer induction overcomes this repression and, in addition, provides cues that organize the neural plate with anteroposterior polarity and elicit regionally diverse differentiations. This new and, in historical terms, heterodox view of vertebrate nervous system ontogeny has important parallels in the well-understood origin of the nervous system in *Drosophila* [reviewed by Campos-Ortega, 1990].

C. The Relationship Between Dorsal and Anterior Determination

During gastrulation, the leading dorsal mesoderm cells come to occupy anterior positions [see Keller, 1991]. These movements are most likely responsible for the observations that interventions that enhance the dorsal nature of mesoderm enhance anterior development (generally, but most obviously in the nervous system), and interventions that ventralize lead to anterior deficiencies. Examples include (i) the ventralizing effect of inhibition of the cortical rotation, which leads to a series of

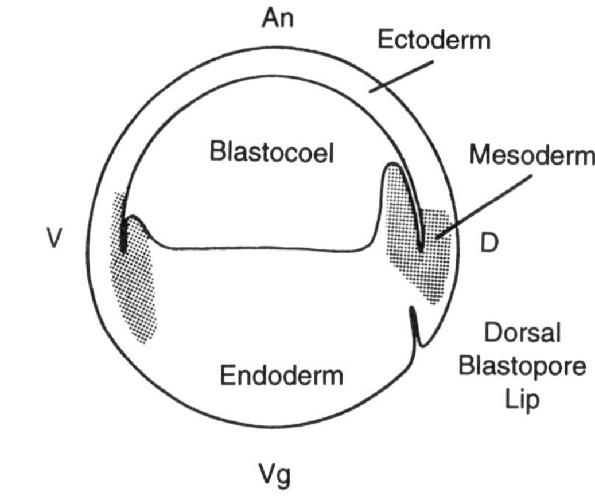
An
Ectoderm
Blastocoel
Mesoderm
V
D
Dorsal
Blastopore
Lip
Endoderm
Vg

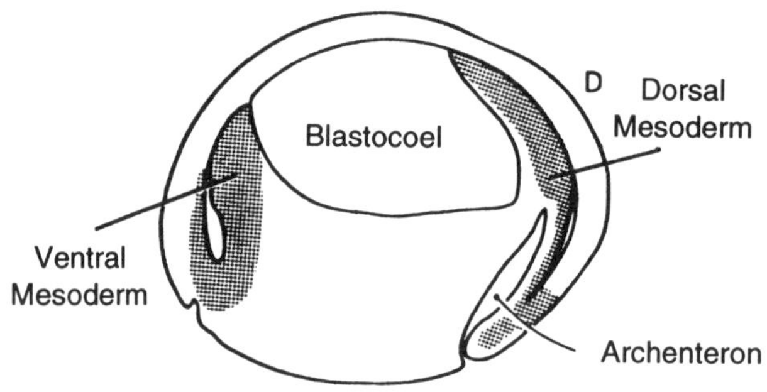
D
Dorsal
Mesoderm
Blastocoel
Ventral
Mesoderm
Archenteron

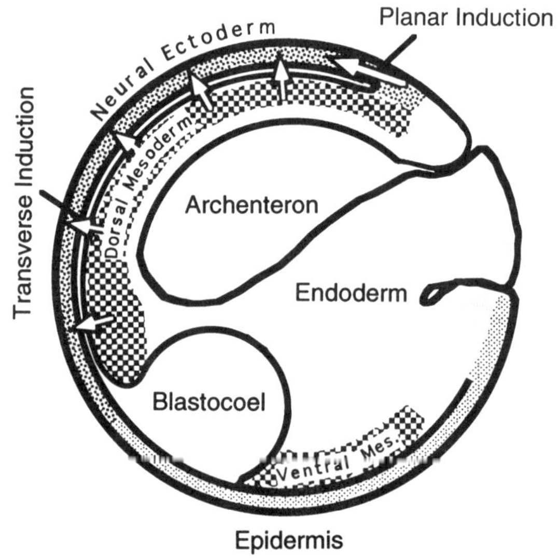
Planar Induction
Neural Ectoderm
Dorsal Mesoderm
Transverse Induction
Archenteron
Endoderm
Blastocoel
Ventral Mes.
Epidermis

defects from microcephaly to absence of dorsal tissues [Grant and Wacaster, 1972; Malacinski et al., 1975; Scharf and Gerhart, 1983; Gerhart et al., 1989]; (ii) treatment of blastula embryos with LiCl, which hyperdorsalizes and leads to excessive head development (Kao et al., 1986; Kao and Elinson, 1988, 1989]; and (iii) retinoic acid (RA), which both ventralizes mesoderm and leads to anterior deficiencies in the nervous system and the entire embryo [Durston et al., 1989; Ruiz i Altaba and Jessel 1991a,b; Sive et al., 1990; Sive and Cheng, 1991; Sharpe, 1991]. As a result of this relationship, Kao and Elinson [1988] have generated a dorsoanterior index (DAI), a standardized scale of developmental states that includes normal embryos as the midpoint.

IV. MESODERM INDUCTION ASSAYS

A. The Nieuwkoop Experiment and the Animal Cap Assay

The animal region of the blastula is fated to become ectoderm (Figs. 1,2), and will develop into what is called "atypical" epidermis when excised and cultured in amphibian salt solution. Nieuwkoop and his colleagues [Nieuwkoop, 1969a,b, 1973, 1977; Sudarwati and Nieuwkoop, 1971; Boterenbrood and Nieuwkoop, 1973] have shown that such animal explants ("animal caps") will form mesodermal tissues when cultured in contact with vegetal cells. This explant system has been applied more recently [Smith, 1987; Slack et al., 1987; Kimelman and Kirschner, 1987; Rosa et al., 1988] as an assay for soluble mesoderm-inducing factors, and led to the identification of such factors and to the characterization of cellular responses that include shape changes, gene activation and repression, and differentiation of various cell types (Fig. 4). In the animal cap assay we use a group of cells that normally do not form mesoderm as a test material because they are capable of reacting to inducing signals. While this assay involves an artificial situation, there is much evidence that inductive interactions are required for mesoderm specification *in vivo*. Examples of such evidence are (i) the inability of embryonic cells, prevented from interacting by culture in dispersion, to differentiate into muscle [Sargent et al., 1986], and (ii) the results of negative dominant receptor studies discussed later.

Application of the animal cap assay has led to the identification of two classes of mesoderm-inducing factors: members of the fibroblast growth factor (FGF) family; and activins, as well as other members of the transforming growth factor beta (TGF-β) superfamily. These are discussed below.

B. RNA Injection as an Assay for Dorsalization and Mesoderm Induction

The animal cap assay has proved very useful, but it has its limitations. First, the study of an explant is necessarily an artificial situation;

Fig. 3. *Some aspects of gastrulation in* Xenopus *[for general review see Hausen and Riebesell, 1991; Keller, 1991]. The drawings show, top to bottom, early, mid, and late gastrula. Gastrulation begins with formation of the dorsal blastopore lip and invagination of dorsal mesoderm. By mid gastrula the dorsal mesoderm has moved well into the blastocoel, and the archenteron has begun to open. Ventral mesoderm also invaginates, but to a lesser extent. The late gastrula, which has rotated by 90° so that its dorsal side is now facing up, is labeled more completely to show the major events. The enlarging archenteron now exceeds in volume the shrinking blastocoel. Dorsal mesoderm has moved along the blastocoel roof beyond the original location of the animal pole. Dorsal ectoderm has become determined as the neural plate, while ventrolateral ectoderm is differentiating to epidermis. The formation of the neural plate depends on interactions of ectoderm with the blastopore lip and the dorsal mesoderm [reviewed in Sharpe, 1990; Yamada, 1990; Slack and Tannahill, 1992; Kintner, 1992]. These interactions appear to take place in two distinct directions: planar induction which moves through the plane of the ectoderm from the lip towards the animal pole, and transverse (or vertical) induction which projects from the invaginated mesoderm radially outward (see text).*

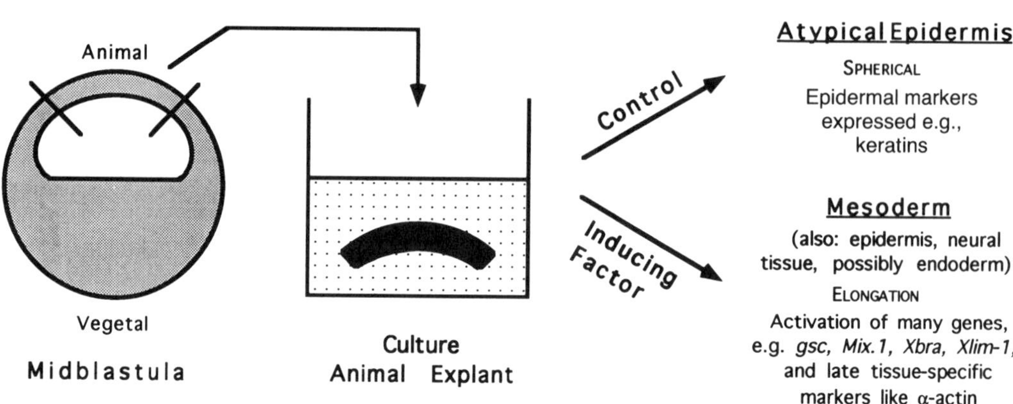

Fig. 4. *The animal cap induction assay. Animal tissue, fated to become ectoderm, can be induced towards mesoderm by vegetal tissue (not illustrated in the figure), or by soluble growth factors, thus providing an assay system for inducing substances.*

second, it requires the availability of a soluble factor, which has proven impractical when studying the Wnt proteins (discussed in Section VI). Fortunately, a different assay system has contributed valuable information. When mRNA, isolated from tissues or synthesized *in vitro* from cloned genes, is injected into an early frog embryo, where it is stable at least up to gastrula stages, it produces its cognate protein *in situ* and potentially elicits phenotypes that provide important clues to its function. The approach is illustrated in Figure 5, and the specific actions of various injected RNAs will be discussed below.

V. MESODERM-INDUCING FACTORS

A. FGF Is Involved in Ventroposterior Mesoderm Induction

1. Functions of FGF in the animal cap assay. Explants exposed to FGF differentiate into structures that include mesodermal tissues of a ventral to lateral character like mesenchyme, mesothelium, and skeletal muscle [Slack et al., 1987]. Different isoforms of FGF, including basic FGF (bFGF), acidic FGF (aFGF), kFGF (FGF-4), and a *Xenopus*-derived form, XeFGF, have similar effects [Slack et al., 1987; Kimelman and Kirschner, 1987; Paterno et al., 1989; Isaacs et al., 1992]. The RNA for XeFGF [Isaacs et al., 1992], and bFGF protein have been detected in early embryos [Kimelman et al., 1988; Slack and Isaacs, 1989; Godsave and Shiurba, 1992], so that the action of an FGF isoform in mesoderm induction is plausible.

2. FGF overexpression phenotypes. In *Xenopus* it is impractical to generate mutants that inactivate a gene, and antisense approaches have not been easily applicable. Thus, functional studies depend primarily on overexpression and ectopic expression of the gene of interest (Fig. 5), and on the exploitation of dominant-negative mutants as described in the next section. Overexpression of FGF has been induced by injection of mRNA into fertilized eggs [Kimelman and Maas, 1992; Thompson and Slack, 1992]. The resulting embryos were surprisingly normal, although injection of high levels of FGF mRNA resulted in abnormalities. Abnormal development was pronounced when a construct of bFGF was injected in which a signal sequence had been added to the protein; likewise, kFGF, which (unlike bFGF) carries a signal sequence, was quite effective [Thompson and Slack, 1992].

A different assay for the action of FGF combines the mRNA injection and animal cap assays, as illustrated in Figure 6. Embryos that

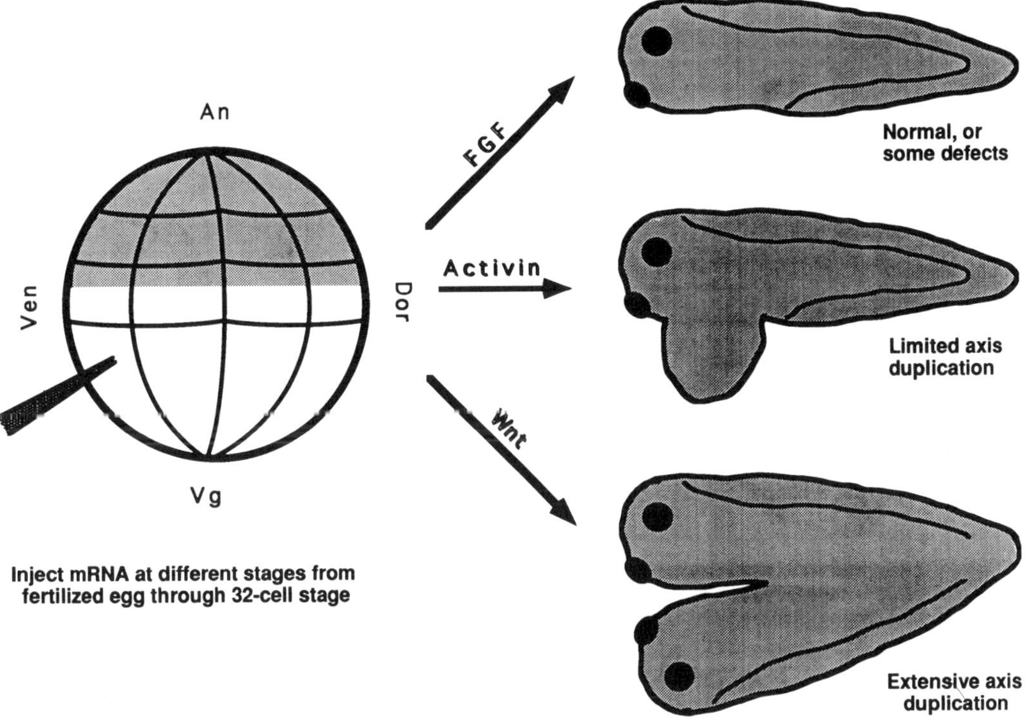

Fig. 5. *Injection of mRNA into the early embryo (the 32-cell stage being illustrated) may lead to developmental consequences, and thus provides an alternate assay method for inducing factors affecting embryogenesis. As discussed in the text, injection of FGF mRNA results in normal or partially deformed embryos [Kimelman and Maas, 1992; Thompson and Slack, 1992]; activin mRNA results in partial axis duplication [Thomsen et al., 1990]; Wnt-1 or Xwnt-8 mRNAs induce extensive axis duplications [Smith and Harland 1991; Sokol et al., 1991]. In addition, noggin mRNA has dorsalizing ability, as shown in an injection assay that involved UV ventralized embryos [Smith and Harland, 1992].*

had been injected with FGF mRNA develop largely normally. When animal caps were explanted from such embryos, the isolated animal caps differentiated into mesodermal derivatives without addition of exogenous factors [Kimelman and Maas, 1992; Thompson and Slack, 1992]. This result makes two points: (i) FGF provided *in situ* is capable, like FGF added to explants, of inducing mesoderm differentiation; (ii) the frog embryo displays its well-known regulative powers, through which the whole embryo may overcome inappropriate inductive influences in its animal region.

3. FGF receptors are expressed in early embryos and are required for trunk and tail development. The presence of an FGF receptor (FGFR) in early *Xenopus* embryos was indicated by ligand-binding studies [Gillespie et al., 1989]. Cloning of cDNAs for two types of FGFR showed that RNA for a receptor homologous to the human *flg* gene is expressed in the oocyte and throughout early embryogenesis (Musci et al., 1990; Friesel and Dawid, 1991], while a second type of receptor is expressed later [Friesel and Brown, 1992]. Immunocytochemical analysis of the *flg*-like FGFR suggests a wide distribution with-

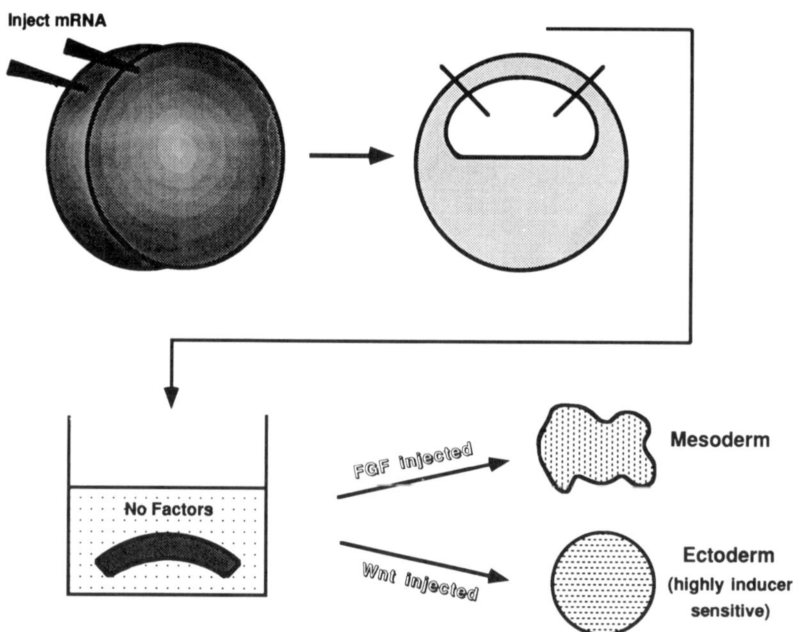

Fig. 6. *Assay combining mRNA injection into early embryos (2-cell stage illustrated) with animal cap explantation. Even though injection of FGF mRNA has comparatively mild consequences in the whole embryo (Fig. 5), animal caps isolated from such embryos spontaneously differentiate mesoderm [Kimelman and Maas, 1992; Thompson and Slack, 1992]. In contrast, caps from Xwnt-8 mRNA-injected embryos do not form mesoderm, in spite of the axis-inducing ability of this RNA (Fig. 5), but are highly sensitive to inducers and differentiate into dorsal mesoderm when exposed to FGF or low levels of activin [Christian et al., 1992; Sokol and Melton, 1992; Sokol, 1993].*

in the embryo and a temporal pattern that is constant through blastula stages, decreases slightly during gastrulation, and increases later [Ding et al., 1992]. Thus, neither the spatial regulation of patterning nor the loss of competence to respond to FGF are likely to be caused directly by the expression pattern of the FGFR.

Compelling evidence for the involvement of at least one form of FGF in mesoderm induction during normal embryogenesis comes from the work of Amaya et al. [1991, 1993]. These authors made a receptor construct which lacks the intracellular kinase domain, and injected the corresponding mRNA into fertilized eggs (Fig. 7). Expression of the truncated receptor had the effect of blocking FGF signal transduction, as seen by the fact that animal caps excised from such embryos did not respond to FGF but still responded to the structurally different inducer, activin. This blockage is believed to be due to obligatory receptor dimer formation during activation [see Schlessinger, 1988]. Embryos expressing the FGFR mutant showed defects mostly in trunk and tail development, while heads were largely normal (Fig. 7). This implies a function for FGF in the formation of lateral and posterior mesoderm [Amaya et al., 1991; 1993].

B. Activins Induce Different Types of Mesoderm in a Concentration-Dependent Manner

1. Activin induces dorsal mesoderm in animal caps. The differentiation of all types of mesoderm, in particular including dorsal structures like notochord, elicited by the culture fluid of XTC cells [Smith, 1987] was a

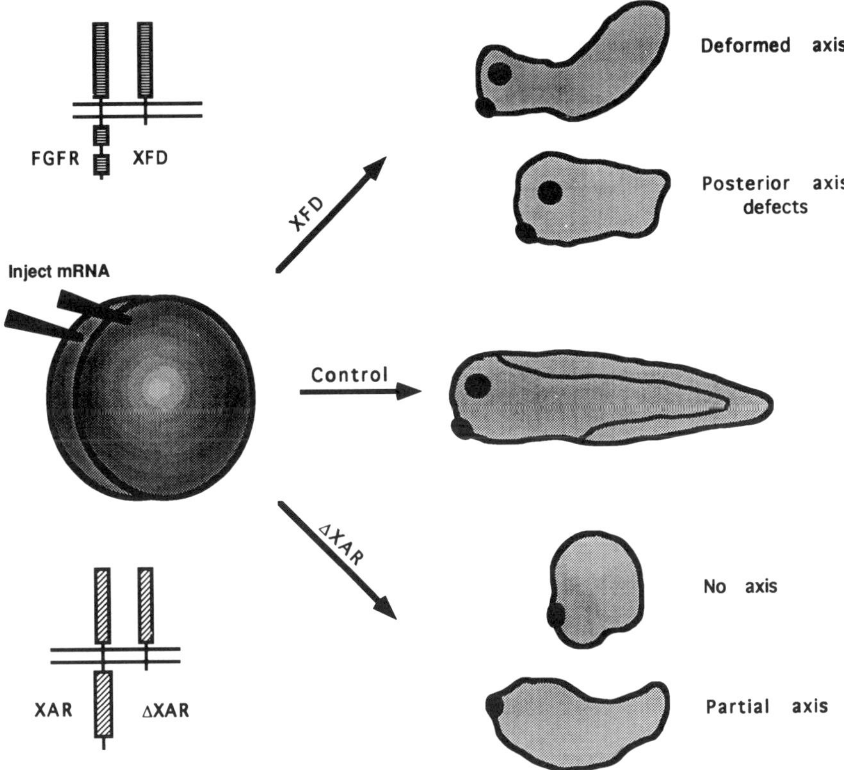

Fig. 7. *Schematic representation of FGF receptor (FGFR), activin receptor (XAR), and their truncated forms XFD and ΔXAR; and the phenotypic consequences of injection of the mRNAs for truncated receptors into early embryos [Amaya et al., 1991; Hemmati-Brivanlou and Melton, 1992].*

major stimulus that helped propel the field of embryonic induction into the high state of activity of the past few years. Observations of the chemical properties of the XTC factor [Smith et al., 1988] and other experiments [Kimelman and Kirschner, 1987; Rosa et al., 1988] suggested that the factor was related to the TGF-β family. This proved to be the case: activins A and B were identified as powerful mesoderm-inducing factors, with activin A being the major inducing component in XTC medium [Asashima et al., 1990; Smith et al., 1990b; Chertov et al., 1990; Thomsen et al., 1990]. Activins may have a role in all vertebrate embryos since they have been shown to affect gene expression in gastrula embryos of the mouse [Blum et al., 1992] and zebrafish [Schulte-Merker et al., 1992].

Injection of activin mRNA into ventral blastomeres of cleavage embryos (see Fig. 5) led to ectopic induction of mesoderm but not to extensive axis duplications [Thomsen et al., 1990].

A mesoderm-inducing factor isolated from chick embryos called vegetalizing factor, studied extensively by Tiedemann and his colleagues [reviewed in Tiedemann, 1990], has been identified by several criteria as chicken activin [Asashima et al., 1991b].

2. Gradients and thresholds in activin action. In the animal cap assay, low concentrations of activin induce more ventral cell types

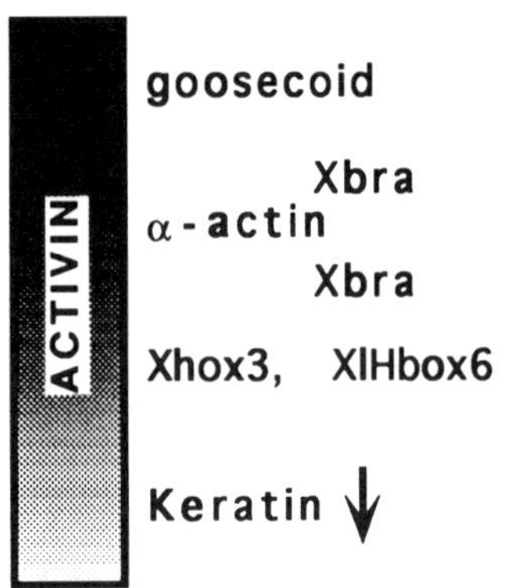

Fig. 8. *Schematic presentation of the results from experiments showing that different concentrations of activin induce different molecular responses in animal cells [Green and Smith, 1990; Green et al., 1992]. Keratin genes are turned off by low activin levels, the other genes listed are turned on in narrow concentration ranges. Ventroposterior genes are affected at low, and dorsoanterior genes at high, activin levels. The following markers are used: Keratins, ectoderm specific [Jamrich et al., 1987];* Xhox3, *expressed in the posterior mesoderm [Ruiz i Altaba and Melton, 1989a,b,c];* XlHbox6, *expressed in posterior neural plate [Sharpe et al., 1987];* Xbra, *expressed in ventrolateral mesoderm and, somewhat later, in the notochord [Smith et al., 1991]; α-actin, specific for skeletal muscle [Mohun et al., 1984]; and* goosecoid, *organizer specific [Blumberg et al., 1991; Cho et al., 1991]. See also Figure 11.*

and molecular markers; high concentrations induce more dorsal cell types and molecular markers, but the ranges of response overlap [Green et al., 1990; Tadano et al., 1993]. This overlap is due to varying accessibility of cells in the tissue to the inducer. When animal cap cells were dissociated by removal of Ca^{2+} from the medium, so that all cells had equal exposure to activin, Green and Smith [1990] and Green et al. [1992] found sharp thresholds and narrow response ranges for the expression of different marker genes (Fig. 8). Induction by FGF also shows some threshold effects [Green et al., 1992]. The expression of marker genes for a series of differentiation states ranging from epidermis to ventroposterior, lateral, and dorsal mesoderm corresponds to a gradient of low-to-high levels of activin. These results might suggest that a dorsoventral gradient of activin could explain the patterning of the mesoderm along the dorsoventral axis. While the threshold effects are impressive and surely have a biological consequence, it is uncertain whether an activin gradient is involved in establishing dorsoventral polarity, because (i) no evidence for an activin gradient in the blastula embryo is available, and (ii) molecules with intense dorsal polarizing effects have been identified, as discussed in Section VI.

3. The expression of activins and activin mRNAs in oocytes and embryos. Asashima et al. [1991a] isolated activin-like proteins with inducing activity from frog eggs, although the particular isoform(s) that are present are not known. Activins are homodimers of the inhibin β chains, two forms of which are known, βA and βB; in *Xenopus,* cDNAs for βA and two closely related forms, βB1 and βB2, have been cloned [Thomsen et al., 1990]. No activin mRNA was found in the egg; βB mRNA arises only in the late blastula and βA mRNA in the late gastrula [Thomsen et al., 1990]. The origin of the activin-like protein in the egg was studied by Rebagliati and Dawid [1993] by considering the possibility that activin mRNA is present transiently in oocytes and leads to stored protein in the egg. No βA and βB1 mRNA could be found by polymerase chain reaction (PCR) methods, but a very low level of βB2 mRNA was detected. In contrast, activin mRNAs (mostly βA) are quite abundant in follicle cells. Thus a consistent hypothesis suggests the synthesis of activins by follicle cells and their import into the oocyte, where they would be stored to become functional in embryogenesis [see also Dohrmann et al., 1993].

4. Activin receptors: expression and function. The first receptor for any TGF-β family member to be isolated was an activin receptor; it proved to be the prototype of a new class of molecules, transmembrane serine/threonine kinases [Mathews and Vale, 1991; reviewed in Lin and Lodish, 1993]. In *Xenopus,* related receptor cDNAs have been cloned and their properties characterized. The receptor mRNA is ubiquitously distributed in the egg and early embryo. Ectopic overexpression of receptor by injection of mRNA leads to developmental malformations, including ectopic mesoderm differentiation and axis duplication, a result which implies that activin is involved in mesoderm specification [Kondo et al., 1991; Hemmati-Brivanlou et al., 1992; Mathews et al., 1992; Nishimatsu et al., 1992a,b].

Strong evidence for a requirement for activin signalling in mesoderm development comes from the dominant-negative receptor experiments of Hemmati-Brivanlou and Melton [1992]. The RNA for truncated activin receptor (ΔXAR) was injected into the 2-cell stage (Fig. 7), and animal caps from the resulting blastulae were tested. They proved to be insensitive to activin but still responsive to FGF, the opposite of what is observed with truncated FGFR [Amaya et al., 1991]. The effect of ΔXAR on development was particularly striking: many of the embryos had no axial structures whatever, and indeed had no mesodermal tissues. Truncated activin receptor also led to strong inhibition of several mesodermal marker genes like *brachyury* and muscle actin. Inhibition by ΔXAR suggests that the activin receptor dimerizes, just as tyrosine receptor kinases do, but this point is not yet explicitly demonstrated. Another unresolved issue relates to the role of distinct activin receptors, i.e., whether the truncated receptor molecule that is capable of abrogating mesoderm induction can associate with all activin receptors that are expressed in the embryo.

A naturally truncated activin receptor mRNA was identified by Nishimatsu et al. [1992a] in the *Xenopus* embryo. In this form, XSTK2, only a portion of the kinase domain is missing. Injection of XSTK2 mRNA into the embryo lead to posterior axis duplications, similar to injection of full-length receptor mRNA. This result suggests that the partially truncated form is capable of transmitting an activin signal. The precise structural attributes of activin receptors that behave as active, inactive, or negative dominant forms remain to be elucidated.

C. Additional Mesoderm Inducers in the TGF-β Family

Early reports on the inducing activities of TGF-β1 [Kimelman and Kirschner, 1987], TGF-β2 [Rosa et al., 1988] and TGF-β3 [Roberts et al., 1990a] served to focus attention on this family of factors, but their low specific activity questions their role in early embryogenesis. TGF-β5, a form whose mRNA is expressed in the *Xenopus* egg [Kondaiah et al., 1990], does not induce mesoderm in animal caps [Roberts et al., 1990b].

An intriguing molecule is Vg1, a TGF-β family member whose mRNA is localized to the vegetal pole of the unfertilized frog egg [Rebagliati et al., 1985; Melton, 1987; Weeks and Melton, 1987]. Localization and sequence thus suggest that Vg1 has a role in mesoderm induction or axis formation, but direct evidence for such an idea has not been obtained to date [see Thomsen and Melton, 1993; Dale et al., 1993].

A new member of the TGF-β family, nodal, has been isolated recently on the basis of a mouse mutant in which no mesoderm forms. The wild-type gene is expressed at the gastrula stage in the node, the mouse equivalent of the frog organizer region [Zhou et al., 1993]. Nodal overexpression has a profound influence on frog development (C. V. E. Wright, personal communication). This implies that a homologous factor occurs in frogs and may be an important player in mesoderm formation in this animal, as it clearly is in the mouse. For further discussion of the TGF-β family in murine development, see the chapter by Akhurst (this volume).

1. Ventralizing mesoderm-inducing factors. Factors belonging to the BMP (bone morphogenetic protein) family, a subgroup of the TGF-β superfamily, are expressed in frog embryos and show biological activities in the system. BMP-2, BMP-4, and BMP-7 mRNAs have been detected in the egg and early embryo [Köster et al., 1991; Dale et al., 1992; Jones et al., 1992; Nishimatsu et al., 1992c], as have the BMP-2 and BMP-4 proteins [Ueno et al., 1992; Shoda et al., 1992; Nishimatsu et al., 1993]. BMP-4 (also called DVR-4) has the ability to induce ventral mesoderm in animal caps [Köster et al., 1991; Jones et al., 1992]. Even more interesting were the results of assaying the function of BMP-4 by injection of mRNA into different regions of the early embryo (see Fig. 5): ventral mesoderm was induced but expression of dorsal mesoderm was inhibited [Dale et al., 1992; Jones et al., 1992]. This could be seen by injecting BMP-4 RNA into the dorsal side of early embryos, an operation which resulted in ventralized embryos that mimick the effect of UV suppression of the cortical rotation, and by the ventralization of animal caps exposed to both activin and BMP-4. *In vivo* BMP-4 may induce ventral mesoderm and prevent the "excessive" dorsalization of the VMZ (ventral marginal zone) by dorsal inducers.

VI. DORSALIZING FACTORS

A. The Relationship Between Mesoderm Formation and Establishment of the Dorsoventral Axis

In an earlier section the evidence was reviewed for the hypothesis that cortical rotation imparts special properties to dorsal blastomeres; that all dorsal blastomeres have some ability of self differentiation as dorsal structures; and that, in spite of this ability, inductive interactions are required for the development of dorsal structure in embryogenesis. The induction of mesoderm in general and of dorsal structures is connected in time and place. Also, experiments with activin in which a low level induces ventral mesoderm and a high level induces dorsal mesoderm, might suggest that dorsal induction is simply more intense mesoderm induction.

Several observations argue against such a view: (i) Mesoderm formation and dorsalization can be dissociated. Inhibition of cortical rotation by UV light leads to the development of embryos that have ventral mesoderm but no axis [Grant and Wacaster, 1972; Malacinski et al., 1975; Scharf and Gerhart, 1983; Gerhart et al., 1989]. An opposite effect can be obtained by treatment of embryos with LiCl: such treatment dorsalizes the entire mesoderm [Kao et al., 1986; Kao and Elinson, 1988, 1989], but Li^+ is not a mesoderm inducer [Cooke et al., 1989]. (ii) Animal cap explants have a "prepattern," i.e., a latent dorsoventral polarity that can be seen only by exposing them to inducing factors [Sokol and Melton, 1991; Dawid et al., 1992; see Fig. 9]. This polarity in animal tissue is probably due to differential competence (or sensitivity) to respond to inducers. (iii) Most importantly, recent results indicate that mesoderm is patterned by an interaction between dorsalizing factors and inducing factors.

B. Dorsalizing Effects Can Be Transferred by Injection of Cytoplasmic Components

The early dorsal axis-inducing ability can be transferred to ventral blastomeres by injecting them with dorsal cytoplasm [Yuge et al., 1990; Hainsky and Moody, 1992], or with RNA isolated from dorsal, but not from ventral, blastomeres [Hainsky and Moody, 1992]. These experiments provide direct evidence for the long-held conviction that cytoplasmic rearrangements associated with cortical rotation [Danilchik and Denegre, 1991] move specific components to the dorsal side of the egg, although specific activation of dorsal components could also play a role in rotation-induced dorsalization. While the inducing RNA (or RNAs) in dorsal blastomeres has yet to be identified, the early dorsalizing effect can be elicited by two types of factors, wnt proteins and noggin.

C. The Striking Effects of Wnt Proteins

1. Wnt proteins as dorsalizing factors. The *wnt* genes encode secreted proteins with many interesting functions in different tissues and developmental stages [reviewed by Nusse and Varmus, 1992; Moon, 1993; Christian and Moon, 1993a]. Several members of the *wnt* family have been isolated from *Xenopus* and their expression pattern has been studied [Noordemeer et al., 1989; Christian et al., 1991a,b; McGrew et al., 1992; Wolda and

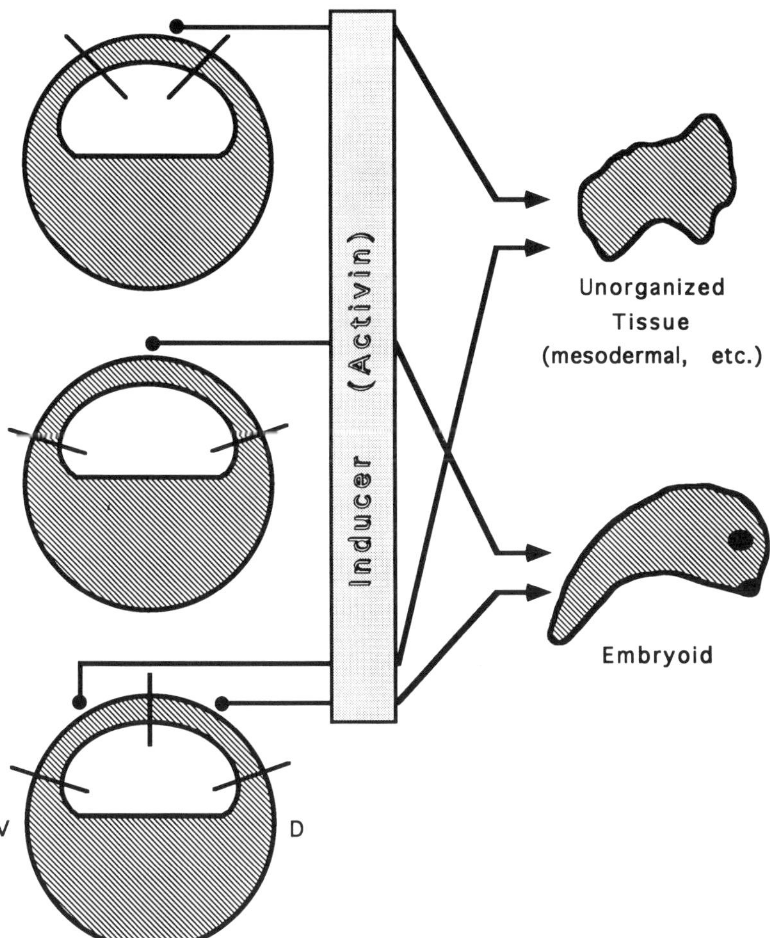

Fig. 9. *The animal hemisphere of the blastula carries a prepattern. While small polar or ventral explants yield only disorganized mesodermal tissues after activin treatment, large caps or their dorsal halves are capable of forming structures with embryo-like polarity ("embryoids") after induction [Sokol and Melton, 1991; Dawid et al., 1992]. The prepattern is latent, however, because even large dorsal explants form atypical epidermis when cultured without factors.*

Moon, 1992; Wolda et al., 1993]. The main focus of interest is the ability of mouse Wnt-1, and *Xenopus* Xwnt-8 mRNA to induce axis duplications when injected into the early embryo (Fig. 5). Duplications of the neural plate were first observed after Wnt-1 RNA injection by McMahon and Moon [1989], but it was not clear whether the effect is directly on the nervous system or mediated by other cell types. The issue was resolved [Sokol et al., 1991; Smith and Harland, 1991] by showing that Xwnt-8 (and Wnt-1) RNA generates a secondary dorsal axis when injected into the ventral, but not the dorsal, blastomeres of normal embryos, and is also capable of rescuing UV-ventralized embryos when injected into any vegetal cell. Such injected cells behave like dorsovegetal blastomeres: they induce an axis without contributing to it; in other words, they form a Nieuwkoop center [Smith and Harland, 1991]. Family members other than *wnt-1*, *Xwnt-8*, and *Drosophila wingless* [Chakrabarti et al., 1992], do not have these effects [see Moon, 1993].

2. Wnts are not mesoderm inducers. It has not been possible to obtain soluble Wnt proteins for a direct test of their function, presumably because these proteins adhere to the cell surface or extracellular matrix [Papkoff and Schryver, 1990; Bradley and Brown, 1990]; *in vivo* expression was used instead. When Wnt RNA was injected into early embryos, and animal caps were explanted from the resulting blatulae, no mesoderm differentiation was observed; yet the explants showed a changed sensitivity to inducers, so that FGF or a low level of activin, which elicited ventral mesoderm in control explants, led to the differentiation of dorsal mesoderm in Xwnt-8-injected explants [Christian et al., 1992; Sokol and Melton, 1992; see Fig. 6]. Wnt expression thus changes the competence of the cells to respond to induction. There are good indications that dorsalizing (Wnt-like) and mesoderm inducing (activin-like) signals cooperate *in vivo* [Sokol, 1993]. FGF and Wnt have quite different effects in the context of the whole embryo (Fig. 5) and in animal caps (Fig. 6), exemplifying the regulative ability of the embryo.

The molecular mechanism of action of the Wnt proteins in the embryo is unknown, but one potentially important effect is a change in gap junctional communication between cells. Olson et al., [1991] and Olson and Moon [1992] have shown that dorsal cells are more coupled than ventral cells in the normal embryo, and that ectopic expression of Wnt and of activin, but not of FGF, increases coupling in ventral cells, making them more like dorsal cells in this, as in other, respects.

3. Xwnt-8 is a ventralizing factor during gastrulation. For all its dramatic dorsalizing effects, Xwnt-8 is not the natural axis-inducing agent. Its normal expression beings in the late blastula when mesoderm induction and axis determination have largely taken place, and Xwnt-8 RNA is localized in the ventral and lateral mesoderm of the gastrula embryo [Christian et al., 1991b]. Why, then, does Xwnt-8 RNA induce a dorsal axis? The answer is not known, but it is possible that injected Xwnt-8 mimicks the function of a different, hypothetical wnt family member that is dorsally expressed in the fertilized egg. It is also possible that Xwnt-8 can stimulate a signal transduction pathway that is present for a different purpose but normally does not have a Wnt effector. A possible model for such an effect is the action of Li^+ on the embryo. The dorsalizing effects of this ion [Kao et al., 1986; Kao and Elinson, 1988, 1989] appear to be mediated by inhibition of the phosphoinositide cycle [Busa and Gimlich, 1989; Maslanski et al., 1992]. Of course, Li^+ is not a natural dorsalizing agent, but it appears to be able to take control of a signal transduction pathway whose components are in place.

Elegant experiments by Christian and Moon [1993b] have shed light on the probable natural function of Xwnt-8 in the embryo. To avoid the early axis-duplicating effect, these authors injected an expression construct rather than mRNA into the embryo. Xwnt-8 expression was then initiated at the midblastula stage, in concert with its natural expression period. By ectopic expression of *Xwnt-8* in the DMZ a profound ventralizing effect became apparent, consistent with the fact that this gene is normally expressed in the entire marginal zone except the dorsal quadrant (i.e., the organizer). The different outcomes of Xwnt-8 expression before and after the midblastula stage illustrate that signalling molecules can have very different effects, depending on the properties of the responding cell.

D. Noggin, a Newly Discovered Secreted Protein With Dorsalizing Function

By combining a highly effective expression cloning method and injection assay, Smith and Harland [1991, 1992] isolated the *Xwnt-8* gene that had also been obtained by sequence homology, and a new factor, which they named noggin. Noggin mRNA shows the same remarkable ability as Xwnt-8 to rescue UV-ventralized embryos, by way of generating a Nieuwkoop center in the injected blastomere [Smith and Harland, 1992]. A low level of noggin mRNA is expressed maternally but is not localized in the egg; much higher expression begins at the midblastula transition (MBT), and the RNA is localized in the organizer region of the gastrula. The isolation of noggin by Smith and Harland [1992] has the exciting promise of leading to a new class of growth factors with multiple potential roles in vertebrate biology.

Is noggin the natural early dorsalizing agent? At this time, the answer must be "maybe." The early dorsalizing activity is localized, and is associated with localized RNA [Hainsky and Moody, 1992], but maternal noggin RNA is not localized. This point argues against noggin as the "Nieuwkoop factor." However, noggin protein might be localized even if the RNA is not, and the issue remains open for now.

While its role in the early blastula is uncertain, a dorsalizing function for noggin during gastrulation is strongly supported. In the gastrula, an influence from the dorsal mesoderm (the DMZ, or the Spemann organizer) affects lateral and ventral mesoderm (the LMZ and VMZ), partially dorsalizing them and inducing the differentiation of axial structures like somites (skeletal muscle). This dorsalizing effect is temporally separated from the earlier effects of the Nieuwkoop center. In the view of Slack [Lettice and Slack, 1993], the late dorsalizing effect of the DMZ is qualitatively distinct from the early effect of the Nieuwkoop center. Smith et al. [1993] have provided evidence that the noggin protein can dorsalize the VMZ, and that it induces the formation of muscle but not of notochord. Activin was unable to dorsalize the VMZ. This effect of noggin mimicks closely the action of DMZ (organizer) on VMZ. Since noggin is secreted and its RNA is highly localized in the organizer during gastrulation, it fulfills all the requirements for a molecule that mediates the dorsalizing effect of the organizer on ventral mesoderm. Of course, its action as mesoderm dorsalizer during gastrulation does not preclude the possible role of noggin in other interactions, e.g., in Nieuwkoop center signalling or in neural induction.

VII. MESODERM INDUCTION AND THE ESTABLISHMENT OF THE DORSOVENTRAL AXIS: MODELS

A. The Three-Signal Model

Transplantation and implantation experiments have led Slack and his colleagues to propose a three-signal model (Fig. 10) that has been heuristically very useful [Smith and Slack, 1983; Dale and Slack, 1987b; Slack, 1991]. The three signals include a vegetal signal inducing ventrolateral mesoderm; a dorsal vegetal signal that, loosely, may correspond to what other authors call the Nieuwkoop center; and a later-acting dorsalizing signal that emanates from the DMZ or organizer during gastrulation and induces somites and other tissues in ventrolateral mesoderm. In terms of general interactive phenomena this model continues to be appropriate, but it does not appear that each of the three signals simply corresponds to one growth factor–like molecule. Variant models are therefore emerging to account for newly discovered interactions [Kimelman et al., 1992; Sive, 1993; Moon, 1993; Christian and Moon, 1993a].

B. A Snapshot of Current Knowledge

The drawing at the right of Figure 10 is not so much meant as a model of early *Xenopus* development as a graphic summary of known and implied interactive events in the embryo. The drawing, just like the three-signal model, is something of a time composite, showing events during blastula and gastrula stages that have complex relationships. As has been outlined above, we now know about three types of signals, each represented by more than one molecule. References for the statements in the following discussion are found in the preceding sections.

Mesoderm inducers are represented by FGF (possibly various isoforms), and TGF-β family members, i.e., activin A and B, and, possibly, *Xenopus* homologues of nodal. At least one member of each group is undoubtedly involved, as shown by the dominant-negative receptor experiments. Nieuwkoop-type induction experiments indicate that the

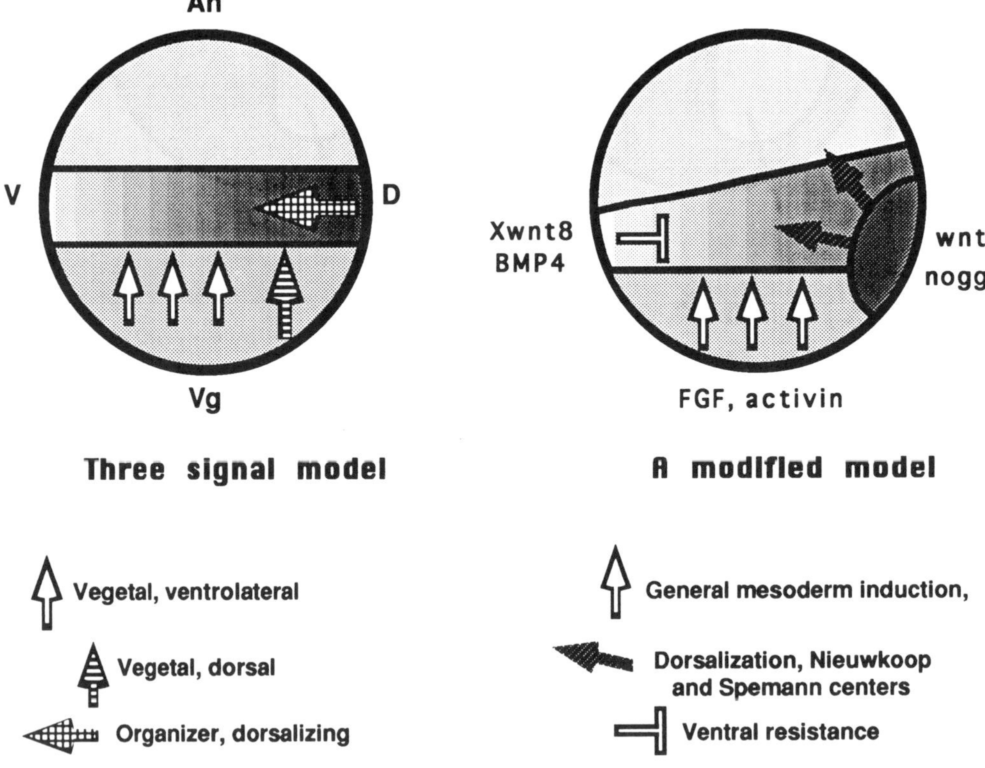

Fig. 10. *Induction models. The three-signal model (left) is based on biological tests that indicated that action of at least three signals in* Xenopus *embryogenesis: a signal by ventrolateral vegetal tissue, inducing mesoderm in the marginal zone; a dorsal vegetal signal, inducing the organizer; and organizer action which dorsalizes lateral and ventral mesoderm at later stages [Smith and Slack, 1983; Dale and Slack, 1987b; Slack, 1991]. The model on the right incorporates present knowledge about biological events and factors known to be effective in mesoderm induction and dorsal axis determination, although not all of these factors are known to act* in vivo. *See text for further discussion.*

mesoderm-inducing signal derives from the vegetal region *in vivo,* and acts from shortly after cleavage begins until late blastula stages. Distribution studies for inducers or their receptors are incomplete, particularly since most studies look at mRNA, although protein is the relevant molecule. Nevertheless there is no evidence for a nonuniform distribution of FGF, activins, or their receptors that could account for the induction of dorsoventral polarity.

Dorsalizing factors are clearly involved in the establishment of the dorsoventral axis, in several ways: (i) as dorsal inducers in early development, accounting for the Nieuwkoop center; (ii) as factors responsible for the self-differentiation ability of dorsal blastomeres into dorsal structures; (iii) as dorsal inducers during gastrulation. In gastrulation, signals must account for (a) the dorsalization of ventrolateral mesoderm, and (b) neural induction. Xwnt-8 (and Wnt-1, but not several other wnt family members) and noggin can elicit the formation of a Nieuwkoop center. Whether they do so *in vivo* is unclear for noggin. The Nieuwkoop center is not elicited by Xwnt-8, although it may be by another wnt protein. Noggin is capable of dorsalizing gastrula VMZ, and thus accounts for one of the dorsalizing functions listed. The second function, self-differentiation, could be due to a secreted factor (autocrine, or short-range paracrine, effect) or to a cell-autonomous factor, but no information is available on this point.

One mechanism of action of dorsalizing factors is to sensitize cells and to change their response to mesoderm inducers; by themselves, dorsalizers are not inducers. Whether this competence-modifying effect is the only mode of action of dorsalizers is not known, but it is surely important in their function.

To obtain mesoderm patterned with dorsoventral polarity the interaction of dorsalizing and inducing factors is undoubtedly required. Yet, the concentration-dependent effects of activin (Fig. 8) suggest that varying levels of inducer could establish polarity. The relevance of inducer gradients as distinct from interactions among factors remains to be established *in vivo.*

No evidence has been obtained in transplantation experiments for a ventralizing effect, e.g., VMZ is dorsalized by DMZ, but the reverse does not happen. Yet two substances with ventralizing activity are expressed in the embryo, BMP-4 and Xwnt-8. A role is implied for these substances *in vivo,* especially for Xwnt-8, which is localized specifically to the ventrolateral marginal zone of the gastrula. Since ventralization *per se* does not appear to occur, one may imagine that such factors are "resisters," whose function may be to prevent overly effective dorsalizing influence from recruiting more than their fair share of the marginal zone.

VIII. CONSEQUENCES OF MESODERM INDUCTION: ACTIVATION OF REGULATORY GENES

A. Genes Encoding Putative Transcription Factors Are Rapidly Induced by Growth Factors in Regionally-Specific Ways

The analysis of genes responding to inducing factors is a useful approach to the study of the mechanism by which these factors execute their developmental role. By using the criterion of resistance of induction to cycloheximide (CHX; see Rosa, 1989), activated genes can be classified as part of initial and subsequent waves of responses. Activin, and in some cases FGF, activates genes that are normally expressed in various distinct patterns in the marginal zone or the entire vegetal hemisphere, while suppressing certain ectoderm-specific genes.

Figure 11 shows schematically some examples of early response genes encoding putative transcription factors. The *Mix.1* [Rosa, 1989], *gsc* [Cho et al., 1991], and *Xlim-1* [Taira et al, 1992] genes are homeobox genes; *XFKH-1* [Dirksen and Jamrich, 1992, also called *pintallavis,* Ruiz i Altaba and Jessel, 1992; and XFD-1, Knöchel et al., 1992] belongs to the forkhead class of factors; and *Xbra* [Smith et al., 1991] is the frog homologue of the mouse *brachyury* gene. It appears that these (and additional) regulatory genes divide the area of the gastrula embryo into distinct regions, thereby controlling pattern formation.

1. The organizer-generating ability of *goosecoid* RNA. As discussed above, two types of secreted molecules, Wnts and noggin, are capable of generating a secondary embryo when injected into early ventral blastomeres. These molecules achieve this effect by eliciting the formation of an ectopic Nieuwkoop center. Injection of *gsc* RNA can also generate a secondary axis, as shown by Cho et al. [1991] and confirmed in our laboratory [M. Taira, H. Otani, and I. B. Dawid, unpublished]. Injection of *gsc* mRNA usually leads to secondary trunk formation (see Fig. 12), although secondary heads have been observed [Cho et al., 1991]. As might be expected from a homeodomain protein with a presumed nuclear site of action, *gsc* acts within the cell into which it is injected and within its progeny, changing their fate to a dorsoanterior one [Niehrs et al., 1993]. However, *gsc* also has non-cell-autonomous effects, in that injected cells recruit other ventral cells into the dorsal axis, presumably by generating signals that affect the fate of their neighbors. In addition, gsc alters the adhesive and motile properties of the cells in which it is expressed. These ele-

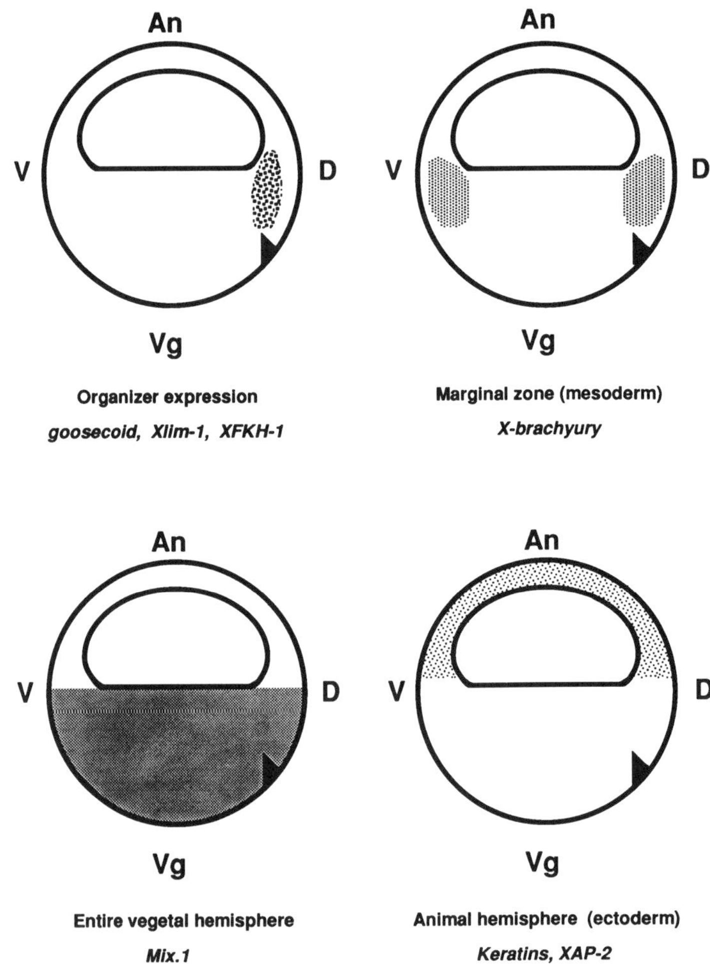

Fig. 11. *A schematic representation of the expression patterns of some genes that respond rapidly to mesoderm induction. The expression patterns of gsc [Cho et al., 1991], Xlim-1 [Taira et al., 1992], and XFKH-1 [Dirksen and Jamrich, 1992; also called pintallavis, Ruiz i Altaba and Jessel, 1992; and XFD1, Knöchel et al., 1992] are similar though not necessarily identical. The Xbra gene is expressed in the marginal zone [Smith et al., 199a], and the Mix.1 gene in the entire vegetal hemisphere [Rosa, 1989]. Ectoderm-specific genes, keratins [Jamrich et al., 1987], and XAP-2 [Winning et al., 1991] are suppressed by mesoderm induction [Symes et al., 1988; Dawid et al., 1988; P. Mathers and T. D. Sargent, personal communication]. See also legend to Figure 8.*

gant results of Niehrs and colleagues [1993] provide insights into some of the molecular mechanisms that are involved in the formation and function of the organizer.

2. More organizer-specific genes. In addition to *noggin* and *gsc,* the transcription factor–encoding genes *Xlim-1* and *XFKH-1* (Fig. 11), and a recently discovered homeobox gene *Xnot* [von Dassow et al., 1993], are expressed in the organizer. The *Xnot* gene shares with *gsc* the distinctive behavior that their expression is superinduced by CHX in animal caps [von Dassow et al., 1993; Tadano et al., 1993]. Because ectopic expression of *Xlim-1* does not (so far) elicit a secondary axis [M. Taira, H. Otani, and I. B. Dawid, unpub-

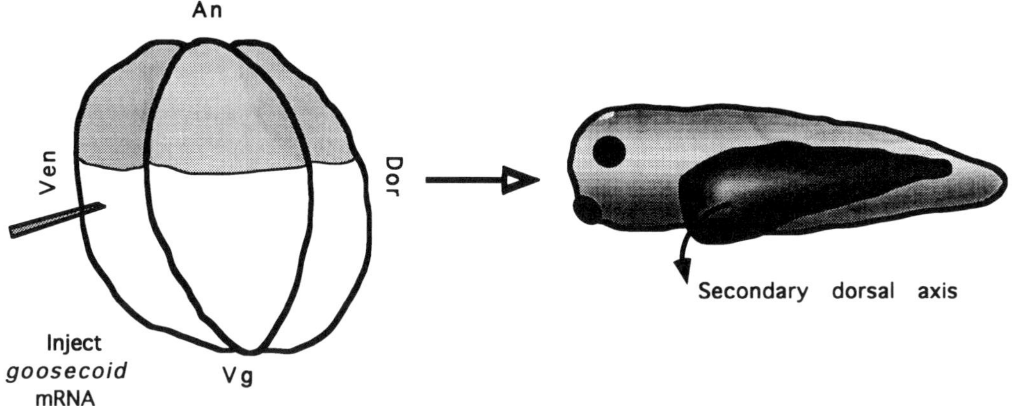

Fig. 12. *Injection of Gsc mRNA into ventral blastomeres leads to axis duplication [Cho et al., 1991]. Although secondary heads can form, trunk duplications are the usual outcome.*

lished] it appears that not every organizer-specific transcription factor will have such an effect. This is not surprising, because one may expect only molecules rather high in a unique line of a regulatory hierarchy to be able to set in motion the many steps required for axis duplication. Other molecules involved in axis formation may depend on the cooperation of different factors, and thus may not be able to start the process by themselves in an ectopic location.

3. Xbra elicits mesoderm formation. Injection of Xbra mRNA into the animal region of early embryos leads to ectopic mesoderm formation [Cunliffe and Smith, 1992]. In addition, animal caps explanted from injected embryos express marker genes characteristic for lateral and posterior mesoderm such as muscle actin, *Xhox3* (see also below), and *Xsna* (The *Xenopus snail* homologue), but not the dorsoanterior marker *gsc*. Cunliffe and Smith [1992] thus suggest that *Xbra* is involved in the specification of posterior mesoderm. This suggestion is consistent with the phenotype of the mouse *brachyury* mutant [Willison, 1990].

B. Subsequent Waves of Gene Activation in Induced Mesoderm

Further specification of regional identity and tissue specificity is provided by the regulated expression of additional genes, some of which are expressed in different regions of the mesoderm and are likely to be involved in the specification of mesodermal fates. An example is provided by the expression of the *Xenopus twist* gene in the entire mesoderm except the muscle-forming region, i.e., in the notochord and ventrolateral mesoderm [Hopwood et al., 1989b]. This pattern suggests that twist is involved in subdividing the mesoderm, possibly in interaction with Xsna, the *Xenopus* homologue of snail [Sargent and Bennet, 1990].

1. Formation of skeletal muscle. Most studied among mesodermal tissues is perhaps the formation of somites and differentiation of muscle which is associated with the expression of MyoD and Myf5 [Hopwood et al., 1989a; Harvey, 1990; Hopwood et al., 1991]. While there is good reason to believe that MyoD and Myf5 regulate muscle-specific differentiation in the embryo, ectopic expression of either gene does not lead to ectopic myogenesis [Hopwood and Gurdon, 1990; Hopwood et al., 1991a].

It is of substantial interest to note that muscle induction proceeds effectively only in groups of cells rather than in single cells, a phenomenon named the community effect [Gurdon 1988], it is indicative of cell interactions whose molecular basis is not understood. Whether other tissues show a community effect in their specification is not known. Only at a later stage in their development are individual muscle cells capable of executing muscle differentiation in an ectopic site, as shown by single-cell transplantation experiments [Kato and Gurdon, 1993].

2. *Xhox3,* an inducible gene involved in establishing the anteroposterior axis. The homeobox gene *Xhox3* is inducible in the animal cap assay, but it is not known whether it is an immediate early response gene. It has several interesting features. In the normal embryo it is expressed in the posterior mesoderm [Ruiz i Altaba and Melton, 1989a]. In explants, it responds to FGF better than to activin, consistent with a function of FGF in ventroposterior induction [Ruiz i Altaba and Melton, 1989b]. Lastly, ectopic expression of Xhox3 mRNA inhibits anterior development [Ruiz i Altaba and Melton, 1989c]. These results implicate *Xhox3* in the specification of the anteroposterior axis and suggest that growth factors influence axis polarization by differentially inducing transcription factor encoding genes.

IX. CONCLUSION

Secreted peptide factors play a major and early role in amphibian and, most likely, all vertebrate embryogenesis. In the best-studied case of the frog *Xenopus laevis,* peptide growth factors regulate the specification of the mesoderm and the establishment of the dorsoventral axis. A major consequence of growth factor action in the embryo is the spatially specific activation of transcriptional regulatory genes that are likely to serve in specifying regional identities in the embryo.

ACKNOWLEDGMENT

I thank Mike Rebagliati for many thoughtful suggestions for changes and additions to the manuscript, and Jeanne-Pierre Saint-Jeannet and Masanori Taira for additional helpful comments.

REFERENCES

Amaya E, Musci TJ, Kirschner MW (1991): Expression of a dominant negative mutant of FGF receptor disrupts mesoderm formation in Xenopus embryos. Cell 66:257–270.

Amaya E, Stein PA, Kirschner MW (1993): FGF signalling in the early specification of mesoderm in *Xenopus*. Development 118:477–487.

Asashima M, Nakano H, Shimada K, Kinoshita K, Ishii K, Shibai H, Ueno N (1990): Mesodermal induction in early amphibian embryos by activin A (erythroid differentiation factor). Roux's Arch Dev Biol 198:330–335.

Asashima M, Nakano H, Uchiyama H, Sugino H, Nakamura T, Eto Y, Ejima D, Nishimatsu S, Ueno N, Kinoshita K (1991a): Presence of activin (erythroid differentiation factor) in unfertilized eggs and blastulae of *Xenopus laevis*. Proc Natl Acad Sci USA 88:6511–6514.

Asashima M, Uchiyama H, Nakano H, Eto Y, Ejima D, Sugino H, Davids M, Plessow S, Born J, Hoppe P, Tiedemann H, Tiedemann H (1991b): The vegetalizing factor from chicken embryos: its EDF (activin-A)-like activity. Mech Dev 34:135–141.

Blum M, Gaunt SJ, Cho KWY, Steinbeisser H, Blumberg B, Bittner D, De Robertis EM (1992): Gastrulation in the mouse: the role of the homeobox gene *goosecoid*. Cell 69:1097–1106.

Blumberg B, Wright CVE, De Robertis EM, Cho KWY (1991): Organizer-specific homeobox genes in *Xenopus laevis* embryos. Science 253:194–196.

Boterenbrood EC, Nieuwkoop PD (1973): The formation of mesoderm in urodelan amphibians; V. Its regional induction by the endoderm. Wilh Roux's Arch 173:319–332.

Boucaut JC, Darribère T, Boulekbache H, Thiery JP (1984): Prevention of gastrulation but not neurulation by antibodies to fibronectin in amphibian embryos. Nature 307:364–367.

Boucaut JC, Darribère T, Shi DL, Boulekbache H, Yamada KM, Thiery JP (1985): Evidence for the role of fibronectin in amphibian gastrulation. J Embryol Exp Morphol 89 (Suppl.):211–227.

Bradley RS, Brown AM (1990): The proto-oncogene *int-1* encodes a secreted protein associated with the extracellular matrix. EMBO J 9:1569–1575.

Busa WB, Gimlich RL (1989): Lithium-induced teratogenesis in frog embryos prevented by a polyphosphoinositide cycle intermediate or a diacylglycerol analog. Dev Biol 132:315–324.

Campos-Ortega JA (1990): Mechanisms of a cellular decision during embryonic development of *Drosophila melanogaster:* epidermogenesis or neurogenesis. Adv Genet 27:403–53.

Chakrabarti A, Matthews G, Colman A, Dale L (1992): Secretory and inductive properties of *Drosophila wingless* protein in *Xenopus* oocytes and embryos. Development 115:355–369.

Chertov OY, Krasnoselsky AL, Bogdanov ME, Hoperskaya OA (1990): Mesoderm-inducing factor from bovine amniotic fluid: purification and N-terminal sequence. Biomed Sci 1:499–506.

Cho KWY, Blumberg B, Steinbeisser H, De Robertis EM (1991): Molecular nature of Spemann's organizer: the role of the Xenopus homeobox gene *goosecoid.* Cell 67:1111–1120.

Christian JL, Moon RT (1993a): When cells take fate into their own hands: differential competence to respond to inducing signal generates diversity in the embryonic mesoderm. Bioessays 15:135–140.

Christian JL, Moon RT (1993b): Interactions between Xwnt-8 and Spemann organizer signaling pathways generate dorsoventral pattern in the embryonic mesoderm of *Xenopus*. Genes Dev 7:1–16.

Christian JL, Gavin BJ, McMahon AP, Moon RT (1991a): Isolation of cDNAs partially encoding four Xenopus Wnt-1/int-1 related proteins and characterization of their transient expression during embryonic development. Dev Biol 143:230–234.

Christian JL, McMahon JA, McMahon AP, Moon RT (1991b): *Xwnt-8, a Xenopus Wnt-1/int-1* related gene responsive to mesoderm inducing factors, may play a role in ventral mesodermal patterning during embryogenesis. Development 111:1045–1055.

Christian JL, Olson DJ, Moon RT (1992): *Xwnt-8* modifies the character of mesoderm induced by bFGF in isolated *Xenopus* ectoderm. EMBO J 11:33–41.

Cooke J, Symes K, Smith EJ (1989): Potentiation by the lithium ion of morphogenetic responses to a *Xenopus* induction factor. Development 105:549–558.

Cunliffe V, Smith JC (1992): Ectopic mesoderm formation in *Xenopus* embryos caused by widespread expression of a Brachyury homologue. Nature 358:427–430.

Dale L, Slack JMW (1987a): Fate map for the 32-cell stage of Xenopus laevis. Development 99:527–551.

Dale L, Slack JMW (1987b): Regional specification within the mesoderm of early embryos of Xenopus laevis Development 100:279–295.

Dale L, Howes G, Price BMJ, Smith JC, (1992): Bone morphogenetic protein 4: a ventralizing factor in early Cenopus development. Development 115:573–585.

Dale L, Matthews G, Colman A (1993): Secretion and mesoderm-inducing activity of the TGF-β-related domain of *Xenopus* Vg1. EMBO J 12:4471–4480.

Danilchik MV, Denegre JM (1991): Deep cytoplasmic rearrangements during early development in Xenopus laevis. Development 111:845–856.

Danilchik MV, Gerhart JC (1987): Differentiation of the animal-vegetal axis in *Xenopus laevis* oocytes, I. Polarized intracellular translocation of platelets establishes the yolk gradient. Dev Biol 122:101–112.

Darribére T, Yamada KM, Johnson KE, Boucaut JC (1988): The 140-kDa fibronectin receptor complex is required for mesodermal cell adhesion during gastrulation in the amphibian Pleurodeles walti. Dev Biol 126:182–194.

Davids M, Loppnow B, Tiedemann H, Tiedemann H (1987): Neural differentiation of amphibian gastrula ectoderm exposed to phorbol ester. Roux's Arch Dev Biol 196:137–140.

Davidson, EH (1986): Gene Activity in Early Development, 3rd edn. Orlando, FL: Academic Press.

Dawid IB (1991): Mesoderm induction. Methods Cell Biol 36:311–328.

Dawid IB (1992): Mesoderm induction and axis formation in *Xenopus laevis.* Bioessays 10:687–691.

Dawid IB, Sargent TD (1988): Xenopus laevis in developmental and molecular biology. Science 240:1443–1448.

Dawid IB, Rebbert ML, Rosa F, Jamrich J, Sargent TD (1988): Gene expression in amphibian embryogenesis. In G Eguchi, TS Okada, L Saxén (eds): Regulatory Mechanisms in Developmental Processes. Dublin: Elsevier, pp 67–74.

Dawid IB, Sargent TD, Rosa F (1990): The role of growth factors in embryonic induction in amphibians. Curr Top Dev Biol 24:261–288.

Dawid IB, Taira M, Good PJ, Rebagliati MR (1992): The role of growth factors in embryonic induction in Xenopus laevis. Mol Reprod Dev 32:136–144.

Ding X-y, McKeehan WL, Xu J, Grunz H (1992): Spatial and temporal localization of FGF receptors in Xenopus laevis. Roux's Arch Dev Biol 201:334–339.

Dirksen ML, Jamrich M (1992): A novel, activin inducible, blastopore lip specific gene of Xenopus laevis contains a fork head DNA binding domain. Genes Dev 6:599–608.

Dixon JE, Kintner CR (1989): Cellular contacts rquired for neural induction in Xenopus embryos: evidence for two signals. Development 106:749–757.

Dohrmann CE, Hemmati-Brivanlou A, Thomsen GH, Fields A, Woolf TM, Melton DA (1993): Expression of activin mRNA during early development in *Xenopus laevis*. Dev Biol 157:474–483.

Doniach T, Phillips CR, Gerhart JC (1992): Planar induction of anteroposterior pattern in the developing central nervous system of Xenopus laevis. Science 257:542–545.

Durston AJ, Timmermans JPM, Hage WJ, Hendriks HFJ, de Vries NJ, Heideveld M, Nieuwkoop PD (1989): Retinoic acid causes an anteroposterior transformation in the developing central nervous system. Nature 340:140–144.

Friesel R, Brown SAN (1992): Spatially restricted expression of fibroblast growth factor receptor-2 during *Xenopus* development. Development 116:1051–1058.

Friesel R, Dawid IB (1991): cDNA cloning and developmental expression of fibroblast growth factor receptors from *Xenopus laevis*. Mol Cell Biol 11:2481–2488.

Gallagher BC, Hainski AM, Moody SA (1991): Autonomous differentiation of dorsal axial structures from an animal cap cleavage stage blastomere in *Xenopus*. Development 112:1103–1114.

Gerhart J, Danilchik M, Doniach T, Roberts S, Rowning B, Stewart R (1989): Cortical rotation of the Xenopus egg: consequence for the anteroposterior pattern of embryonic dorsal development. Development 107(Suppl.):37–51.

Gillespie LL, Paterno GD, Slack JMW (1989): Analysis of competence: receptors for fibroblast growth factor in early Xenopus embryos. Development 106:203–208.

Gimlich RL (1986): Acquisition of developmental autonomy in the equatorial region of the Xenopus embryo. Dev Biol 115:340–352.

Gimlich RL, Gerhart JC (1984): Early cellular interactions promote embryonic axis formation in Xenopus laevis. Dev Biol 104:117–130.

Godsave SF, Shiurba RA (1992): Xenopus blastulae show regional differences in competence for mesoderm induction: correlation with endogenous basic fibroblast growth factor levels. Dev Biol 151:506–515.

Godsave SF, Slack JMW (1989): Clonal analysis of mesoderm induction in *Xenopus laevis*. Dev Biol 134:486–490.

Godsave SF, Slack JMW (1991): Single cell analysis of mesoderm formation in the *Xenopus* embryo. Development 111:523–530.

Grant P, Wacaster JF (1972): The amphibian grey crescent—a site of developmental information? Dev Biol 28:454–471.

Green JB, Smith JC (1990): Graded changes in dose of a Xenopus activin homologue elicit stepwise transitions in embryonic cell fate. Nature 347:391–394.

Green JBA, Howes G, Symes K, Smith JC (1990): The biological effects of XTC-MIF: quantitative comparison with *Xenopus* FGF. Development 108:173–183.

Green JBA, New HV, Smith JC (1992): Responses of embryonic Xenopus cells to activin and FGF are separated by multiple thresholds and correspond to distinct axes of the mesoderm. Cell 71:731-739.

Grunz H, Tacke L (1989): Neural differentiation of Xenopus laevis ectoderm takes place after disaggregation and delayed reaggregation without inducer. Cell Differ Dev 28:211–218.

Gurdon JB (1992): The generation of diversity and pattern in animal development. Cell 68:185–199.

Gurdon JB (1988): A community effect in animal development. Nature 336:772–774.

Hainski AM, Moody SA (1992): *Xenopus* maternal RNAs from a dorsal animal blastomere induce a secondary axis in host embryos. Development 116:347–355.

Harvey RP (1990): The *Xenopus MyoD* gene: an unlocalized maternal RNA predates lineage-restricted expression in the early embryo. Development 108:669–680.

Hausen P, Riebesell M (1991): The Early Development of Xenopus laevis: An Atlas of the Histology. Berlin Heidelberg New York: Springer-Verlag.

Hemmati-Brivanlou A, Melton DA (1992): A truncated activin receptor inhibits mesoderm induction and formation of axial structures in *Xenopus* embryos. Nature 359:609–614.

Hemmati-Brivanlou A, Stewart RM, Harland RM (1990): Region-specific induction of an *engrailed* protein by anterior notochord in *Xenopus*. Science 250:800–802.

Hemmati-Brivanlou A, Wright DA, Melton DA (1992): Embryonic expression and functional analysis of a *Xenopus* activin receptor. Dev Dyn 194:1–11.

Hopwood ND, Gurdon JB (1990): Activation of muscle genes without myogenesis by ectopic expression of MyoD in frog embryos cells. Nature 347:197–200.

Hopwood ND, Pluck A, Gurdon JB (1989a): MyoD expression in the forming somites is an early response to mesoderm induction in *Xenopus* embryos. EMBO J 8:3409–3417.

Hopwood ND, Pluck A, Gurdon JB (1989b): A Xenopus mRNA related to Drosophila twist is expressed in response to induction in the mesoderm and the neural crest. Cell 59:893–903.

Hopwood ND, Pluck A, Gurdon JB (1991): Xenopus Myf-5 marks early muscle cells and can activate muscle genes ectopically in early embryos. Development 111:551–560.

Howard JE, Hirst EMA, Smith JC (1992): Are β_1 integrins involved in *Xenopus* gastrulation? Mech Dev 38:109–120.

Isaacs HV, Tannahill D, Slack JMW (1992): Expression of a novel FGF in the *Xenopus* embryo. A new candidate inducing factor for mesoderm formation and anteroposterior specification. Development 114:711–720.

Jamrich M, Sargent TD, Dawid IB (1987): Cell-type specific expression of epidermal cytokeratin genes during gastrulation of Xenopus laevis. Genes Dev 1:124–132.

Janeczek J, Born J, Hoppe P, Tiedemann H (1992): Partial characterization of neural-inducing factors from *Xenopus* gastulae. Evidence for a larger protein complex containing the factor. Roux's Arch Dev Biol 201:30–35.

Jessel TM, Melton DA (1992): Diffusible factors in vertebrate embryonic induction. Cell 68:257–270.

Jones CM, Lyons KM, Lapan PM, Wright CVE, Hogan BLM (1992): DRV-4 (bone morphogenetic protein-4) as a posterior-ventralizing factor in *Xenopus* mesoderm induction. Development 115:639–647.

Johnson KE, Darribère T, Boucaut J-C (1993): Mesodermal cell adhesion to fibronectin-rich fibrillar extracellular matrix is required for normal *Rana pipiens* gastrulation. J Exp Zool 265:40–53.

Kageura H (1990): Spatial distribution of the capacity to initiate a secondary embryo in the 32-cell embryo of *Xenopus laevis*. Dev Biol 142:432–438.

Kageura H, Yamana K (1984): Pattern regulation in defect embryos of *Xenopus laevis*. Dev Biol 101:410–415.

Kageura H, Yamana K (1986): Pattern formation in 8-cell composite embryos of *Xenopus laevis*. J Embryol Exp Morph 91:79–100.

Kao KR, Elinson RP (1988): The entire mesodermal mantle behaves as Spemann's organizer in dorsoanterior enhanced *Xenopus laevis* embryos. Dev Biol 127:64–77.

Kao KR, Elinson RP (1989): Dorsalization of mesoderm induction by lithium. Dev Biol 132:81–90.

Kao KR, Masui Y, Elinson RP (1986): Lithium-induced respecification of pattern in Xenopus laevis embryos. Nature 322:371–373.

Kato K, Gurdon JB (1993): Single-cell transplantation determines the time when *Xenopus* muscle precursor cells acquire a capacity for autonomous differentiation. Proc Natl Acad Sci USA 90:1310–1314.

Kay BK, Peng HB (eds) (1991): *Xenopus laevis;* Practical Uses in Cell and Molecular Biology. Methods Cell Biol 36.

Keller R (1991): Early embryonic development of *Xenopus laevis*. Methods Cell Biol 36:61–113.

Kimelman D, Kirschner M (1987): Synergistic induction of mesoderm by FGF and TGF-beta and the identification of an mRNA coding for FGF in the early Xenopus embryo. Cell 51:869–877.

Kimelman D, Maas A (1992): Induction of dorsal and ventral mesoderm by ectopically expressed Xenopus basic fibroblast growth factor. Development 114:261–269.

Kimelman D, Abraham JA, Haaparanta T, Palisi TM, Kirschner MW (1988): The presence of FGF in the frog egg: its role as natural mesoderm inducer. Science 242:1053–1056.

Kimelman D, Christian JL, Moon RT (1992): Synergistic principles of development: overlapping patterning systems in Xenopus mesoderm induction. Development 116:1–9.

Kintner C (1992): Molecular basis of early neural development in Xenopus embryos. Annu Rev Neurosci 15:251–284.

Kintner CR, Dodd J (1991): Hensen's node induces neural tissue in Xenopus ectoderm. Implications for the action of the organizer in neural induction. Development 113:1495–1505.

Knöchel S, Lef J, Clement J, Klocke B, Hille S, Köster M, Knöchel W (1992): Activin A induced expression of a *fork head* related gene in posterior chordamesoderm of *Xenopus laevis* embryos. Mech Dev 38:157–165.

Kondaiah P, Sands MJ, Smith JM, Fields A, Roberts AB, Sporn MB, Melton, DA (1990): Identification of a novel transforming growth factor-β (TGF-β5) mRNA in *Xenopus laevis*. J Biol Chem 265:1089–1093.

Kondo M, Tashiro K, Fujii G, Asano M, Miyoshi R, Ryutaro Y, Muramatsu M, Shiokawa K (1991): Activin receptor mRNA is expressed in *Xenopus* embryogenesis and the level of the expression affects the body axis formation. Biochem Biophys Res Commun 181:684–690.

Köster M, Plessow S, Clement JH, Lorenz A, Tiedemann H, Knöchel W (1991): Bone morphogenetic protein 4 (BMP-4), a member of the TGF-β family, in early embryos of *Xenopus laevis:* an analysis of mesoderm inducing activity. Mech Dev 33:191–200.

Lettice LA, Slack JMW (1993): Properties of the dorsalizing signal in gastrulae of Xenopus laevis. Development 117:263–271.

Lin HY, Lodish HF (1993): Receptors for the TGF-β superfamily: multiple polypeptides and serine/threonine kinases. Trends Cell Biol (in press).

Malacinski GM, Benford H, Chung H-M (1975): Association of an ultraviolet irradiation sensitive cytoplasmic localization with the future dorsal side of the amphibian egg. J Exp Zool 191:97–110.

Maslanski JA, Leshko L, Busa WB (1992): Lithium-sensitive production of inositol phosphates during amphibian embryonic mesodem induction. Science 256:243–245.

Mathews LS, Vale WW (1991): Expression cloning of an activin receptor, a predicted transmembrane serine kinase. Cell 65:973–982.

Mathews LS, Vale WW, Kintner CR (1992) Cloning of a second type of activin receptor and functional characterization in *Xenopus* embryos. Science 255:1702–1705.

McGrew LL, Otte AP, Moon RT (1992): Analysis of *Xwnt-4* in embryos of *Xenopus laevis:* A *Wnt* family member expressed in the brain and floor plate. Development 115:463–473.

McMahon AP, Moon RT (1989): Ectopic expression of the proto-oncogene *int-1* in *Xenopus* embryos leads to duplication of the embryonic axis. Cell 58:1075–1084.

Melton DA (1987): Translocation of a localized maternal mRNA to the vegetal pole of *Xenopus* oocytes. Nature 328:80–82.

Mohun TJ, Brennan S, Dathan N, Fairman S, Gurdon JB (1984): Cell type-specific activation of actin genes in the early amphibian embryo. Nature 311:716–721.

Moody SA (1987a): Fates of the blastomeres of the 16-cell stage *Xenopus* embryo. Dev Biol. 119:560–578.

Moody SA (1987b): Fates of the blastomeres of the 32-cell-stage *Xenopus* embryo. Dev Biol 122:300–319.

Moon RT (1993): In pursuit of the functions of the Wnt family of developmental regulators: insights from *Xenopus laevis*. Bioessays 15:91–97.

Musci TJ, Amaya E, Kirschner MW (1990): Regulation of the fibroblast growth factor receptor in early *Xenopus embryos*. Proc Natl Acad Sci USA 87:8365–8369.

Nakatsuji N (1984): Cell locomotion and contact guidance in amphibian gastrulation. Am Zool 24:615–627.

Nakatsuji N, Johnson KE (1983): Comparative study of extracellular fibrils on the ectodermal layer in gastrulae of five amphibian species. J Cell Sci 59:61–70.

Niehrs C, Keller R, Cho KWY, De Robertis EM (1993): The homeobox gene *goosecoid* controls cell migration in Xenopus embryos. Cell 72:491–503.

Nieuwkoop PD (1969a): The formation of the mesoderm in urodelean amphibians, I. Induction by the endoderm. Wilhelm Roux' Arch. 162:341–373.

Nieuwkoop PD (1969b): The formation of the mesoderm in urodelean amphibians, II. The origin of the dorsoventral polarity of the mesoderm. Wilhelm Roux' Arch. 163:295–315.

Nieuwkoop PD (1973): The "organisation center" of the amphibian embryo: its origin, spatial organization, and morphogenetic action. Adv Morphogenet 10:1–39.

Nieuwkoop PD (1977): Origin and establishment of embryonic polar axes in amphibian development. Curr Top Dev Biol 11:115–132.

Nishimatsu S, Iwao M, Nagai T, Oda S, Suzuki A. Asashima M, Murakami K, Ueno N (1992a): A carboxyl-terminal truncated version of the activin receptor mediates activin signals in early *Xenopus* embryos. FEBS Lett 312:169–173.

Nishimatsu S-I, Oda S, Murakami K, Ueno N (1992b): Multiple genes for *Xenopus* activin receptor expressed during early embryogenesis. FEBS Lett 303:81–84.

Nishimatsu S, Suzuki A, Shoda A, Murakami K, Ueno N (1992c): Genes for bone morphogenetic proteins are differentially transcribed in early amphibian embryos. Biochem Biophys Res Commun 186:1487–1495.

Nishimatsu S-I, Takebayashi K, Suzuki A, Murakami K, Ueno N (1992d): Immunodetection of *Xenopus* bone morphogenetic protein-4 in early embryos. Growth Factors (in press).

Noordemeer J, Meijlink F, Verrijzer P, Rijsewijk F, Destree O (1989). Isolation of the *Xenopus* homolog of *int-1/wingless* and expression during neurula stages of early development. Nucl Acids Res 17:11–18.

Nusse R, Varmus H (1992): *Wnt* genes. Cell 69:1073–1087.

Olson DJ, Moon RT (1992): Distinct effects of ectopic expression of *Wnt*-1, activin B, and FGF on gap junctional permeability in 32-cell *Xenopus* embryos. Dev Biol 151:204–212.

Olson DJ, Christian JL, Moon RT (1991). Effect of *Wnt-1* and related proteins on gap junctional communication in *Xenopus laevis* embryos. Science 252:1173–1176.

Otte AP, Moon RT (1992): Protein kinase c isozymes have distinct roles in neutral induction and competence in Xenopus. Cell 68:1021–1029.

Otte AP, Koster CH, Snoek GT, Durston AJ (1988): Protein kinase C mediates neural induction in *Xenopus laevis*. Nature 334:618–620.

Otte AP, Van Run P, Heideveld M, Van Driel R, Durston AJ (1989): Neural induction is mediated by cross-talk between the protein kinase C and cyclic AMP pathways. Cell 58:641–648.

Otte AP, Kramer IM, Durston AJ (1991): Protein kinase C and regulation of the local competence of *Xenopus* ectoderm. Science 251:570–573.

Papkoff J, Schryver B (1990): Secreted *int-1* protein is associated with the cell surface. Mol Cell Biol 10:2723–2730.

Paterno GD, Gillespie LL, Dixon MS, Slack JMW, Heath JK (1989): Mesoderm-inducing properties of Int-2 and kFGF: two oncogene encoded growth factors related to FGF. Development 106:79–83.

Rebagliati MR, Dawid IB (1993). Expression of activin transcripts in follicle cells and oocytes of *Xenopus laevis*. Dev Biol 159:574–580.

Rebagliati MR, Weeks DL, Harvey RP, Melton DA (1985): Identification and cloning of localized maternal RNAs from Xenopus eggs. Cell 42:769–777.

Roberts AB, Kondaiah P, Rosa F, Watanabe S, Good P, Danielpour D, Roche NS, Rebbert ML, Dawid IB, Sporn MB (1990a): Mesoderm induction in *Xenopus laevis* distinguishes between the various TGF-β isoforms. Growth Factors 3:277–286.

Roberts AB, Rosa F, Roche NS, Coligan JE, Garfield M, Rebbert ML, Kondaiah P, Danielpour D, Kehrl JH, Wahl SM, Dawid IB, Sporn MB (1990b): Isolation and characterization of TGF-α2 and TGF-α5 from medium conditioned by *Xenopus* XTC cells. Growth Factors 2:135–147.

Rosa FM (1989): Mix.1, a homeobox mRNA inducible by mesoderm inducers, is expressed mostly in the presumptive endodermal cells of Xenopus embryos. Cell 57:965–974.

Rosa F, Roberts AB, Danielpour D, Dart LL, Sporn MB, Dawid IB (1988): Mesoderm induction in amphibians: the role of TGF-β2-like factors. Science 239:783–785.

Ruiz i Altaba A (1992): Planar and vertical signals in the induction and patterning of the *Xenopus* nervous system. Development 116:67–80.

Ruiz i Altaba A, Jessell T (1991a): Retinoic acid modifies mesodermal patterning in early *Xenopus embryos*. Genes Dev 5:175-187.

Ruiz I Altaba A, Jessell TM (1991b): Retinoic acid modifies the pattern of cell differentiation in the central nervous system of neurula stage Xenopus embryos. Development 112:945–958.

Ruiz i Altaba A, Jessel TM (1992): *Pintallavis,* a gene expressed in the organizer and midline cells of frog embryos: involvement in the development of the neural axis. Development 116:81–93.

Ruiz i Altaba A, Melton DA (1989a): Bimodal and graded expression of the *Xenopus* homeobox gene Xhox3 during embryonic development. Development 106:173–183.

Ruiz i Altaba A, Melton DA (1989b): Interaction between peptide growth factors and homeobox genes in the establishment of antero-posterior polarity in frog embryos. Nature 341:33–38.

Ruiz i Altaba A, Melton DA (1989c): Involvement of the Xenopus homeobox gene *Xhox3* in pattern formation along the anterior-posterior axis. Cell 57:317–326.

Saha MS, Grainger RM (1992): A labile period in the determination of the anterior-posterior axis during early neural development in Xenopus. Neuron 8:1003–1014.

Saint-Jeannet J-P, Huang S, Duprat A-M (1990): Modulation of neural committment by changes in target cell contacts in *Pleurodeles waltl.* Dev Biol 141:93–103.

Saint-Jeannet J-P, Pituello F, Huang S, Foulquier F, Duprat A-M (1993): Experimentally provoked neural induction results in an incomplete expression of neuronal traits. Exp Cell Res 207:383–387.

Sargent MG, Bennett MF (1990): Identification in *Xenopus* of a structural homologue of the Drosophila gene snail. Development 109:967–973.

Sargent TD, Jamrich M, Dawid IB (1986): Cell interactions and the control of gene activity during early development of Xenopus laevis. Dev Biol 114:238–246.

Sato SM, Sargent TD (1989): Development of neural inducing capacity in dissociated Xenopus embryos. Dev Biol 134:263–266.

Saxén L (1961): Transfilter neural induction of amphibian ectoderm. Dev Biol 3:140–152.

Scharf SR, Gerhart JC (1983): Axis determination in eggs of *Xenopus laevis:* a critical period before first cleavage, identified by the common effects of cold, pressure and ultraviolet irradiation. Dev Biol 99:75–87.

Schlessinger J (1988): Signal transduction by allosteric receptor oligomerization. Trends Biochem Sci 13:443–447.

Schulte-Merker S, Ho RK, Herrmann BG, Nüsslein-Volhard C (1992): The protein product of the zebrafish homologue of the mouse *T* gene is expressed in nuclei of the germ ring and the notochord of the early embryo. Development 116:1021–1032.

Sharpe CR (1990): Regional neural induction in *Xenopus laevis*. Bioessays 12:591–596.

Sharpe CR (1991): Retinoic acid can mimic endogenous signals involved in transformation of the Xenopus nervous system. Neuron 7:239–247.

Sharpe CR, Gurdon JB (1990): The induction of anterior and posterior neural genes in Xenopus laevis. Development 109:765–774.

Sharpe CR, Fritz A, De Robertis EM, Gurdon JB (1987): A homeobox-containing marker of posterior neural differentiation shows the importance of predetermination in neural induction. Cell 50:749–758.

Shoda A, Murakami K, Ueno N (1992): Presence of high molecular weight forms of BMP-2 in early Xenopus embryos. Growth Factors (in press).

Sive HL (1993): The frog prince-ss: a molecular formula for dorsoventral patterning in *Xenopus*. Genes Dev 7:1–12.

Sive HL, Cheng PF (1991): Retinoic acid perturbs the expression of *Xhox.lab* genes and alters mesodermal determination in Xenopus laevis. Genes Dev 5:1321–1332.

Sive HL, Draper BW, Harland RM, Weintraub H (1990): Identification of a retinoic acid-sensitive period during primary axis formation in Xenopus laevis. Genes Dev 4:932–942.

Sive HL, Hattori K, Weintraub H (1989): Progressive determination during formation of the anteroposterior axis in Xenopus laevis. Cell 58:171–180.

Slack JMW (1991): From Egg to Embryo, 2nd edn. Cambridge, UK: Cambridge University Press.

Slack JMW, Isaacs H (1989): Presence of basic fibroblast growth factor in the early *Xenopus* embryo. Development 105;147–153.

Slack JMW, Tannahill D (1992): Mechanism of anteroposterior axis specification in vertebrates. Lessons from the amphibians. Development 114:285–302.

Slack JMW, Darlington BG, Heath JK, Godsave SF (1987): Mesoderm induction in early Xenopus embryos by heparin-binding growth factors. Nature 326:197–200.

Smith JC (1987): A mesoderm-inducing factor is produced by a Xenopus cell line. Development 99:3–14.

Smith JC (1989): Mesoderm induction and mesoderm-inducing factors in early amphibian development. Development 105:665–677.

Smith JC, Slack JMW (1983): Dorsalization and neural induction: properties of the organizer in Xenopus laevis. J Embryol Exp Morphol 78:299–317.

Smith JC, Yaqoob Y, Symes K (1988): Purification, partial characterization and biological effects of the XTC mesoderm-inducing factor. Development 103:591–600.

Smith JC, Price BMJ, Van Nimmen K, Huylebroeck D (1990a): Identification of a potent Xenopus mesoderm inducing factor as a homologue of activin A. Nature 345:729–731.

Smith JC, Symes K, Hynes RO, DeSimone D (1990b): Mesoderm induction and the control of gastrulation in Xenopus laevis: the roles of fibronectin and integrins. Development 108:229–238.

Smith JC, Price BMJ, Green JBA, Weigel D, Herrmann BG (1991): Expression of a Xenopus homolog of Brachyury (T) is an immediate early response to mesoderm induction. Cell 67:79–87.

Smith WC, Harland RM (1991): Injected Xwnt-8 RNA acts early in Xenopus embryos to promote formation of a vegetal dorsalizing center. Cell 67:753–765.

Smith WC, Harland RM (1992): Expression cloning of noggin, a new dorsalizing factor localized to the Spemann organizer in Xenopus embryos. Cell 70:829–840.

Smith WC, Knecht AK, Wu M, Harland RM (1993): Secreted noggin protein mimics the Spemann organizer in the dorsalizing Xenopus mesoderm. Nature 361:547–549.

Sokol SY (1993): Mesoderm formation in *Xenopus* ectodermal explants overexpressing Xwnt8: Evidence for a cooperating signal reaching the animal pole by gastrulation. Development 118:1335–1342.

Sokol S, Melton DA (1991): Pre-existent pattern in *Xenopus* animal pole cells revealed by induction with activin. Nature 351:409–411.

Sokol SY, Melton DA (1992): Interaction of Wnt and activin in dorsal mesoderm induction in *Xenopus*. Dev Biol 154:348–355.

Sokol S, Christian JL, Moon RT, Melton DA (1991): Injected Wnt RNA induces a complete body axis in Xenopus embryos. Cell 67:741–752.

Spemann H, Mangold H (1924): Über Induktion von Embryonalanlagen durch Implantation artfremder Organisatoren. Wilh Roux' Arch 100:599–638.

Sudarwati S, Nieuwkoop PD (1971): Mesoderm formation in the Anuran *Xenopus laevis* (Daudin). Wilh Roux's Arch Dev Biol 166:189–204.

Symes K, Yaqoob M, Smith JC (1988): Mesoderm induction in Xenopus laevis: responding cells must be in contact for mesoderm formation but suppression of epidermal differentiation can occur in single cells. Development 104:609–618.

Tadano T, Otani H, Taira M, Dawid IB (1993): Differential induction of regulatory genes during mesoderm formation in Xenopus laevis embryos. Dev Genet 14:204–211.

Taira M, Jamrich M, Good PJ, Dawid IB (1992): The LIM domain-containing homeobox gene Xlim-1 is expressed specifically in the organizer region of Xenopus gastrula embryos. Genes Dev 6:356–366.

Takasaki H, Konishi H (1989): Dorsal blastomeres in the equatorial region of the 32-cell Xenopus embryo autonomously produce progeny committed to the organizer. Dev Growth Differ 31:147–156.

Thompson J, Slack JMW (1992): Over-expression of fibroblast growth factors in *Xenopus* embryos. Mech Dev 38:175–182.

Thomsen G, Melton DA (1993): Processed Vg1 protein is an axial mesoderm inducer in *Xenopus*. Cell 74:433–441.

Thomsen G. Woolf T, Whitman M, Sokol S, Vaughan J, Vale W, Melton DA (1990): Activins are expressed early in Xenopus embryogenesis and can induce axial mesoderm and anterior structures. Cell 63:485–493.

Tiedemann H (1990): Cellular and molecular aspects of embryonic induction. Zoological Science 7:171-186.

Toivonen S, Tarin D, Saxén L, Tarin PJ, Wartiovaara J (1975): Transfilter studies on neural induction in the newt. Differentiation 4:1–7.

Ueno N, Shoda A, Takebayashi K, Suzuki A, Nishimatsu S-I, Kikuchi T, Wakimasu M, Fujino M, Murakami K (1992): Identification of bone morphogenetic protein-2 in early *Xenopus laevis* embryos. Growth Factors 7:233–240.

Vincent J-P, Oster GF, Gerhart JC (1986): Kinematics of gray crescent formation in *Xenopus* eggs: the displacement of subcortical cytoplasm relative to the egg surface. Dev Biol 113:484–500.

Vincent J-P, Gerhart JC (1987): Subcortical rotation in Xenopus eggs: an early step in embryonic axis specification. Dev Biol 123:526–539.

Von Dassow G, Schmidt JE, Kimelman D (1993): Induction of the *Xenopus* organizer: expression and regulation of *Xnot*, a novel FGF and activin-regulated homeobox gene. Genes Dev 7:355–366.

Weeks DL, Melton DA (1987): A maternal mRNA localized to the vegetal hemisphere in *Xenopus* eggs codes for a growth factor related to TGFβ. Cell 51:861–867.

Wilson EB (1925): The Cell in Development and Heredity, 3rd edn. New York: MacMillan.

Willison K (1990): The mouse Brachyury gene and mesoderm formation. Trends Genet 6:104–105.

Winklbauer R (1990): Mesodermal cell migration during *Xenopus* gastrulation. Dev Biol 142:155–168.

Winning RS, Shea LJ, Marcus SJ, Sargent TD (1991): Developmental regulation of transcription factor AP-2 during *Xenopus laevis* embryogenesis. Nucl Acids Res 19:3709–3714.

Wolda SL, Moon RT (1992): Cloning and developmental expression in *Xenopus laevis* of seven additional members of the *Wnt* family. Oncogene 7:1941–1947.

Wolda SL, Moody CJ, Moon RT (1993): Expression of *Xwnt-3A* in neutral tissue of *Xenopus laevis* embryos. Dev Biol 155:46–57.

Yamada T (1990): Regulations in the induction of the organized neural system in amphibian embryos. Development 110:653–659.

Yuge M, Kobayakawa Y, Fujisue M, Yamana K (1990): A cytoplasmic determinant for dorsal axis formation in an early embryo of *Xenopus laevis*. Development 110:1051–1056.

Zhou X, Sasaki H, Lowe L, Hogan BLM, Kuehn MR (1993): Nodal is a novel TGF-β like gene expressed in the mouse node during gastrulation. Nature 361:543–547.

ABOUT THE AUTHOR

IGOR B. DAWID is Chief of the Laboratory of Molecular Genetics, National Institute of Child Health and Human Development, National Institutes of Health, in Bethesda, Maryland. He received his Ph.D. in 1960 from the University of Vienna, and then did postdoctoral work with John Buchanan at MIT. He subsequently spent 15 years on the staff of the Department of Embryology of the Carnegie Institution at Baltimore before moving to the NIH in 1978. He has worked on molecular aspects of development throughout his career, mostly using *Xenopus,* but also studying *Drosophila* and, recently, the zebrafish. His earlier interests focused on mitochondrial and ribosomal DNA, while recent work concerns embryonic induction and the role of LIM class homeobox genes. Dr. Dawid has been Editor-in-Chief of *Developmental Biology* (1975–80), Chairman of the Editorial Board of *The Proceedings of the National Academy of Sciences* (1988–91), and has been a member of the editorial boards of *Cell, Development Growth and Differentiation, Nucleic Acids Research, Annual Reviews of Biochemistry,* and *Genes and Development*.

Acronyms Used in This Volume

AC, anchor cell
ADP, adenosine diphosphate
AER, apical ectodermal ridge
ARAM, antigen recognition activation motifs
ARH1, antigen receptor homology 1 motifs (same as TAM)
ATP, adenosine triphosphate
bFGF, basic fibroblast growth factor
βLAP, TGFβ amino terminal latency-associated peptide
BMP, bone morphogenetic protein
CAK, cyclin-dependent kinase activating kinase
CHX, cycloheximide
CNS, central nervous system
CRE, cAMP-responsive element
CREB, cAMP-responsive element binding protein
CsA, cyclosporin A
CSF, colony stimulating factor
CSP, cell surface protein
cAMP, cyclic adenosine 3′,5′-monophosphate
DAG, diacylglycerol
DARPP-32, dopamine and cAMP-regulated phosphoprotein
DAI, dorsoanterior axis index
ΔXAR, truncated activin receptor
DMZ, dorsal marginal zone
DNA, deoxyribonucleic acid
DTC, distal tip cell
EC, embryonal carcinoma cells
ECM, extracellular matrix
EGF, epidermal growth factor
EGFR, EGF receptor
ENU, ethylnitrosourea
ER, endoplasmic reticulum
Erk, extracellular regulated protein kinase
ES, embryonal stem cell
FGF, fibroblast growth factor
FGFR, FGF receptor
FKBP, FK506-binding protein
G-protein, GTP-binding protein
G-subunit, glycogen-targeting subunit of PP-1
GAG, glycosaminoglycan
GAP, GTPase-activating protein
GDI, guanine nucleotide diphosphate dissociation inhibitor
GDS, guanine nucleotide diphosphate dissociation stimulator
GDP, guanosine diphosphate
GMCSF, granulocyte-macrophage colony stimulating factor
GNEF, guanine nucleotide exchange factor; same as GNRP and GRF
GNRP, guanine nucleotide release protein; same as GNEF and GRF
GRF, guanine nucleotide releasing factor; same as GNEF and GNRP
GTP, guanosine triphosphate
GTPase, guanosine-triphosphatase
GTPγS, guanosine-5′-O-[3-thio]-triphosphate
GVBD, germinal vesicle breakdown
HBGF, heparin-binding growth factor
HNF3, hepatocyte nuclear factor-3
hnRNP, heteronuclear ribonuclear proteins
HSPG, heparan sulfate proteoglycan
HTLV-1, human T cell leukemia virus
I-1, inhibitor 1 of protein phosphatase
IGF, insulin-like growth factor
IGFBP, insulin-like growth factor binding protein
IGF-II/Man-6-P receptor, insulin-like growth factor-II/mannose-6-phosphate receptor
IL-1, interleukin-1

IL-2, interleukin-2; also called TCGF, TMF, and TSF
IP3, inositol triphosphate
INH, MPF inhibitory activity in *Xenopus*; same as PP-2A (see PP)
IRS1, insulin receptor substrate-1
LAK, lymphocyte activated killer cells
LAR, leukocyte common antigen related protein
LTBP, latent TGFβ-binding protein
Man6P, mannose-6-phosphate
MCSF, macrophage colony stimulating factor
MIF, mesoderm-inducing factor
MPF, maturation (M-phase) promoting factor composed of $p34^{cdc2}$ (see cdc2 in Glossary) and cyclin
mRNA, messenger ribonucleic acid
NAD, nicotinamide-adenine diphosphate
NCAM, neural cell adhesion molecule
NF-AT, nuclear factor of activated T cells
NF-κB, nuclear transcription factor-κB
NGF, nerve growth factor
NK, natural killer cell
NLS, nuclear localization signal
NMR, nuclear magnetic resonance
PAI-1, plasminogen activator inhibitor, type I
PCR, polymerase chain reaction
PCNA, proliferating cell nuclear antigen
PDGF, platelet-derived growth factor
PDGFR, platelet-derived growth factor receptor
PHA, phytohemagglutinin
PI, phosphatidylinositol
PI3K, phosphatidylinositol-3-kinase
PKA, protein kinase A
PKC, protein kinase C
PLC, phospholipase C
PNS, peripheral nervous system
PP, protein phosphatase (four PPs are identified: PP-1, PP-2A, PP-2B, and PP-2C)
PPIase, peptidyl-prolyl isomerase
PTP, protein tyrosine phosphatase
PTX, pertussis toxin
RA, retinoic acid
RGD, arg-gly-asp sequence
RNA, ribonucleic acid
RT-PCR, reverse transcription polymerase chain reaction
SDS-PAGE, sodium dodecyl sulfate polyacrylamide gel electrophoresis
SH2, Src homology region 2
SH3, Src homology region 3
SM, sex myoblast
SPARC, secreted protein, acidic and rich in cysteine
TAM, tyrosine-based activation motifs
TCGF, T-cell growth factor; same as IL-2
TCR, T-cell receptor
TGF-α, transforming growth factor alpha
TGF-β, transforming growth factor beta
TIMP, tissue inhibitor of metalloproteases
TMF, T-cell mitogenic factor; same as IL-2
TPA, 12-O-Tetradecanoyl phorbol-13-acetate (also called PMA, phorbol myristate acetate)
TRE, TPA-responsive element
TSF, thymocyte-stimulating factor; same as IL-2
uPA, urokinase plasminogen activator
UV, ultraviolet
VMZ, ventral marginal zone (giving rise to ventral mesoderm)
VPC, vulval precursor cells
VU, ventral uterine precursor cell

Glossary of Proteins and Their Genes and Gene Families

In this book we have adopted the following convention which has been adhered to as closely as possible. Acronyms for genes and their primary transcripts are italicized and in lower case (e.g., *wng*) as are the full names for genes (e.g., *wingless*). Acronyms for proteins and the processed mRNAs that encode them are in lower case with the first letter capitalized. Also used is the alternative format for proteins in which the apparent molecular weight of the protein is represented in the name as the letter "p" followed by the number of kilodaltons, and the acronym attached as a superscript (e.g., $p60^{src}$). Protein families are referred to in lower case (e.g., src family). Standard exceptions to this convention are that for yeast genes and their protein products, those from *S. cerevisiae* are capitalized whereas those from *S. pombe* are in lower case. Italics and standard letters still designate yeast genes and proteins respectively. Additionally gene names in which the first letter of the acronym refers to the genus are italicized and with the first letter capitalized.

l(1)ph: a *Drosophila* gene that encodes a Raf1 homologue

3BP-1: a protein that binds to the SH3 domain of the tyrosine protein kinase, Abl. The sequence of 3BP-1 is homologous to that of a C-terminal segment of Bcr and of Rho-GAP.

3CH34: a mammalian gene that encodes a VH1-type PTP.

aar: abnormal anaphase resolution. A *Drosophila* gene that encodes a regulatory subunit of PP-1.

actin: a muscle protein which, along with myosin, forms fibers that are involved in contraction and which also forms fibers in nonmuscle cells. These latter fibers participate in cell movement and in maintaining cell shape.

aFGF: acidic fibroblast growth factor; also called FGF-1. A growth factor originally identified for its ability to stimulate fibroblast proliferation but which affects many other target cells.

AP-1: activator protein-1. A Jun/Fos heterodimer which was originally described as a protein that activates transcription by binding to its consensus binding sequence (TGACTCA) in the SV40 enhancer and MTIIA gene. The AP-1 transcription factor has been shown to mediate transcriptional regulation by PKC.

armadillo: a *Drosophila* gene encoding a protein with homology to β-catenin.

bek: bacterially expressed kinase. A murine gene that encodes an FGFR-2.

bFGF: basic fibroblast growth factor. It is also called FGF-2. A growth factor for many cell types.

bimG: blocked in mitosis G. An *Aspergillus* gene that encodes a phosphatase related to PP-1.

Bithorax: a gene cluster in *Drosophila* that functions to assign unique identities to body segments in the abdomen and posterior thorax.

boss: bride of sevenless. A *Drosophila* gene encoding a putative ligand for the sevenless receptor.

brachyury: a mesodermal determinant gene from the T-locus on chromosome 17 in the mouse. It is also called the T gene.

BWS1: a gene in *S. pombe* that encodes a phosphatase related to PP-1. It is also known as *dis2*.

cactus: a *Drosophila* gene that is homologus to IκB.

crkle: a *Drosophila* gene that performs a function similar to *Drk*.

CD3: cluster of differentiation #3. A human leukocyte differentiation antigen; an invariant component of the multisubunit T-cell receptor (TCR) complex.

CD4: cluster of differentiation #4. A human leukocyte differentiation antigen; a transmembrane glycoprotein and a coreceptor with the TCR in a subpopulation of T cells that recognize the antigen bound to the major histocompatability complex, class II (MHC class II) molecules.

CD8: cluster of differentiation #8. A human leukocyte differentiation antigen; a transmembrane glycoprotein and a coreceptor with the TCR in a subpopulation of T cells that recognize the antigen bound to the major histocompatability complex, class I (MHC class I) molecules.

CD45: cluster of differentiation #45. Also called human leukocyte common antigen (LCA); a polymorphic transmembrane glycoprotein with a cytoplasmic region composed of tandemly duplicated domains with PTP activity.

CDC #: cell division cycle genes of *S. cerevisiae*.

cdc #: cell division cycle genes of *S. pombe*.

cdc2: an *S. pombe* gene encoding a serine/threonine kinase, Cdc2 which is also called $p34^{cdc2}$. The equivalent gene in vertebrates is also called *cdc2* whereas the *S. cerevisiae* gene is called CDC28. Cdc2 is regulated by its interaction with cyclin. The *Xenopus* Cdc2-cyclin complex is also called MPF.

cdc25: an *S. pombe* gene that encodes a protein tyrosine phosphatase.

CDC55: an *S. cerevisiae* gene that encodes a protein homologous to the regulatory subunit of PP-2A.

cdk: cyclin-dependent kinase.

cek: chicken embryo kinase. A chicken gene that encodes an FGFR.

clr: genes that produce the Clear phenotype in *C. elegans*.

CLN1 and *CLN2: S. cerevisiae* genes encoding G1 cyclins.

CNA1 and *CNA2: S. cerevisiae* genes encoding the catalytic subunit A of the protein phosphatase, PP-2B.

CNB: an *S. cerevisiae* gene encoding the regulatory B subunit of the protein phosphatase, PP-2B.

csw: corkscrew. A *Drosophila* gene that encodes a phosphotyrosine phosphatase.

cyclins: regulatory subunits of the cdc and cdk kinases.

daf-1: Dauer activating factor-1. A *C. elegans* gene which is involved in Dauer larval differentiation.

DARPP-32: dopamine and cAMP-regulated phosphoprotein; an inhibitor of protein phosphatase-1

Der: the *Drosophila* EGFR.

*dis*2: same as *bws1*.

disheveled: a *Drosophila* gene that has not yet been cloned.

DLAR: a *Drosophila* gene that encodes a transmembrane PTP related to mammalian LAR.

DMZ: the dorsal marginal zone in *Xenopus* which gives rise to dorsal mesoderm.

DPTP99A: a transmembrane PTP in *Drosophila*.

DPTP10D: a transmembrane PTP in *Drosophila*.

drk: downstream of receptor kinase. A *Drosophila* gene that encodes a protein made of SH2 and SH3 domains which is homologous to mammalian GRB2 and *C. elegans sem-5*.

dorsal: a *Drosophila* gene encoding a protein that is homologous to transcription factors such as NF-κB and *c-rel*.

dpp: the *decapentaplegic* homeobox gene in *Drosophila* which encodes a protein related to the mammalian TGFβ and bone morphogenetic proteins (BMPs).

Drafl: Drosophila rafl. A *Drosophila* gene which encodes a kinase homologous to the mammalian Raf1 kinase.

Dsor: Drosophila suppressor of *Raf*. A *Dro-*

sophila gene which encodes a kinase homologous to the MEK1 gene product.

Dtrk: a *Drosophila* gene encoding a transmembrane tyrosine kinase and related to the *trk* gene in mammals which encodes one of the receptors for nerve growth factor.

DVR: decapentaplegic-Vg-related genes. An alternate name for the subfamily of TGF-β-related proteins that includes BMPs, VGRs (Vg1-related proteins) and *dpp*.

easter: a *Drosophila* gene encoding a serine protease-like protein.

egl: egg laying defective. A *C. elegans* gene.

ellipse: a *Drosophila* gene which is an allele of the gene *Der* which encodes a transmembrane receptor related to the mammalian EGFR.

emt: a human gene encoding a 72-kD tyrosine kinase that is homologous to the *itk* gene in mouse.

engrailed: a *Drosophila* gene that encodes a homeodomain transcription factor.

Erk: extracelluar-regulated protein kinase. A serine/threonine kinase which is part of the signal transduction pathway initiated by growth factors and that involves Ras. There at least two Erks, designated 1 and 2. They are also called the MAP kinases.

fem: feminized. A *C. elegans* gene required for proper sex determination.

FKBP: the FK506 (an immunosuppressive drug) binding protein which has been found to have *cis, trans*-proline isomerase activity.

flg: the *fms*-like gene in mammals which encodes an FGFR.

fork head: a *Drosophila* gene that encodes a transcription factor related to XFKH in *Xenopus* and HNF3 in mammals.

c-fos: the cellular homologue of the transforming gene of the Finkel-Biskis-Jinkins (FBJ) murine osteosarcoma virus. A subunit of the AP-1 transcription factor which recognizes the TRE sequence.

fyn: a gene encoding p59^{fyn}, a protein tyrosine kinase of the src family.

GAP: GTPase-activating protein. A regulator of Ras activity and a potential effector of the Ras signal. The *Drosophila* gene, *gap1*, encodes a GAP which is homologous with mammalian RasGAP. There are also GAPs which are specific for other members of the RAS superfamily such as for Rap, Rac, and Rho.

gastrulation defective: a *Drosophila* gene encoding a serine protease-like protein.

glp: germ-line proliferation defective; *C. elegans* genes.

glp-1: related to *lin-12* and to Notch.

GMCSF: granulocyte-macrophage colony stimulating factor

G-protein: GTP-binding protein. A protein that binds GTP and hydrolyzes it to form GDP and orthophosphate. G-proteins are used as "molecular switches" in molecular pathways such as signal transduction (e.g., Ras) and protein synthesis (e.g., eIF-2). The G-protein adopts an active conformation when bound to GTP and an inactive conformation when bound to GDP. There are two types of G-protein: monomeric and heterotrimeric.

GRB2: a mammalian protein consisting of two SH3 domains flanking an SH2 domain. Homologous proteins are Sem-5 in *C. elegans* and Drk in *Drosophila*.

GRP33: see p62.

gsc: goosecoid. A *Xenopus* homeobox gene. The name is derived from the names of the *Drosophila* genes *gooseberry* and *bicoid*.

G-subunit: a glycogen-targeting subunit of PP-1.

HCS26: an *S. cerevisiae* gene encoding a cyclin.

hedgehog: a *Drosophila* gene encoding a putative transmembrane protein that regulates *wingless* expression.

her: her (rather than him). *C. elegans* genes required for sex determination.

HOX: Homeobox regulatory genes originally identified in *Drosophila* and encoding transcription factors that determine developmental patterns of body segmentation.

huckebein: a *Drosophila* gene which encodes a putative transcription factor.

KC: a PDGF-inducible early response gene and a cell-cycle-regulated gene in the mouse.

I-1: the inhibitor 1 of protein phosphatases.

Id: inhibitor of differentiation. A myogenic

inhibitory factor which is a member of the MyoD-related helix-loop-helix transcription factors but which lacks the basic DNA-binding domain.

IGF: insulin-like growth factor. A growth factor that is induced by growth hormone and is also called somatomedin C.

IGFBP: insulin-like growth factor binding protein. A family of soluble, secreted proteins which are specific for IGF-1 and are believed to regulate the interaction of IGF-1 with its cell surface receptor.

IGF-II/Man-6-P receptor: the insulin-like growth factor-II/mannose-6-phosphate receptor. In mammals, this transmembrane protein binds both IGF-II and mannose-6-phosphate via two different binding sites. This protein is also referred to as the cation-independent mannose-6-phosphate receptor.

IκB: an inhibitor of the mammalian transcription factor NF-κB.

INH: an MPF inhibitory activity in *Xenopus* which is identical with PP-2A.

int: mouse genomic sequences adjacent to the integration site (*int*-1) or activated by the integrated mouse mammary tumor provirus (*int*-2, *int*-3) prototype in mouse. *Int-1, -2,* and *-3* are unrelated genes.

int-1: a mouse gene that encodes a secreted protein similar to the *Drosophila* wingless protein. These genes are in the *wnt* gene family which also includes *irp.*

int-2: a mouse gene that encodes a secreted protein in the FGF gene family.

int-3: a mouse gene which encodes a transmembrane protein that is similar to *Drosophila* Notch and *C. elegans* Lin12.

IL-1: interleukin I

IL-2: interleukin-2; also called TCGF, TMF, and TSF

IRS1: insulin receptor substrate-1; a protein which, once phosphorylated by the insulin receptor kinase, interacts with phospholipase Cγ via its SH2 domain and activates PLCγ.

itk: a mouse gene encoding a 72-kD tyrosine kinase that is homologous to the *emt* gene in humans.

JE: a PDGF-inducible early response gene in the mouse that is related to platelet factor 4.

jun: c-jun, a cellular homologue of the transforming gene of the avian sarcoma virus 17 (ASV-17). A subunit of the AP-1 transcription factor which recognizes the TRE sequence.

kfgf: an FGF gene which was originally isolated from human stomach tumor DNA. This gene is also called *hst* and *fgf-4.*

kit: a gene that encodes a tyrosine kinase receptor which is related to the PDGF and CSF-1 receptors. The prototype, *v-kit,* is the oncogene of the Hardy-Zuckerman 4 feline sarcoma virus (HZ4-FeSV). The *c-kit* gene is located in the W-locus on the mouse chromosome. The ligand for Kit is Steel.

k-sam: kATO-III cell–derived stomach cancer amplified gene. A human gene that encodes a receptor homologous with FGFR-2.

lag: C. elegans genes that can mutate to cause the same phenotype as a *lin-12 glp-1* double mutant.

lck: a mouse gene that encodes a protein tyrosine kinase, $p56^{lck}$, of the src family which is found almost excusively in cells of the T-cell lineage.

let: lethal genes in *C. elegans.*

Let-23: a *C. elegans* tyrosine kinase receptor of the EGF receptor subfamily.

Let-60: the *C. elegans* Ras protein which acts downstream of Let-23.

Let-341: a *C. elegans* putative guanine nucleotide exchange factor.

lin: lineage abnormal genes in *C. elegans.*

Lin-3: a *C. elegans* protein containing an EGF-like motif which is probably a ligand for Let-23.

Lin-12: a *C. elegans* transmembrane protein in the Lin-12/Notch family of receptors for intercellular signals.

Lin-31: a *C. elegans* transcription factor of the HNF3/Forkhead family.

M-subunit: myofibril-targeting subunit of PP-1.

MAP kinase: mitogen-activated protein kinase. A serine/threonine kinase which is activated upon cell stimulation by growth factors and other mitogens and which is part of the signal transduction pathway that includes Ras. Also called ERK (extracellular regulated kinase).

MCSF: macrophage colony stimulating factor.

MEK kinase: MAP kinase kinase, or ERK kinase kinase. A serine/threonine kinase which activates the MAP (ERK) kinases by phosphorylating them.

MIF: mesoderm-inducing factor.

mik1: a gene in *S. cerevisiae* which encodes a dual specificity protein kinase.

mil (or *mht*): a homologue of the transforming gene of the avian retrovirus, Mill Hill 2.

mlx: mesoderm inducible homeobox gene expressed in the *Xenopus* gastrula stage.

MPF: maturation (or M-phase) promoting factor; composed of $p34^{cdc2}$ and cyclin and promotes maturation of *Xenopus* oocytes in response to the appropriate hormonal stimulus.

M-subunit: the myofibril-targeting subunit of PP-1.

myc: homologue of the oncogene of the avian myelocytometosis virus (*v-myc:* the viral oncogene; *c-myc:* the cellular protooncogene) which encodes a nuclear transcription factor with a helix-loop-helix motif.

myf4: myogenic factor 4. A human gene that encodes a member of the MyoD-related helix-loop-helix transcription factor family which is the human homologue of myogenin.

myf5: myogenic factor 5. A mammalian gene that promotes myogenesis.

myoD: a gene identified from its cDNA cloned from a mouse aza-cytidine-treated myoblast-myotube library. It is a myogenic determination gene that encodes a helix-loop-helix transcription factor.

myogenin: a rat gene that stimulates expression of muscle genes and encodes a member of the MyoD-related helix-loop-helix transcription factors.

neu: a protooncogene that encodes a tyrosine kinase transmembrane receptor that is related to the EGFR.

NF-AT: the nuclear factor of activated T cells; a transcription factor comprised of two components; NF-ATc and NF-ATn.

NF-κB: a mammalian transcription factor that regulates expression of the immunoglobulin κ light chain gene amongst others. It is homologous with the *c-rel* oncogene product and the dorsal protein in *Drosophila*.

noggin: a *Xenopus* dorsalizing gene.

norpA: a *Drosophila* gene which encodes a phospholipase C.

Notch: a *Drosophila* gene which encodes a transmembrane protein that contains EGF-like repeats in the extracellular domain and which influences neural development.

nudel: a *Drosophila* gene that has not yet been cloned.

p53: a tumor suppressor gene that encodes a nuclear protein which is found inactivated in a variety of human neoplasms.

p62: a GAP-associated tyrosine phosphoprotein of unknown function which has extensive homology with a putative hnRNP protein, GRP33.

p140: Ras-GRF.

pac-1: a mammalian gene which encodes a nuclear localized PTP.

Patch: a murine embryonic lethal dominant mutation with partial penetration in the heterozygote.

patched: a *Drosophila* gene encoding a transmembrane protein which regulates *wingless* expression.

PCNA: proliferating cell nuclear antigen which is found in abundance in the nuclei of proliferating cells and is a subunit of DNA polymerase δ.

PDGFRα: the α subunit of the platelet-derived growth factor receptor.

PDGFRβ: the β subunit of the platelet-derived growth factor receptor.

pelle: a *Drosophila* gene encoding a serine/threonine kinase.

pim-1: a gene that encodes a protein tyrosine kinase and which was identified as an oncogene that is activated by insertional mutagenesis in T-cell lymphomas.

pintallavis: a *Xenopus* gene, encoding a transcription factor, which is also called *XFD1* or *XFKH*.

pipe: a *Drosophila* gene that has not yet been cloned.

PP2B: a Ca^{2+}-dependent serine/threonine phosphatase which is called calcineurin.

pole hole: a *Drosophila* gene encoding the Raf protein kinase.

porcupine: a *Drosophila* gene that has not yet been cloned.

PP: a general term for serine/threonine protein phosphatase; includes PP-1, PP-2A, PP-2B, PP-V, PP-X, PP-y, PP-Z, PP-2BW, PPHA1, PPHA2, PPH3, PPH22.

pRB: the retinoblastoma susceptibility gene product encoded by the *Rb* gene which is inactivated in familial retinoblastoma.

PTP: a general term for protein tyrosine phosphatase. For the names of the members of this superfmaily see Table I, Ingebritsen et al. (this volume).

pyp1, pyp2, and *pyp3: S. pombe* genes that encode PTPs.

rac: Ras-related C3 botulinum toxin substrate. Originally identified as a cDNA from a differentiated HL-60 library, *rac* encodes a protein homologous to Rho and Ras.

raf-1: a homologue of the transforming gene of the murine sarcoma virus MSV-3611. *v-mil* is the avian homologue. The protein product, Raf-1, is a serine/threonine protein kinase.

ras: the cellular homologue of the rat sarcoma virus oncogene, *v-ras*. The *c-ras* gene encodes a monomeric G-protein which is involved in signal transduction from protein tyrosine kinase transmembrane receptors.

RasGAP: Ras GTPase-activating protein which stimulates the GTPase activity of Ras and also may act as a downstream effector of Ras.

Rb: 'Retinoblastoma' gene in humans. A gene which was first discovered as the mutated gene in familial retinoblastosis. Unlike most genes which are associated with specific cancers, the *Rb* gene encodes a protein that suppresses cell proliferation. Only when both alleles of this gene are mutated to forms that do not produce active proteins does cancer ensue. Because of these properties, this and other genes have been referred to as antioncogenes or tumor suppressor genes.

rdgc: retinal degeneration C. A *Drosophila* gene encoding a serine/threonine protein phosphatase.

rel: a mammalian gene encoding a transcription factor. Rel is a member of the rel family which includes c-Rel, v-Rel, NF-κB, and Dorsal. c-Rel is the cellular homolog of the v-Rel oncoprotein which is the product of the avian Rev-T retroviral oncogene.

RGD: the arg-gly-asp sequence which is a recognition site for the integrin receptors.

rho: Ras homologous open reading frame. The gene encodes a G-protein related to Ras.

rsk: a gene which encodes $p90^{rsk}$, the S6 protein kinase, which is a serine/threonine kinase which phosphorylates the S6 protein of the 4OS eukaryotic ribosomal subunit.

s6k: A gene which encodes $p70^{s6k}$, the S6 protein kinase, which is a serine/threonine kinase.

sdc: sex determination and dosage compensation genes in *C. elegans.*

sds21: an *S. pombe* gene encoding a phosphatase of the PP-1 family.

sds22: an *S. pombe* gene that encodes what may be a nuclear targeting subunit of the protein phosphatase PP-1.

sem: sex myoblast migration genes in *C. elegans.*

sem-5: a *C. elegans* gene encoding a protein related in structure and function to mammalian GRB2 and *Drosophila drk.*

sev: The *sevenless* gene in *Drosophila* which encodes a tyrosine kinase receptor.

shaggy: see *zeste-white-3.*

shc: a human gene encoding src homology region 2–containing protein.

shortened gastrulation: a *Drosophila* gene that has not yet been cloned.

sina: the *seven in abstentia* gene in *Drosophila* which encodes a nuclear protein.

sis: c-sis, the mammalian cellular homologue of the transforming gene of the Simian sarcoma virus which encodes the B-chain of platelet-derived growth factor.

SIT4: an *S. cerevisiae* gene which encodes a phosphatase related to the catalytic subunit of PP-2A.

snake: a *Drosophila* gene encoding a serine protease-like protein.

sos: the *son of sevenless* gene in *Drosophila* which encodes a gene related to CDC25, the yeast GTP exchange factor.

spatzle: a *Drosophila* gene that encodes a protein which is probably the ligand for the Toll protein.

src: the *c-src* gene is the cellular homologue of v-*src* the transforming gene of Rous sarcomavirus transforming gene, v-src. These genes encode soluble, cytoplasmic tyrosine protein kinases which interact with many transmembrane PTP receptors that are able to recognize the SH2 domain on the Src protein for which this protein holds the prototype.

Steel: the ligand for the *c-kit* gene product which is encoded in the *Sl* locus of the mouse and is identified by the Steel-Dickie mutation. The protein product is variously referred to as the kit ligand (KL), stem-cell factor (SCF), and mast cell growth factor (MSGF).

SWI: the mating type switching abnormal gene in *S. cerevisiae.*

SW14: an *S. cerevesiae* gene that encodes a transcriptional regulator.

syn muv: the synthetic multivulva genes in *C. elegans.*

tailless: a *Drosophila* gene that encodes a putative nuclear transcription factor.

TAN-1: the Translocation Associated Notch homologue in humans.

TGF-α: transforming growth factor-α, a growth factor related to EGF.

TGF-β: transforming growth factor-β, a member of the TGF_{β} superfamily that also includes the bone morphogenetic proteins (BMPs).

TK-14: a human gene that encodes a homologue of FGFR-2 which was isolated from a human tumor cDNA library.

toll: a *Drosophila* gene that encodes a transmembrane receptor protein.

tolloid: a *Drosophila* gene encoding a protein homologous to the mammalian bone morphogenetic protein 1 (BMP1).

torso: a *Drosophila* gene that encodes a tyrosine kinase receptor.

torpedo: a *Drosophila* allele fo the *Der* gene that encodes a transmembrane tyrosine kinase receptor related to the EGFR.

tra: the transformed genes in *C. elegans* which are required for sex determination.

Tra-1: a putative DNA binding protein from *C. elegans.*

Tra-2: a transmembrane protein in *C. elegans* that is downstream from Her-1 in signal transduction.

TRE: the TPA-responsive element which is the recognition site for the transcription factor, AP-1.

trk: a mammalian gene which is selectively expressed in the developing nervous system and that encodes a transmembrane tyrosine protein kinase, a receptor for nerve growth factor (NGF). Coexpression of Trk and the p75 NGF receptor is required for high affinity binding of NGF.

twi: twist. A *Drosophila* gene encoding a helix-loop-helix transcription factor that is involved in mesoderm determination. A *Xenopus* homolog is also identified.

vav: a mammalian gene encoding a protein with an SH2-domain that may be a nuclear transcription factor.

vg1: vegetal zone gene 1. A *Xenopus* gene encoding an mRNA localized at the vegetal pole of the egg.

wee1: an *S. cerevisiae* gene that encodes a dual specificity protein kinase.

windbeutel: A *Drosophila* gene that has not yet been cloned.

wng: wingless. a *Drosophila* gene encoding a protein related to the *int-1* gene product in mammals. The wingless protein regulates the activity of the zeste-white-3 protein kinase.

wnt: A family of related genes that includes *wingless* and *Xwnt* (*Drosophila*) and *int-1* (mammals). The name is derived from the *wingless* gene and the *int-1* protooncogene.

WT-1: a recessive gene which, when mutated, is responsible for the occurrence of Wilm's tumor in humans (a childhood nephroblastoma).

xbra: the *Xenopus* homolog of the mouse *brachyury* gene.

XeFGF: an isoform of FGF isolated from *Xenopus.*

XFD1: another name for *XFKH*.

XFKH: a *Xenopus* member of *fork head* class of transcription factor genes. The family is identified by the *Drosophila* gene *fork head* and the mammalian factor HNF3.

Xhox-3: the *Xenopus* homeobox gene-3, which is expressed in the posterior of the embryo.

Xlim: members of the LIM class of homeobox genes which are defined by the association of two copies of a cysteine-rich domain with a homeodomain.

Xnot: a *Xenopus* homeobox gene expressed in notochord.

xol: the XO lethal genes required for sex determination and dosage compensation in *C. elegans*.

Xsna: the *Xenopus* homolog of *snail* (*sna*), a gene that is involved in mesoderm determination in *Drosophila*.

XTC cells: a cultured cell line from *Xenopus*.

xwnt: a *Xenopus* member of the *wnt* gene family.

YOP-51: a plasmid-borne gene of *Yersinia enterolitica* that encodes a PTP.

ZAP-70: a soluble protein tyrosine kinase that associates with the TCR.

zen: the *Drosophila zerknullt* gene that encodes a homeodomain protein.

zeste-white-3: a *Drosophila* gene encoding a serine/threonine kinase related to the mammalian glycogen synthase kinase-3. The zeste-white-3 kinase regulates engrailed activity and is regulated by wingless.

Index